图书在版编目(CIP)数据

海洋碳中和前沿进展.2022:中文、英文/李建平主编.--青岛:中国海洋大学出版社,2023.5
ISBN 978-7-5670-3499-0

Ⅰ.①海… Ⅱ.①李… Ⅲ.①海洋-二氧化碳-节能减排-研究报告-2022-汉、英 Ⅳ.①X511

中国国家版本馆CIP数据核字(2023)第082194号

海洋碳中和前沿进展(2022)
HAIYANG TANZHONGHE QIANYAN JINZHAN (2022)

出版发行	中国海洋大学出版社		
社　　址	青岛市香港东路23号	邮政编码	266071
网　　址	http://pub.ouc.edu.cn		
出 版 人	刘文菁		
责任编辑	姜佳君	电　　话	0532-85901040
网　　址	j.jiajun@outlook.com		
印　　制	青岛海蓝印刷有限责任公司		
版　　次	2023年5月第1版		
印　　次	2023年5月第1次印刷		
成品尺寸	185 mm×260 mm		
印　　张	19.75		
字　　数	550千		
印　　数	1—1 300		
审 图 号	GS鲁(2023)0121号		
定　　价	175.00元		
订购电话	0532-82032573		

发现印装质量问题,请致电0532-88785354,由印刷厂负责调换。

海洋碳中和前沿进展（2022）

Advances in Ocean Carbon Neutrality (2022)

李建平　主编

中国海洋大学出版社
·青岛·

编委会

主　编：李建平

副主编：（按姓氏笔画排序）

　　　　马　崑　卢　昆　包　锐　冯玉铭
　　　　刘　臻　李　岩　徐　胜　梁生康
　　　　董云伟　魏俊峰

成　员：（按姓氏笔画排序）

　　　　丁黎黎　于　蒙　于通顺　王　栋
　　　　许博超　余　静　张　潮　徐晓峰
　　　　高　阳　高会旺　郭金家　曹飞飞
　　　　董双林　谢素娟

前　言

全球气候变化是全人类共同面临的重大挑战。2015年12月,《巴黎气候协定》正式签署,其核心目标是将全球气温上升控制在远低于工业革命前水平的2 ℃以内,并努力控制在1.5 ℃以内。要实现这一目标,全球温室气体排放需要在2030年之前减少一半,在2050年左右达到净零排放,即碳中和。2020年12月,联合国秘书长安东尼奥·古特雷斯指出,2050年实现碳中和是当今世界最为紧迫的使命。

应对气候变化是全人类的共同事业。习近平主席在第七十五届联合国大会一般性辩论上郑重宣布:中国将提高国家自主贡献力度,采取更加有力的政策和措施,二氧化碳排放力争于2030年前达到峰值,努力争取2060年前实现碳中和。这是中国应对全球气候问题作出的庄严承诺。2022年5月16日,习近平总书记在《求是》杂志上发表文章指出,碳达峰碳中和是我国进入新发展阶段需要正确认识和把握的重大问题之一。党的二十大报告提出,"积极稳妥推进碳达峰碳中和"。这是以习近平同志为核心的党中央统筹国内国际两个大局作出的重大决策部署,为推进碳达峰碳中和工作提供了根本遵循,对于全面建设社会主义现代化国家、促进中华民族永续发展和构建人类命运共同体都具有重要意义。

海洋作为全球最大、最重要的生态系统,在应对全球气候变化、保护生物多样性和实现可持续发展等方面发挥着至关重要的作用。海洋碳储量是陆地碳库的20倍、大气碳库的50倍,海洋碳汇稳定了大气中的二氧化碳含量,缓解了全球变暖的压力。

为积极响应国家"双碳"目标,中国海洋大学于2022年6月正式成立海洋碳中和创新研究中心(下称"中心")。中心以服务国家"双碳"目标重大需求为宗旨,充分发挥海洋领域"文理工农"等多学科交叉融合优势,以"深耕海洋、服务国家、世界一流"为建设目标,提供"蓝碳"中国方案,打造具有鲜明海洋特色的"政产学研金服用"综合创新平台。中心开展有组织科研,在重大科学问题、工程技术难题和产业技术问题领域凝聚资源,形成科研集群力量,长时间持续攻关,吸引了全校百余位相关专业的专家学者。

作为一个新兴的研究领域,海洋碳中和的相关成果和技术不断涌现,亟待总结和交流。因此,中心精选了17位成员在2022年发表的高水平学术成果,分为"海洋碳增汇关键过程与理论创新""非碳能源替代方案与关键技术""海洋碳增汇应用典型示范与推广""海洋碳交易市场标准与气候变化评估"四大版块,分别由梁生康教授、刘臻教授、董云伟教授和徐胜教授组稿形成了《海洋碳中和前沿进展(2022)》,期望能为关注全球气候变化问题的业内人士、专家学者提供一些新的视角,带来一些新的启示,同时也让更多的人关注与了解海洋碳中和。

未来,海洋碳中和领域新的挑战和问题也将不断出现,仍需面对和解决。后续中心将深化有组织科研,每年出版《海洋碳中和前沿进展》,形成系列论文集,期望为社会各界的志同道合者提供长期学习交流平台,助力国家实现碳达峰碳中和目标,提升服务国家重大需求和区域经济社会发展的能力。敬请读者对该系列论文集批评指正。海洋碳中和事业任重道远,让我们齐心协力,相互支持。本书的出版得到了中国海洋大学科技处等职能部门的大力支持,也得到了中央高校基本科研业务费专项"碳中和交叉团队"项目(202242001)的资助。

志合者,不以山海为远。我们期望,本书作为中心成员们的智慧结晶,可以在国内乃至国际社会广泛传播,吸引国内外更多专家学者交流合作,共同携手,为应对全球气候变化贡献力量!

感谢为本书付出心血的每一位同志!

2023年5月1日

目 录 CONTENTS

第一篇　海洋碳增汇关键过程与理论创新
Part Ⅰ　Key Processes and Theoretical Innovations in Ocean Carbon Sequestration

01 A global assessment of the mixed layer in coastal sediments and implications for carbon storage ……… 003
02 Threat by marine heatwaves to adaptive large marine ecosystems in an eddy-resolving model ………… 017
03 Differential mobilization and sequestration of sedimentary black carbon in the East China Sea ……… 032
04 Persistently high efficiencies of terrestrial organic carbon burial in Chinese marginal sea sediments over the last 200 years ……………………………………………………………………………… 048

第二篇　非碳能源替代方案与关键技术
Part Ⅱ　Emission-free Energy Solutions and Key Technologies

01 Semitransparent polymer solar cell/triboelectric nanogenerator hybrid systems: Synergistic solar and raindrop energy conversion for window-integrated applications ……………………………… 069
02 Experimental investigation of a novel OWC wave energy converter …………………………………… 090
03 Numerical investigation of spudcan penetration under partially drained conditions ………………… 106
04 Comparative study on metaheuristic algorithms for optimizing wave energy converters …………… 121

第三篇　海洋碳增汇应用典型示范与推广
Part Ⅲ　Typical Demonstrations and Promotion of Implemented Ocean Carbon Sequestration

01 Optimization of aquaculture sustainability through ecological intensification in China ……………… 153
02 Weakened fertilization impact of anthropogenic aerosols on marine phytoplankton— a comparative analysis of dust and haze particles ……………………………………………………… 168
03 Effect of anthropogenic aerosol addition on phytoplankton growth in coastal waters: Role of enhanced phosphorus bioavailability ……………………………………………………………………… 185
04 A low-cost in-situ CO_2 sensor based on a membrane and NDIR for long-term measurement in seawater …………………………………………………………………………………………………… 202

第四篇　海洋碳交易市场标准与气候变化评估

Part Ⅳ　Standardization of Ocean Carbon Trading and Climate Change Assessment

- ① Examining the social pressures on voluntary CSR reporting: The roles of interlocking directors ········ 219
- ② Carbon emission intensity and biased technical change in China's different regions: A novel multidimensional decomposition approach ··· 245
- ③ Green finance and high-quality development of marine economy ································ 263
- ④ Global trends and prospects of blue carbon sinks: A bibliometric analysis ···················· 277

主编简介 ··· 295
主要作者简介 ··· 297

第一篇
海洋碳增汇关键过程与理论创新

Part I
Key Processes and Theoretical Innovations in Ocean Carbon Sequestration

A global assessment of the mixed layer in coastal sediments and implications for carbon storage[①]

Shasha Song[1,2,3], Isaac R. Santos[4,5], Huaming Yu[6,7], Faming Wang[8], William C. Burnett[9], Thomas S. Bianchi[10], Junyu Dong[11], Ergang Lian[12], Bin Zhao[1,2], Lawrence Mayer[13], Qingzhen Yao[1,2], Zhigang Yu[1,2], Bochao Xu[1,2*]

1 Frontiers Science Center for Deep Ocean Multispheres and Earth System, Key Laboratory of Marine Chemistry Theory and Technology, Ministry of Education, Ocean University of China, 266100 Qingdao, China
2 Laboratory for Marine Ecology and Environmental Science, Qingdao National Laboratory for Marine Science and Technology, 266100 Qingdao, China
3 College of Chemistry and Chemical Engineering, Ocean University of China, 266100 Qingdao, China
4 Department of Marine Sciences, University of Gothenburg, Gothenburg, Sweden
5 National Marine Science Centre, School of Environment, Science and Engineering, Southern Cross University, Coffs Harbour, 2450 NSW, Australia
6 College of Oceanic and Atmospheric Sciences, Ocean University of China, 266100 Qingdao, China
7 Sanya Oceanographic Institution, Ocean University of China, 572000 Sanya, China
8 Xiaoliang Research Station for Tropical Coastal Ecosystems, Key Laboratory of Vegetation Restoration and Management of Degraded Ecosystems, and the CAS engineering Laboratory for Ecological Restoration of Island and Coastal Ecosystems, South China Botanical Garden, Chinese Academy of Sciences, 510650 Guangzhou, China
9 Department of Earth, Ocean, and Atmospheric Science, Florida State University, Tallahassee, 32306 Florida, USA
10 Department of Geological Sciences, University of Florida, Gainesville, 32611-2120 Florida, USA
11 School of Computer Science and Technology, Ocean University of China, 266100 Qingdao, China
12 State Key Laboratory of Marine Geology, Tongji University, 200092 Shanghai, China
13 School of Marine Sciences, University of Maine, Walpole, Maine 04573, USA
* Corresponding Author: Bochao Xu (xubc@ouc.edu.cn)

Abstract

The sediment-water interface in the coastal ocean is a highly dynamic zone controlling biogeochemical fluxes of greenhouse gases, nutrients, and metals. Processes in the sediment mixed layer (SML) control the transfer and reactivity of both particulate and dissolved matter in coastal interfaces. Here we map the global distribution of the coastal SML based on excess ^{210}Pb (^{210}Pb$_{ex}$) profiles and then use a neural network model to upscale these observations. We show that highly dynamic regions such as large estuaries have thicker SMLs than most oceanic sediments. Organic carbon preservation and SMLs are inversely related as mixing stimulates oxidation in sediments which enhances organic matter decomposition. Sites with SML thickness > 60 cm usually have lower organic carbon accumulation rates (<50 g·m^{-2}·a^{-1}) and total organic carbon/specific surface area ratios (<0.4 mg·m^{-2}). Our global scale observations reveal that reworking can accelerate organic matter degradation and reduce carbon storage in coastal sediments.

---

① 本文于2022年8月发表在 Nature Communications 第13卷，https://doi.org/10.1038/s41467-022-32650-0。

Introduction

Coastal sediments record detailed historical changes of land-use and climate, which can impact source-to-sink particle dynamics across the land-ocean boundary[1]. These sediment records can be altered by physical and/or biological mixing which can modify sedimentary structures and obscure record interpretations[2]. Constraining the thickness and location of the sediment mixed layer (SML) is essential for resolving key pathways in marine biogeochemical cycles[3,4]. For instance, the SML is an important driver of the exchange of nutrients, organic carbon, redox-sensitive elements and greenhouse gases between the seafloor and the overlying seawater[5,6].

Quantifying the thickness of the SML is challenging, particularly in highly dynamic coastal sediments. Atmospherically derived ^{210}Pb (half-life of 22.3 years) is a natural "tracer" for sediments accumulating over time. Pioneering work using ^{210}Pb for dating marine sediments[7], established chronologies in estuarine and coastal sediments[8-10]. In rapidly changing coastal environments, sediments are commonly remobilized as stationary fluid muds, and/or resuspended and laterally transported as dense suspensions[11,12]. Particle reworking and remobilization can result in loss of chronological information[13-15] and is highly heterogeneous spatially and temporally. Thus, the distribution and thickness of the SML are important to improve global flux estimates of dissolved and particulate constituents of key biogeochemical cycles in Earth System Models[16].

Neural networks have emerged as a powerful tool to resolve complex spatial and temporal patterns that are common in large datasets in the geosciences[17,18]. Some examples of neural networks applications in the earth sciences include: hydrology, including flooding forecasts, and water quality modeling[19,20]; geophysics/geomorphology, including earthquake predictions, and simulating land-use change[21,22]; and atmospheric sciences, by modeling cloud formation and temperatures[23,24]. Recently, machine learning (e.g., K-nearest neighbor, random forest and neural networks) has been applied to resolve important questions in oceanography, including the global distribution of seafloor total organic carbon, benthic properties, sediment porosity/density and sediment accumulation rates[25-30].

While the SML has been assessed on local and regional scales[12,15], global-scale datasets remain sparse. A tracer-identified surface mixed-depth global mean value of 9.8 cm ± 4.5 cm was obtained in the 1990's[31,32] and later updated to 5.75 cm ± 5.67 cm[33]. Nevertheless, there remains a significant gap in our knowledge of the distribution patterns of SMLs in the global coastal ocean. Hence, we posit that a more accurate estimate of SMLs is needed for better incorporation into global biogeochemical ocean models.

Here, we define the SML as the sediment thickness captured by ^{210}Pb$_{ex}$ profiles that has been reworked over time scales of months[34,35]. We estimated SML thicknesses in the global coastal ocean by compiling data from 742 globally distributed sediment cores. First, we evaluated the spatial patterns and drivers of sediment mixing. We then used a neural network model to estimate the thickness of SMLs in the entire global ocean shallower than 200 m, linked it to organic carbon burial in marine sediments. This global assessment of coastal SMLs improves our ability to estimate and interpret the burial capacity and remineralization of organic matter and nutrients in ocean sediments.

Results and Discussion

▶ *Estimating the global coastal sediment mixed layer*

Profiles of excess ^{210}Pb (^{210}Pb$_{ex}$) in coastal shelf and estuarine sediments, devoid of sediment mixing, typically reflect exponential decay with depth[1]. Vertical ^{210}Pb$_{ex}$ profiles have previously been divided into eight types to cover multiple mixing possibilities[12]. Here, we simplified this classification into five common types (Fig. 1a). A Type Ⅰ profile is produced by constant sediment accumulation

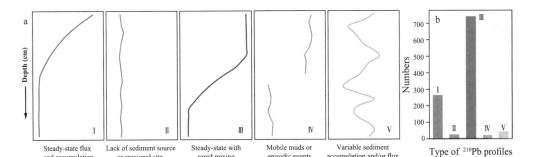

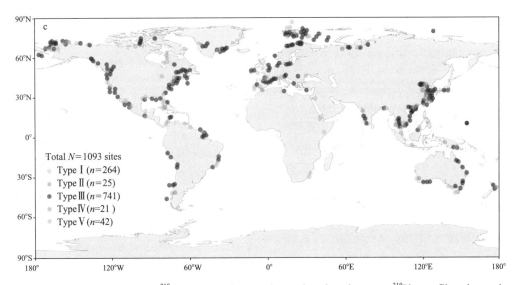

a. Sketches of five sedimentary $^{210}Pb_{ex}$ patterns in coastal areas based on downcore $^{210}Pb_{ex}$ profiles; the x-axis represents $^{210}Pb_{ex}$ activities. The frequency b. and global distribution c. of five types of $^{210}Pb_{ex}$ profiles in the global coastal ocean is based on data from 1093 cores collected from the literature (Supplementary Data 1).

Fig. 1　Profiles of excess ^{210}Pb ($^{210}Pb_{ex}$) in global coastal ocean

under steady-state conditions. In contrast, Type II profiles show no excess ^{210}Pb activities with depth, reflective of scoured, eroded areas with exposures of once deeply buried material[34]. These two profiles are produced by sediment deposition in non-reworked settings; the other three types of $^{210}Pb_{ex}$ profiles reflect disturbances related to sediment mixing (physical and/or biological). A Type III profile reflects constant $^{210}Pb_{ex}$ activities along the upper layers of the core overlaying an exponentially decaying trend, usually attributed to mixing from sediment resuspension or reworking very recently (months to years)[35,36]. A Type IV profile also reveals an intense reworked surficial layer during an episodic disturbance event such as a storm, leading to deep homogenous $^{210}Pb_{ex}$ activities in the upper section, overlying sediments with no excess ^{210}Pb. The Type V profile can be used as evidence of repetitive reworking and deposition in a highly reworked mud stratigraphy[37]. The criteria used here in defining the thickness of the SML is based on the upper homogenized layer of $^{210}Pb_{ex}$ in Type III and Type IV profiles, indicative of mud layers reworked on a timescale of months. Since any exclusion of other $^{210}Pb_{ex}$ profile types can result in an underestimation of mixing depths. Thus, our use of Type III and Type IV profiles represents a conservative estimate of SML thicknesses on a global scale. Furthermore, it is possible that an exponential decrease in the upper section of a Type I profile could originate from slow/deep bioturbation. This kind of reworking is indistinguishable from profiles originating from sediment accumulation and/or a combination of sediment accumulation and bioturbation.

We compiled 1093 published $^{210}Pb_{ex}$ profiles from the global continental shelf (see Methods and Supplementary Data 1). The geographic distribution of $^{210}Pb_{ex}$ profile types of all 1093 sites (Fig. 1b and 1c) revealed a dominance of Type Ⅲ, followed by Type Ⅰ. Studies that used gravity cores to define SML were not applied in our modeling (except for the Amazon Shelf), because gravity cores are well-known to create a bow wave artifact that disrupts near-surface sediments[38]. However, the significant greater thicknesses of the SML in the Amazon Estuary made it difficult to obtain a complete $^{210}Pb_{ex}$ profile when using short box-cores (<1 m). So, gravity-core data were included when analyzing SMLs in Amazon Shelf sediments. Overall, 742 cores had complete $^{210}Pb_{ex}$ profiles and recognizable SMLs to be used in our model.

The overall thickness of SMLs in the coastal ocean proved to be very heterogeneously distributed; approximately 47% of the SMLs were less than 5 cm, 36% had SMLs ranging from 5 cm to 20 cm, and only 3% had SMLs thicker than 30 cm (Fig. 2a). Thick SMLs primarily occurred in large-river delta-front estuaries (LDEs). Deep mixing could be viewed as inconsistent with the notion that these sediments are potential "recorders" of past natural and anthropogenic changes[39]. However, thick SMLs demonstrate the heterogeneity of depositional environments in LDEs, and the need for careful site selection of sampling locations for paleo-reconstruction work, which has proven to be successful in some LDEs (e.g., Mississippi River)[40,41]. These LDEs include the Yangtze (SML usually > 20 cm and reaching more than 100 cm locally, Fig. 2b)[12], Amazon (usually more than 100 cm, Fig. 2c)[42], and Ganges. SMLs in Asia are thicker, with thicknesses of 10–20 cm estimated to be 23% of the total (Supplementary Fig. 1). This finding is not surprising since Asia has the largest number of great rivers in the world[43]. In contrast, much thinner SMLs (0–5 cm, 35%) were found in North America away from large river sources. In some enclosed marginal seas such as the Gulf of Mexico (Fig. 2d), Gulf of California and the Baltic Sea, the SML thicknesses are also significant, reaching up to 40 cm. Areas with negligible SMLs are mainly found in the coastal ocean at high latitudes.

▶ *Drivers of SML*

The SML thickness is controlled by (1) environmental factors such as precipitation, temperature, and water depth; (2) physical factors including winds, tides, waves, and bottom stress; (3) biological activities including feeding and burrowing; and (4) sediment sources. We identified 12 influencing factors (see Methods) and analyzed their relationship to SMLs (Supplementary Fig. 2). The strongest correlation coefficients, in rank order, were river discharge (0.58), bottom stress (0.52), total suspended matter (0.35), primary productivity (0.20), and water depth (0.16). Correlations between SML and other potential drivers such as sediment accumulation rate, relative sea level change, tropical cyclone frequency and mean annual precipitation were non-significant. Bottom stress alone predicts a tight response of SML thickness with a range of 2 m (Supplementary Fig. 3a). Removing values with bottom stress > 1 Pa shows that primary productivity (implying food-driven bioturbation) also predicts SML thickness (Supplementary Fig. 3b), with values consistent with depths mixed by animals[32]. Thus, we infer that physical forces account for the greatest sediment disturbance in coastal oceans, but that bioturbation is a strong control in regions with little physical mixing. Shallow systems with large sediment sources (such as deltas) usually have higher SML thicknesses due to stronger hydrodynamic forcing and larger watersheds with more extensive and varied human disturbances.

We incorporated these five strongest SML drivers into our neural network model to predict global SMLs (see Methods section). Since it is well established that SMLs largely occur in fine-

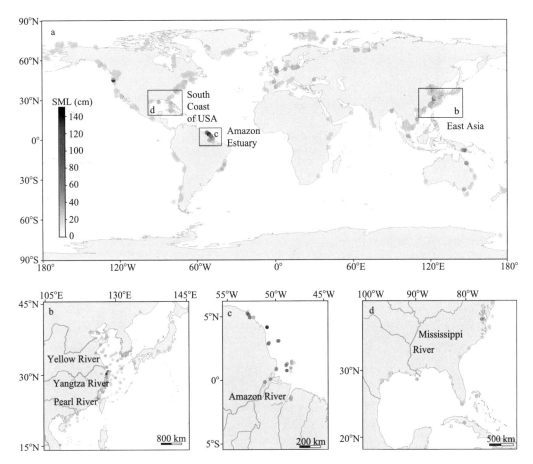

a. The global coastal ocean with 742 sites; b. The coastal ocean of East Asia; c. Amazon Estuary; d. Gulf of Mexico.

Fig. 2　The thickness of the sediment mixed layer (SML) estimated by excess ^{210}Pb (^{210}Pb$_{ex}$) profiles

grained deposits (e.g., both organic and inorganic particles)[44], the ^{210}Pb$_{ex}$ profiles used here are almost exclusively from muddy deposits.

▶ *Global upscaling with a neural network model*

Our neural network simulation of SMLs in the global coastal ocean with water depths shallower than 200 m predicts that more than 50% of the SMLs have a thickness of 0–5 cm, with only 3% thicker than 30 cm (Fig. 3a). The maximum SML thickness of nearly 200 cm was found in the Amazon Delta (Fig. 3b) due to the abundant river particulate sources and high bottom stress. Other areas with thick SMLs include the inner shelf of the Gulf of Mexico (Fig. 3c), and the East China Sea (Fig. 3d).

Areas such as the northwest Australia, the Fly River Estuary, the Bay of Bengal, and the Hudson Bay had SML thicknesses of ca. 30 cm. In high latitude areas such as the Arctic, southern regions of Africa, northern America, and southern Australia, SMLs approached 0 cm. Overall, these model simulation results matched well with empirical observations (Fig. 2) with a correlation coefficient of 0.73 (Supplementary Fig. 4).

Based on a probability distribution of modeled average SML thicknesses (Supplementary Fig. 5), the average SML thickness in fine-grained sediments of the global coastal ocean is estimated to be in the range of 7.0–10.0 cm. Most modeled results occur

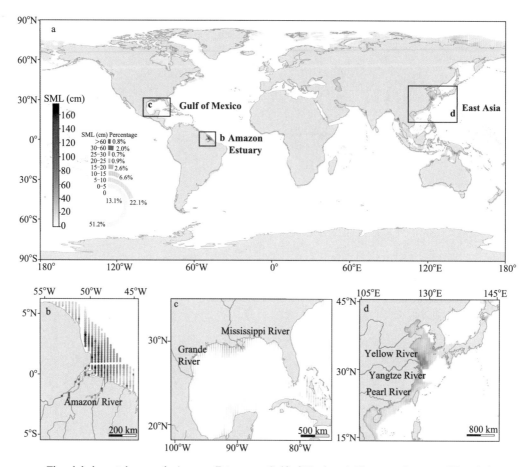

a. The global coastal ocean; b. Amazon Estuary; c. Gulf of Mexico; d. The coastal ocean of East Asia.

Fig. 3   Simulation of sediment mixed layer (SML) thicknesses

between 8.0 cm and 9.5 cm, with a mean value of 8.5 cm ± 0.6 cm, which is between previous estimates of 5.7 cm[33] and 9.8 cm[32]. The timescale of SML formation varies depending upon the environmental conditions. Individual particles within the SML, both organic and inorganic, will range in age depending upon the sediment accumulation rate and intensity of physical and biological mixing[45]. For example, mudflats adjacent to the Amazon Estuary have deposition rates reaching 1 cm per day[46]. Similarly, continental shelf sediments near the mouth of the Yangtze River, have short-term deposition rates of 4.4 cm per month[47]. In the northern Gulf of Mexico, near the Mississippi/Atchafalaya LDE complex, SMLs with thicknesses of 10–20 cm, were formed within one year[48]. Many reported SMLs deposit over seasonal to inter-annual timescales[49].

Marine sediments host significant levels of biogenic materials that drive biogeochemical exchange, carbon storage and regulation of greenhouse gases[50]. Previous studies addressed mixing intensity and depth in marine muds dominated by bioturbation[33,51,52]. We focus here on coastal ocean sediments, where mixing is strongly affected by physical (e.g., waves and/or currents set up by rivers, storms), biological (e.g., bioturbation) and human (e.g., trawling) driving forces. The presence of hypoxia/anoxia will limit the presence of infauna which, in turn, will limit the thickness of SMLs driven by biological factors. For example, in cases where bottom waters are anoxic, such as the basins off California, SMLs are absent[8]. Hypoxia/anoxia is increasing in many areas around the world such as in the northern Gulf of Mexico[53], where SML thicknesses are likely to decline with smaller, more opportunistic benthic species dominating in sediments[54]. Our neural network model allows for a more comprehensive and quantitative understanding of reworked muds, induced by physical, biological, and anthropogenic factors, in the coastal ocean.

▶ *Implications for carbon storage*

As anthropogenic climate change modifies the global ocean, the importance of CO_2 sequestration and carbon storage in sediments has received considerable attention[55]. Coastal margins including mangroves and saltmarshes cover approximately 16% of the global seabed area but account for > 90% of total ocean OC burial[56], thereby playing a central role in the global carbon cycle. Reworked muds, including mobile muds, resuspended sediments, and bio-mixed muds, are subject to long-term hydrodynamic sorting, which has significant impacts on OC transport, degradation, and deposition on ocean margins, as well as interpretations of related climate records[57].

Continuous resuspension and redeposition drive periodic oxidation-reduction cycles that accelerate the degradation of organic matter, making the mixed zones effective "incinerators" of organic matter[58]. Enhanced remineralization of sedimentary OC can reduce OC accumulation rates (OCAR)[59]; in contrast, to more quiescent sedimentary environments, that typically favor OC preservation[6]. Total organic carbon/specific surface area ratios (TOC/SSA), commonly used as an indicator of OC preservation that normalizes grain size effects, are usually lower in mobile mud deposits and higher in high productivity/upwelling regions[60,61].

Sediment mixing, as reflected by thicknesses of SMLs, appears to strongly impact organic carbon (OC) preservation in the coastal ocean. The distribution patterns of OCAR (Fig. 4a) and TOC/SSA ratio (Fig. 4b) indicate that OC preservation in different geomorphic settings is highly related to SML thickness. The global coastal margins can be divided into four major morphotypes: narrow-shallow type (SN), deep-glaciated type (DG), wide-flat type (WF), and shelves having intermediate values to the other three morphotypes (IM)[62]. The average thicknesses of SMLs, in these four morphotypes, are 5.8, 3.6, 6.5 and 9.9 cm, respectively. Intense sediment reworking is mostly found in IM coastal-shelf settings, with more than 22% of SML deeper than 10 cm and about 2% higher than 60 cm (Fig. 4c). Most of the largest, river-dominated coastal margins (e.g., Amazon, Mississippi, and Ganges) are categorized as the IM type (Fig. 4a). An inverse relationship between OCAR and SML thickness (Fig. 4d) implies that OCAR decreases significantly when the SML thickness is thicker than 10 cm.

For regions with OCAR greater than 400 $g \cdot m^{-2} \cdot a^{-1}$, almost all SMLs are thinner than 10 cm. For example, the mangrove-dominated Indonesian coast has OCAR = 1722 $g \cdot m^{-2} \cdot a^{-1}$ with an SML of less than 10 cm (Fig. 4a). In contrast, the SML in the Amazon Shelf is thicker than 60 cm and OCAR is less than 50 $g \cdot m^{-2} \cdot a^{-1}$. In the mobile mud belt of the Amazon Estuary, more than 50% of the OC input from the river is oxidized and decomposed, and only 13%–17% of the overall OC input is stored in the seabed[14]. Frequent resuspension and redeposition of mobile muds enhance OC degradation by increasing exposure to powerful oxidants such as oxygen[14], as well as labile organic matter that acts as primers[47]. Mobile muds are thus important sites for remineralization of OC. Similarly, high TOC/SSA ratios are found at sites with thin SML (e.g., > 1.5 $mg \cdot m^{-2}$ when SMLs < 10 cm), and lower TOC/SSA ratios in areas with thick SMLs (e.g., < 0.4 $mg \cdot m^{-2}$ when SMLs > 60 cm) (Fig. 4e).

In the other three coastal morphotypes, less than 10% of SMLs have thicknesses > 10 cm (Fig. 4c). In the DG coastal-shelf settings, > 30% of the sites have negligible SMLs. At high latitudes, deglaciation is changing coastal sedimentary dynamics, particularly in fjords[63], resulting in local changes in the density of deposits and sources of

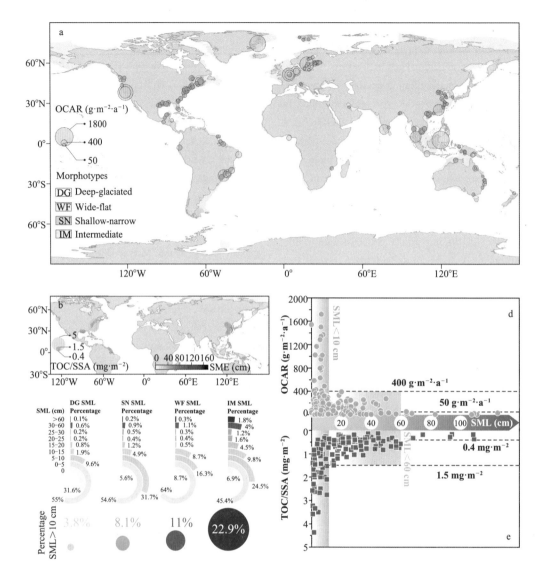

a. Distribution of organic carbon accumulation rates (OCAR) in the global coastal ocean (reproduced with permission of "Harris, P. T. & Macmillan-Lawler, M. Global Overview of Continental Shelf Geomorphology Based on the SRTM30_PLUS 30-Arc Second Database. In: Finkl, C., Makowski, C. (eds) Seafloor Mapping along Continental Shelves. 169-190 (2016)", copyright of © 2016 Springer International Publishing Switzerland); b. Distribution of the total organic carbon/specific surface area (TOC/SSA) ratios in global coastal sediments; c. Sediment mixed layer (SML) in different shelf morphotypes; d. Plot of OCAR versus SML thickness in global coastal ocean sediments; e. Plot of OCAR and TOC/SSA ratios versus SML thickness in global coastal ocean sediments.

Fig. 4 Impacts of sediment mixed layer (SML) on organic carbon storage

OC[64]. Weak sediment reworking is associated with OC preservation in polar continental margins. For example, coastal sediments in north-east Greenland, characterized by sediments with shallow SMLs had OCAR at 1540 g·m^{-2}·a^{-1} (Fig. 4a)[65]. Therefore, identifying the thickness and distribution of SMLs in high latitude systems needs further attention, as these regions experience dramatic alterations in the sedimentary, cryospheric and hydrologic cycles (https://www.ipcc.ch/report/sixth-assessment-report-working-group-i/).

Narrow shelf (SN) coastal oceans minimize

mineralization of carbon along the transmission path[66]. Interestingly, reported TOC/SSA ratios are all > 0.4 mg·m^{-2} (Fig. 4b), consistent with more rapid transport and reduced OC decay in such geomorphic settings. Wider shelves (WF) with widths up to 380 km largely occur in high latitude regions such as the Arctic, the Siberian Shelf and Chukchi Sea (Fig. 3a) and had very thin SMLs (< 5 cm). Lower latitude WF settings (e.g., Yellow Sea and the inner shelf muddy area of the East China Sea[60]) have thicker SMLs. Sedimentary OC in these mobile-mud regions experience frequent sediment reworking with rapid Fe redox cycling and long-distance transport, resulting in low OC burial (Fig. 4a and 4b) and OC preservation efficiency (about 30%)[67-69].

The estimated thicknesses of SMLs here are based on a decadal time scale from ^{210}Pb profiles, provide a strong connection for assessing carbon storage. These correlations provide global-scale evidence that refreshed exposure of sedimentary OC to the overlying water column reduces OC accumulation. Specific mechanisms will vary among sites and will have different impacts among shelf morphotypes. The simulation presented here provides a first step establishing a more global characterization of SMLs in the current global coastal ocean. This characterization can lead to more insights about hotspots of organic matter cycling in marine sediments[56], which can more broadly support Earth System Models. Finally, as many marine macrofaunal benthos in the coastal ocean are undergoing poleward range expansion due to global warming[70], SMLs will likely undergo additional change that will also need to be included in ongoing modeling efforts.

Methods

▶ *Data source*

The Web of Science (Thomson Reuters, New York, NY), Google Scholar, and Bai Du Scholar were utilized to search the literature using the following key words or phrases: ^{210}Pb with coastal and/or estuary; and reworked muds and/or mixed layer. The data repository created by Solan et al.[71] provided an excellent base to show relationships between benthic faunal community distribution and sediment characteristics, and we also obtained additional ^{210}Pb$_{ex}$ profiles from references therein. We compiled 238 studies with 1093 sites having ^{210}Pb$_{ex}$ profiles distributed globally. Data source references are presented in the Supplementary Data 1. About 62% of the sites were located on continental shelves between 0 m and 200 m water depth, and the remaining locations were in intertidal mud/sand flats, wetlands, and the deep ocean.

Environmental factors, physical dynamics and biological activities were collected for each site and presented in Supplementary Data 1, including mean annual precipitation, water depth, total suspended matter, tidal range, relative sea level rise rate, tropical cyclone frequency, bottom stress, primary productivity, river discharge, river sediment load and sediment accumulation rate, to explore for drivers of the SML in the coastal zone.

Tidal ranges were extracted from a global tidal range dataset[72,73] and water depths from the ocean bathymetry database (https://www.ngdc.noaa.gov/mgg/global/global.html). Tropical cyclone effects and hazard risks were based on the Global Cyclone Hazard Frequency and Distribution, v1 dataset (1980–2000), and assessed on a 2.5 minutes' global grid. More than 1600 storm tracks were assembled and modeled, through the period January 1st, 1980, to December 31st, 2000, for the Atlantic, Pacific, and Indian Oceans, at UNEP/GRID-Geneva PreView (https://www.ldeo.columbia.edu/chrr/). Sediment types were collected from seafloor lithology in the GplatesPortal[74]. We added about 7800 points in areas with poor data coverage and reproduced a new digital map of seafloor lithology (Supplementary Fig. 6). We simulated SML only for sites with fine-grained sediment (silt and clay). In coastal wetlands, sediment accumulation rates were obtained from a recent global assessment of saltmarshes and mangroves[75]. Data of mean annual precipitation were acquired from world climate data, European Climate Assessment & Dataset[76]. For relative sea level rise rates, data were collected from the Permanent Service for Mean Sea Level database[77]. Total suspended matter was derived from Medium Resolution Imaging Spectrometer satellite data, processed in the framework of the GlobColour project[73]. The dataset of river discharge comes from "Dai and Trenberth Global River Flow and Continental Discharge Dataset"[78], which were interpolated according to the river discharge and distance to obtain the river discharge influence at each data point. The data of bottom currents are from TPXO global tide models[79]. The sediment load of global large rivers is from the dataset by Milliman and Katherine[80]. Satellite-observed monthly global climatology sea surface chlorophyll-a concentration and primary production with 4 km resolution were downloaded from Copernicus Marine Service (https://resources.marine.copernicus.eu/). Annual average chlorophyll-a concentration and primary production were originally calculated using these climatology data, which were then linearly interpolated for each location. Part (165 sites) of the organic carbon accumulation rates (OCAR) data for global tidal marshes and mangroves was extracted from the study of Wang et al.[75]. The remaining estimates of OCAR and TOC/SSA ratios were collected from the literature and both the actual data and relevant references are listed in the Supplementary Data 1.

▶ *The neural network model*

A supervised multilayer perceptron (MLP), the most commonly applied type of neural networks[81], was employed to assess the distribution of SML in the global coastal ocean. Briefly, a neural network is a set of neurons that can be connected and combined in one or multiple layers. The first layer, called the input layer, consists of the source data. There are then intermediate layers, called hidden layers. The resulting output, in this case the thickness of the SML, is obtained in the last layer. Two key issues in MLP design are the specification of the number of hidden layers and the number of neurons in these layers. Once the number of layers and number of neurons in each layer have been selected, the network's weights and thresholds must be set to minimize the prediction error made by the network; this task is the role of the training algorithms.

Four essential steps were used in designing the neural network: (a) collecting data, (b) preprocessing data, (c) building the network, and (d) training and test performance of the model (Supplementary Fig. 7)[82]. After data collection, correlation analysis and data normalization were used to train the neural network more efficiently. For correlation analysis, several highly correlated factors were selected, and transformed into values between 0 and 1 using data normalization. The selected data were divided into two randomly selected groups, the training group which corresponded to 70% of the patterns, and the test group, which corresponded to the remaining 30% of the patterns.

Training a MLP helps to accurately estimate desired dependent variables or outputs. To do this, multiple calculations are carried out to modify the weights of every one of the connections between neurons. The first step in this process is achieved with an activation function (equation 1), which handles all the data that enters a neuron[83].

$$x_k = \sum_{j=1} w_{jk} y_j \qquad (1)$$

where w_{jk} stands for the weight that represents the connection between layers j and k, y_j is the value of the input, which is introduced in a neuron, and x_k is the solution provided by the activation function.

The accuracy of the various predictions was evaluated using the correlation coefficient (CC), the root mean-square error (RMSE), the mean absolute error (MAE), and percent correct (PC) between the measured values and the predicted values. Here, we used MAE and R^2 score to evaluate the neural network model. There are other algorithms with very different principles that may also be suitable for predicting SMLs, such as K-nearest neighbor (KNN), Random Forest (RF) and support vector machine (SVM). We compared each algorithm based on their performance and found that the MLP was the most applicable algorithm for predicting global SML (Table S1).

▶ *Structure of the neural network*

The optimal architecture of the developed artificial neural network revealed three hidden layers of 20, 40 and 10 neurons (Supplementary Fig. 8). The neurons in the input layer were equal to the 5 predictors with the best statistical correlation results. This nonlinearity in a neural network model presents advantages and disadvantages. For example, it is difficult to determine the values of the parameters in a computationally intensive nonlinear optimization. Thus, a trial-and-error approach was used to determine a network with optimum performance. After each iteration, the network outputs were compared to the actual target values until the performance function was maximized.

Output values were compared with expected values based on the training data, and the errors computed. Through iterative propagation of errors back to the network, the connection weights were automatically adjusted until the target minimum error was attained. To achieve this, different tests were conducted, and the best learning rate and training time (epoch) obtained were constant at a level of 3000. When epochs reached 3000, the MAE decreased to 0.32 (after data normalization) and remained stable, and the R^2 was 0.475.

▶ *Global upscaling based on the neural network model*

A global map was equally divided into 1000×1000 grids using ArcGis (10.2), resulting in about 50,000 sites along the coastal zone and continental shelf to water depths of 200 m. Each grid is 20 arcmin×10 arcmin with an area of ca. 510 km^2. The operational steps in ArcGis included attributes extraction, fishnet creation, location selection, and data export. Finally, we entered the grid into the trained neural network model to predict the thickness of SML with the same grid resolution. The model prediction results are available as a downloadable file (Supplementary Data 2).

Data Availability

The data generated in this study are provided in the Supplementary Information and Supplementary Data 1 and 2, and also deposited in the Zenodo online repository at https://doi.org/10.5281/zenodo.6901752.

Code Availability

The code used in this study are available at https://doi.org/10.5281/zenodo.6890194.

Acknowledgements

We thank to Wenyuan Li for her assistance during the neural network building. We thank to Dr. Xiaoyong Duan, Limin Hu and Guangquan Chen for providing sediment information. Dr. Peter T. Harris and Miles Macmillan-Lawler are acknowledged for providing the global morphotypes data. This work was supported by Natural Science Foundation of China (NSFC grant 42130410 to Zhigang Yu, 41876075 to Bochao Xu, and U1906210 to Qingzhen Yao).

Author Contributions

Shasha Song, Bochao Xu and Zhigang Yu designed the research. Shasha Song collected the ^{210}Pb data in global coastal ocean and wrote the initial draft, with later input from all authors. Isaac R. Santos, William C. Burnett, and Thomas S. Bianchi helped with the further language editing and designing the manuscript. Huaming Yu collected the data of bottom stress, river discharge and primary productivity of all the sites and helped with the model simulation. Faming Wang collected the climate and environmental factors of all sites. Junyu Dong helped with the neural network. Ergang Lian helped with the figures. Bin Zhao and Lawrence Mayer collected the data of OC burial and supported writing of the related section. Bochao Xu, Qingzhen Yao and Zhigang Yu secured funding for the study.

Competing Interests

The authors declare no competing interests.

References

1. Nittrouer C A, Sternberg R W. The formation of sedimentary strata in an allochthonous shelf environment: The Washington continental shelf[J]. Mar Geol, 1981, 42: 201-232.
2. Aller R C. Mobile deltaic and continental shelf muds as suboxic, fluidized bed reactors[J]. Mar Chem,1998, 61: 143-155.
3. Brainerd K E, Gregg M C. Surface mixed and mixing layer depths[J]. Deep-Sea Res, 1995, 42: 1521-1543.
4. Aller R C, Cochran J K. The critical role of bioturbation for particle dynamics, priming potential, and organic C remineralization in marine sediments: Local and basin scales[J]. Front Earth Sci, 2019, 7: 157.
5. Severmann S, McManus J, Berelson W M, et al. The continental shelf benthic iron flux and its isotope composition[J]. Geochim Cosmochim Acta, 2010, 74: 3984-4004.
6. Zhao B, Yao P, Bianchi T S, et al. Controls on organic carbon burial in the eastern China marginal Seas: A regional synthesis[J]. Glob Biogeochem Cycles, 2021, 35: e2020GB006608.
7. Koide M, Bruland K W, Goldberg E D. ^{228}Th/^{232}Th and ^{210}Pb geochronologies in marine and lake sediments[J]. Geochim Cosmochim Acta, 1973, 37: 1171-1187.
8. Bruland K W, Bertine K, Koide M, et al. History of metal pollution in Southern California coastal zone[J]. Environ Sci Technol, 1974, 8: 425-432.
9. Smith J N, Walton A. Sediment accumulation rates and geochronologies measured in the Saguenay Fjord using the ^{210}Pb dating method[J]. Geochim Cosmochim Acta, 1979, 44: 225-240.
10. Sanchez-Cabeza J A, Ruiz-Fernández A C. ^{210}Pb sediment radiochronology: An integrated formulation and classification of dating models[J]. Geochim Cosmochim Acta, 2012, 82: 183-200.
11. Corbett D R, McKee B, Duncan D. An evaluation of mobile mud dynamics in the Mississippi River deltaic region[J]. Mar Geol, 2004, 209: 91-112.
12. Xu B C, Bianchi T S, Allison M A, et al. Using multi-radiotracer techniques to better understand sedimentary dynamics of reworked muds in the Changjiang River estuary and inner shelf of East China Sea[J]. Mar Geol, 2015, 370: 76-86.
13. DeMaster D J, Kuehl S A, Nittrouer C A. Effects of suspended sediments on geochemical processes near the mouth of the Amazon River: Examination of biological silica uptake and the fate of particle-reactive elements[J]. Cont Shelf Res, 1986, 6: 107-125.
14. Aller R C, Blair N E. Carbon remineralization in the Amazon–Guianas tropical mobile mudbelt: A sedimentary incinerator[J]. Cont Shelf Res, 2006, 26: 2241-2259.
15. Aller R C, Madrid V, Chistoserdov A, et al. Unsteady diagenetic processes and sulfur biogeochemistry in tropical deltaic muds: Implications for oceanic isotope cycles and the sedimentary record[J]. Geochim Cosmochim Acta, 2010, 74: 4671-4692.
16. Hedges J I. Global biogeochemical cycles: Progress and problems[J]. Mar Chem, 1992, 39: 67-93.
17. Bishop C M. New neural network for pattern recognition[M]. Oxford University Press, 1996.
18. Reichstein M, Camps-Valls G, Stevens B, et al. Deep learning and process understanding for data-driven Earth system science[J]. Nature, 2019, 566: 195-204.
19. Hong H, Panahi M, Shirzadi A, et al. Flood susceptibility assessment in Hengfeng area coupling adaptive neuro-fuzzy inference system with genetic algorithm and differential evolution[J]. Sci Total Environ, 2018, 621: 1124-1141.
20. Guillod B P, Orlowsky B, Miralles D G, et al. Reconciling spatial and temporal soil moisture effects on afternoon rainfall[J]. Nat Commun, 2015, 6: 6443.
21. Pham B T, Tien Bui D, Pourghasemi H R, et al. Landslide susceptibility assessment in the Uttarakhand area (India) using GIS: A comparison study of prediction capability of naïve bayes, multilayer perceptron neural networks, and functional trees methods[J]. Theor Appl Climatol, 2015, 128: 255-273.
22. Seydoux L, Balestriero R, Poli P, et al. Clustering earthquake signals and background noises in continuous seismic data with unsupervised deep learning[J]. Nat Commun, 2020, 11: 3972.
23. Lops Y, Pouyaei A, Choi Y, et al. Application of a partial convolutional neural network for estimating geostationary aerosol optical depth data[J]. Geophys Res Lett, 2021, 48: e2021GL093096.
24. Madakumbura G D, Thackeray C W, Norris J, et al. Anthropogenic influence on extreme precipitation over global land areas seen in multiple observational datasets[J]. Nat Commun, 2021, 12: 3944.
25. Lee T R, Wood W T, Phrampus B J. A machine learning (kNN) approach to predicting global seafloor total organic carbon[J]. Glob Biogeochem Cycles, 2019, 33: 37-46.
26. Restreppo G A, Wood W T, Graw J H, et al. A machine-learning derived model of seafloor sediment accumulation[J]. Mar Geol, 2021, 440: 106577.
27. Restreppo G A, Wood W T, Phrampus B J. Oceanic sediment accumulation rates predicted via machine learning algorithm: Towards sediment characterization

on a global scale[J]. Geo-Mar Lett, 2020, 40: 755-763.
28. Graw J H, Wood W T, Phrampus B J. Predicting global marine sediment density using the random forest regressor machine learning algorithm[J]. J Geophys Res Solid Earth, 2021, 126: e2020JB020135.
29. Phrampus B J, Lee T R, Wood W T. A global probabilistic prediction of cold seeps and associated seafloor fluid expulsion Anomalies (SEAFLEAs) [J]. Geochem, Geophys, Geosyst, 2020, 21: e2019GC008747.
30. Lee T R, Phrampus B J, Obelcz J, et al. Global marine isochore estimates using machine learning[J]. Geophys Res Lett, 2020, 47: e2020GL088726.
31. Boudreau B P. Is burial velocity a master parameter for bioturbation? [J]. Geochim Cosmochim Acta, 1994, 58: 1243-1249.
32. Boudreau B P. Mean mixed depth of sediments: The wherefore and the why[J]. Limnol Oceanogr, 1998, 43: 524-526.
33. Teal L R, Bulling M T, Parker E R, et al. Global patterns of bioturbation intensity and mixed depth of marine soft sediments[J]. Aquat Biol, 2008, 2: 207-218.
34. Sommerfield C K, Nittrouer C A, Figueiredo A G. Stratigraphic evidence of changes in Amazon shelf sedimentation during the late Holocene[J]. Mar Geol, 1995, 125: 351-371.
35. Jankowska E, Michel L N, Zaborska A, et al. Sediment carbon sink in low-density temperate eelgrass meadows (Baltic Sea) [J]. J Geophys Res Biogeosci, 2016, 121: 2918-2934.
36. Smoak J M, Patchineelam S R. Sediment mixing and accumulation in a mangrove ecosystem: Evidence from ^{210}Pb, ^{234}Th and ^{7}Be[J]. Mangroves & Salt Marshes, 1999, 3: 17-27.
37. Alongi D M, Wattayakorn G, Pfitzner J, et al. Organic carbon accumulation and metabolic pathways in sediments of mangrove forests in southern Thailand[J]. Mar Geol, 2001, 15: 85-103.
38. Nevissi A E, Shott G J, Crecelius E A. Comparison of two gravity coring devices for sedimentation rate measurement by ^{210}Pb dating techniques[J]. Hydrobiologia, 1989, 179: 261-269.
39. Bianchi T S, Allison M A. Large-river delta-front estuaries as natural "recorders" of global environmental change[J]. PNAS, 2009, 106: 8085-8092.
40. McKee B A, Aller R C, Allison M A, et al. Transport and transformation of dissolved and particulate materials on continental margins influenced by major rivers: Benthic boundary layer and seabed processes[J]. Cont Shelf Res, 2004, 24: 899-926.
41. Sampere T P, Bianchi T S, Allison M A. Historical changes in terrestrially-derived organic carbon inputs to Louisiana continental margin sediments over the past 150 years[J]. J Geophys Res Biogeosci, 2011, 116: G01016.
42. Aller R C, Heilbrun C, Panzeca C, et al. Coupling between sedimentary dynamics, early diagenetic processes, and biogeochemical cycling in the Amazon-Guianas mobile mud belt: Coastal French Guiana[J]. Mar Geol, 2004, 208: 331-360.
43. Zhao C H, Zhu Z H, Zhou D Z. World rivers and dams[M]. Beijing: China Water Power Press, 2000.
44. Swarzenski P W. ^{210}Pb Dating[M]//Rink W J, Thompson J. Encyclopedia of Scientific Dating Methods. Dordrecht, Netherlands: Springer Netherlands, 2014: 1-11.
45. Meile C, Van Capellen P. Particle age distributions and O_2 exposure times: Timescales in bioturbated sediments[J]. Glob Biogeochem Cycles, 2005, 19: GB3013.
46. Allison M A, Nittrouer C A, Faria Jr. L E C. Rates and mechanisms of shoreface progradation and retreat downdrift of the Amazon river mouth[J]. Mar Geol, 1995, 125: 373-392.
47. Mckee B A, Nittrouer C A, Demaster D J. Concepts of sediment deposition and accumulation applied to the continental shelf near the mouth of the Yangtze River[J]. Geology, 1983, 11: 1354-1360.
48. Allison M A, Sheremet A, Goni A, et al. Storm layer deposition on the Mississippi–Atchafalaya subaqueous delta generated by Hurricane Lili in 2002[J]. Cont Shelf Res, 2005, 25: 2213-2232.
49. Allison M A, Kineke G C, Gordon E S, et al. Development and reworking of a seasonal flood deposit on the inner continental shelf off the Atchafalaya River[J]. Cont Shelf Res, 2000, 20: 2267-2294.
50. Solan M, Cardinale B J, Downing A L, et al. Extinction and ecosystem function in the marine benthos[J]. Science, 2004, 306: 1177-1180.
51. Emery K O. Some surface features of marine sediments made by animals[J]. J Sediment Petrology, 1956, 23: 202-204.
52. Boudreau B P. What controls the mixed-layer depth in deep-sea sediments? The importance of particulate organic carbon flux[J]. Limnol Oceanogr, 2004, 49: 620-624.
53. Turner R E, Rabalais N N. Linking landscape and water quality in the Mississippi River Basin for 200 years[J]. BioScience, 2003, 53: 563-571.
54. Middelburg J J, Levin L A. Coastal hypoxia and sediment biogeochemistry[J]. Biogeosciences, 2009, 6: 1273-1293.

55. Bauer J, Cai W J, Raymend P A, et al. The changing carbon cycle of the coastal ocean[J]. Nature, 2013, 504: 61-70.
56. Bianchi T S, Cui X, Blair N E, et al. Centers of organic carbon burial and oxidation at the land-ocean interface[J]. Org Geochem, 2018, 115: 138-155.
57. Magill C R, Ausín B, Wenk P, et al. Transient hydrodynamic effects influence organic carbon signatures in marine sediments[J]. Nat Commun, 2018, 9: 4690.
58. Aller R C, Blair N E, Brunskill G J. Early diagenetic cycling, incineration, and burial of sedimentary organic carbon in the central Gulf of Papua (Papua New Guinea)[J]. J Geophys Res, 2008, 113: F01S09.
59. Bartoli M, Nizzoli D, Zilius M, et al. Denitrification, nitrogen uptake, and organic matter quality undergo different seasonality in sandy and muddy sediments of a turbid estuary[J]. Front Microbiol, 2020, 11: 612700.
60. Mayer L M. Relationships between mineral surfaces and organic-carbon concentrations in soils and sediments[J]. Chem Geol, 1994, 114: 347-363.
61. Mayer L M. Surface-area control of organic-carbon accumulation in continental-shelf sediments[J]. Geochim Cosmochim Acta, 1994, 58: 1271-1284.
62. Harris P T, Macmillan-Lawler M. Global overview of continental shelf geomorphology based on the SRTM30_PLUS 30-Arc Second Database[M]// Seafloor Mapping along Continental Shelves. Finkl C, Makowski C. 2016: 169-190.
63. Bianchi T S, Arndt S, Austin W, et al. Fjords as aquatic critical zones (ACZs)[J]. Earth-Sci Rev, 2020, 203: 103145.
64. Cui X, Bianchi T S, Savage C. Erosion of modern terrestrial organic matter as a major component of sediments in fjords[J]. Geophys Res Lett, 2017, 44: 1457-1465.
65. Glud R N, Risgaard-Petersen N, Thamdrup B, et al. Benthic carbon mineralization in a high-Arctic sound (Young Sound, NE Greenland)[J]. Mar Ecol: Prog Ser, 2000, 206: 59-71.
66. Leithold E L, Blair N E, Wegmann K W. Source-to-sink sedimentary systems and global carbon burial: A river runs through it[J]. Earth-Sci Rev, 2016, 153: 30-42.
67. Blair N E, Aller R C. The fate of terrestrial organic carbon in the marine environment[J]. Ann Rev Mar Sci, 2012, 4: 401-423.
68. Zhao B, Yao P, Bianchi T S, et al. The role of reactive iron in the preservation of terrestrial organic carbon in estuarine sediments[J]. J Geophys Res Biogeosci, 2018, 123: 3556-3569.
69. Middelburg J J. Reviews and syntheses: To the bottom of carbon processing at the seafloor[J]. Biogeosciences, 2018, 15: 413-427.
70. Bianchi T S, Aller R C, Atwood T B, et al. What global biogeochemical consequences will marine animal-sediment interactions have during climate change?[J]. Elem Sci Anth, 2021, 9: 001880.
71. Solan M, Ward E R, White E L, et al. Worldwide measurements of bioturbation intensity, ventilation rate, and the mixing depth of marine sediments[J]. Sci Data, 2019, 6: 58.
72. Pickering M D, Horsburgh K J, Blundell J R, et al. The impact of future sea-level rise on the global tides[J]. Cont Shelf Res, 2017, 142: 50-68.
73. Schuerch M, Spencer T, Temmerman S, et al. Future response of global coastal wetlands to sea-level rise[J]. Nature, 2018, 561: 231-234.
74. Dutkiewicz A, Müller R D, O'Callaghan S, et al. Census of seafloor sediments in the world's ocean[J]. Geology, 2015, 43: 795-798.
75. Wang F M, Sanders C J, Santos I R, et al. Global blue carbon accumulation in tidal wetlands increases with climate change[J]. Natl Sci, 2020, 8: nwaa296.
76. Fick S E, Hijmans R J. WorldClim 2: New 1-km spatial resolution climate surfaces for global land areas[J]. Int J Climatol, 2017, 37: 4302-4315.
77. Holgate S J, Matthews A, Woodworth P L, et al. New data systems and products at the permanent service for mean sea level[J]. J Coast Res, 2013, 29: 493-504.
78. Dai A, Qian T, Trenberth K E, et al. Changes in continental freshwater discharge from 1948 to 2004[J]. J Climate, 2009, 22: 2773-2792.
79. Erofeeva S, Padman L, Howard S L. Tide Model Driver (TMD) version 2.5, Toolbox for Matlab[EB/OL]. https://www.github.com/EarthAndSpaceResearch/TMD_Matlab_Toolbox_v2.5, 2020.
80. Milliman J D, Katherine L F. River discharge to the coastal ocean: A global synthesis[M]. Cambridge University Press, 2013.
81. Cancilla J C, Diaz-Rodriguez P, Izquierdo J G, et al. Artificial neural networks applied to fluorescence studies for accurate determination of N-butylpyridinium chloride concentration in aqueous solution[J]. Sens Actuator B-Chem, 2014, 198: 173-179.
82. Ghorbani M A, Khatibi R, Hosseini B, et al. Relative importance of parameters affecting wind speed prediction using artificial neural networks[J]. Theor Appl Climatol, 2013, 114: 107-114.

Threat by marine heatwaves to adaptive large marine ecosystems in an eddy-resolving model[①]

Xiuwen Guo[1], Yang Gao[1,2,3*], Shaoqing Zhang[2,3,4*], Lixin Wu[3,4], Ping Chang[3,5], Wenju Cai[6,7], Jakob Zscheischler[8,9,10], L. Ruby Leung[11], Justin Small[3,12], Gokhan Danabasoglu[3,12], Luanne Thompson[13] and Huiwang Gao[1]

1 Frontiers Science Center for Deep Ocean Multispheres and Earth System, and Key Laboratory of Marine Environmental Science and Ecology, Ministry of Education, Ocean University of China, Qingdao, 266100, China

2 Laboratory for Ocean Dynamics and Climate, Qingdao National Laboratory for Marine Science and Technology, Qingdao, 266237, China

3 International Laboratory for High – Resolution Earth System Prediction (iHESP), College Station, TX USA

4 Key Laboratory of Physical Oceanography, Institute for Advanced Ocean Study, Frontiers Science Center for Deep Ocean Multispheres and Earth System (FDOMES), College of Oceanic and Atmospheric Sciences, Ocean University of China, Qingdao, 266100, China

5 Department of Oceanography, Texas A&M University, College Station, Texas, 77843, USA

6 Physical Oceanography Laboratory/CIMST, Ocean University of China and Qingdao National Laboratory for Marine Science and Technology, Qingdao, 266100, China

7 CSIRO Marine and Atmospheric Research, Aspendale, Victoria, 3195, Australia

8 Department of Computational Hydrosystems, Helmholtz Centre for Environmental Research – UFZ, 04318 Leipzig, Germany

9 Climate and Environmental Physics, University of Bern, Bern, 3012, Switzerland

10 Oeschger Center for Climate Change Research, University of Bern, Bern, 3012, Switzerland

11 Atmospheric Sciences and Global Change Division, Pacific Northwest National Laboratory, Richland, WA, 99354, USA

12 National Center for Atmospheric Research, Boulder, CO, 80305, USA

13 University of Washington, School of Oceanography, Seattle, WA, 98195, USA

* Corresponding author: Yang Gao (yanggao@ouc.edu.cn); Shaoqing Zhang (szhang@ouc.edu.cn)

Abstract

Marine heatwaves (MHWs), episodic periods of abnormally high sea surface temperature (SST), severely affect marine ecosystems. However, how global warming affects MHWs over Large Marine Ecosystems (LMEs) remains unknown, because such LMEs are confined to the coast where low-resolution climate models are known to have biases. LMEs cover about 22% of the global ocean but account for 95% of global fisheries catches. Here, using a high-resolution Earth system model that resolves ocean mesoscale eddies and simulates more realistic MHWs than low-resolution models, and applying a "future threshold" that considers MHWs as anomalous warming above the long-term mean warming of SSTs, we find that future intensity and annual days of MHWs over majority of the LMEs remains higher than in the present-day climate. Thus, the increases in MHWs under global warming poses a serious threat to coastal marine ecosystems, even if organisms in LMEs could adapt fully to the long-term mean warming.

---

① 本文于2022年2月发表在 *Nature Climate Change* 第12卷，https://doi.org/10.1038/s41558-021-01266-5。

Introduction

The ocean has warmed significantly during the past few decades in most parts of the world[1]. With continuous ocean warming, prolonged extreme ocean warming events, known as marine heatwaves (MHWs), have occurred in many parts of the global ocean in the past decades[2-4]. Severe MHWs have caused negative impacts on marine ecosystems and fisheries[5-8], and the ecological responses to MHWs have been observed across a range of processes, scales, taxa and geographic regions[9]. MHWs have broader and more devastating ecological and socio-economic consequences than the impacts of long-term slower changes in the mean warming for which species might possibly adjust through adaptation[10]. Therefore, it is vital to investigate future changes in MHWs under global warming in order to develop potential mitigation strategies to reduce the overall ecological impact of climate change[11].

Both satellite and field observations of sea surface temperature (SST) have demonstrated that over the past few decades, MHWs have become longer-lasting, more frequent and extensive, primarily attributable to the increase in the mean warming of SST[3,11-13]. Both regional[14] and global[9,15] model simulations project MHWs to intensify and their incidence to increase under a warming climate. For instance, under the fossil fuel intensive scenario of Representative Concentration Pathway (RCP) 8.5[16], the majority of ocean areas is projected to experience almost permanent heatwaves with concomitantly stronger intensity by the end of the 21st century, with MHWs defined based on the conditions of the present climate[17], referred to as a mean warming-inclusive threshold. Similarly, many previous studies define MHWs relative to the mean climate over the historical period to investigate the changing characteristics of MHWs and their potential impact on marine life both in the past and in future[9,15]. In contrast, shifting the baseline temperature for future is useful to isolate the influence of mean background warming and higher moments[18] of temperature statistics on MHWs[17], and a moving threshold is suggested to use to attribute MHW changes in the context of long term warming[19]. Therefore, estimating MHW changes using the mean warming-inclusive and future thresholds brackets scenarios are relevant to marine ecosystems with a range of capacity adapting to future warming.

Numerical models are important tools for elucidating the drivers and characteristics of MHWs, but the capability to reproduce MHWs in the historical record differs substantially among models at different resolutions. Low-resolution models, though computationally less intensive and useful for assessing the impact of climate change on MHWs at continental or global scales[17,20], do not resolve small scale physical processes including boundary currents and eddy transport processes[5] associated with MHWs[21,22]. High-resolution regional models with about 10 km ocean grid have much better fidelity in reproducing the magnitude and spatial structure of MHW events observed during the latter half of the 20th century[14,23]. In a comparison of global model simulations forced by atmospheric reanalysis at 1.0°, 0.25°, and 0.1° ocean grid spacing, the simulations at the 0.1° generally yield the realistic results in reproducing the frequency and duration of global MHWs during 1985–2017[24].

While previous studies have mostly focused on the global MHWs characteristics[3,9,17], there is an obvious increase in the frequency and duration of coastal MHWs from 1981 to 2016 based on four satellite datasets[12]; the largest impact on ecosystems is seen in LMEs found mainly in the coastal ocean[25]. The observed SST trends during the historical period in the LMEs are predominantly positive[26], and under a warming climate, the monthly SST warm extremes in LMEs over parts of the northern oceans depicted substantial increase as well using the Coupled Model Intercomparison Project Phase 5 (CMIP5) and Community Earth System Model large ensemble project (CESM-LENS)[27].

However, approximately 1° resolution of the global models may not be able to resolve the crucial processes such as ocean eddies, coastal upwelling, stressing the need of higher resolution models with daily time scale to broadly investigate the changes in ocean variables such as SST due to climate change[27]. Further, there is a built-in increase in MHWs everywhere in the future ocean climate when using a mean warming-inclusive threshold, obscuring changes in characters that are specific to LMEs. Using climate simulations from a mesoscale-eddy-resolving ultra-high resolution Earth system model[28], we show an enhanced intensity and annual days of MHWs over most of the LMEs in future even using a "future threshold" above mean warming.

Need for High-Resolution Model Simulation

Low-resolution (nominal approximately 1°) global models lack the capability of resolving small-scale processes such as boundary currents, coastal processes and ocean eddy fluxes[21] (the red circles in Fig. 1a), making them difficult to realistically simulate the characteristics of MHWs in LMEs (Fig. 1b) and their impact on marine species such as the Atlantic salmon (*Salmo salar*) [29,30]. In contrast to pelagic areas, the biodiversity of coastal areas is far more abundant[6], including many foundation species[7] as well as economically important fish[29,31,32] that are also vulnerable to MHWs. Observed impacts include coral bleaching[33,34], declining seagrass density[35], spawning reduction and distribution shift of marine fish[36,37].

With the ability to resolve small-scale processes (the red circles in Fig. 1a) and their connections to climate modes of variability (the purple and brown circles in Fig. 1a), high-resolution Community Earth System Model version 1.3 (CESM1.3) is used for simulation from 1850 to 2100 (See "Model descriptions" in "Methods"), providing valuable information for risk assessment and adaptation planning for coastal areas. The model is forced by historical forcings before 2005 and the RCP8.5[16] (a high-emission scenario) thereafter. Results from a low-resolution version of the same model and available CMIP5 models, which are also low-resolution, are used for a comparison.

Observed and Simulated MHWs in the Historical Period

The frequency of MHWs (See "Definition of MHW" in "Methods") is qualitatively similar among remotely sensed National Oceanic and Atmospheric Administration (NOAA) Optimum Interpolation Sea Surface Temperature (OISST; Fig. 2a), Group for High Resolution Sea Surface Temperature (GHRSST) Multi-Product Ensemble (GMPE; Supplementary Fig. 1a) and the modeled SST from high-(CESM-HR) and low-(CESM-LR) resolution configuration of CESM1.3[28] and the CMIP5 ensemble (Fig. 2b,c,d; See "Satellite-based observational dataset" and "Model descriptions" in "Methods"). During the historical period (1975–2004), there are one to three MHW events per year occurring over most of the globe. The obvious low frequency of MHWs in the eastern tropical Pacific is owing to El Niño-Southern Oscillation (ENSO) that can result in long period events that occur only every few years[17] (Fig. 2a). The spatial distributions of MHW frequency indicate CESM-HR (Fig. 2b) is closer to that of OISST (Fig. 2a) and GMPE (Supplementary Fig. 1a) than the low-resolution CESM-LR (Fig. 2c) and CMIP5 (Fig. 2d), as clearly delineated by the latitudinal zonal mean variations (Fig. 2e).

For instance, biases of MHW frequency in CESM-HR, CESM-LR and CMIP5 relative to OISST are –0.31, –0.55 and –0.60 (15%, 27% and 30%) times per year (Fig. 2e), respectively, although negative bias still exists across majority of the latitude bands in CESM-HR. The zonal mean of GMPE only includes the latitudes within 55° of the equator due to the diminished consensus among the multiple data sets that are included in GMPE at high latitudes[38-40].

In the following, we classify the LMEs into 6 groups (Fig. 1b and See "LMEs" in "Methods"), and show that CESM-HR (red) more closely captures the frequency of MHWs when compared to OISST/GMPE than its low-resolution counterparts: CESM-

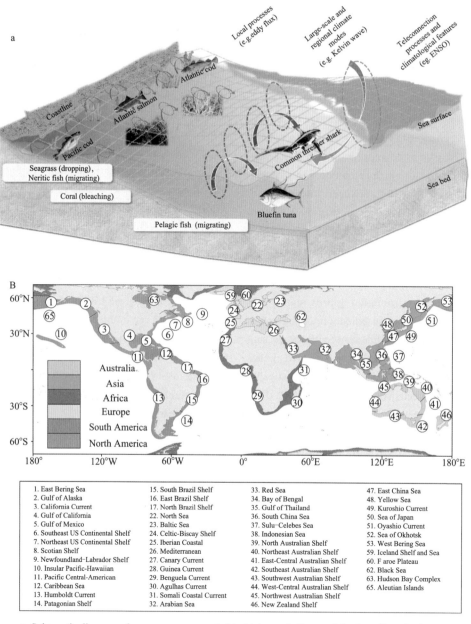

a. Schematic diagram of processes represented in high-resolution models that allow the impact on biodiversity to be evaluated. The red, purple and brown circles indicate local, regional and teleconnection processes, with arrows illustrating the interactions between these processes and the ocean environment.
b. The groups of LMEs (See "LMEs" in "Methods") by continent, including North America, South America, Europe, Africa, Asia and Australia, with a total of 54 LMEs used in this study and the numbers and names of LMEs listed at the bottom.

Fig. 1　The physical processes driving MHWs and locations of LMEs

LR (blue) and the CMIP5 ensemble (orange in Fig. 2f) ($P < 0.05$). The improved simulated global mean SST from CESM-HR relative to CESM-LR[28] is partly explained by the difference in computing eddy vertical heat transport, that is, it is explicitly computed in CESM-HR but parameterized in

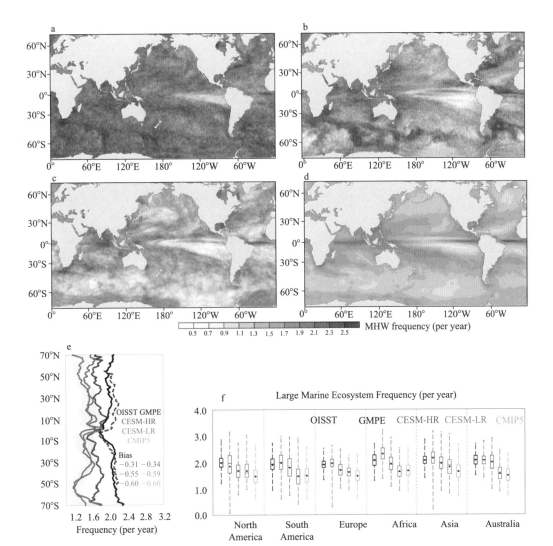

a–d. Spatial distribution of annual MHW frequency based on OISST data (a), as well as the simulation outputs in CESM-HR (b), CESM-LR (c) and CMIP5 (d). e. Zonal mean MHW frequency for OISST (solid black), GMPE (dashed gray), CESM-HR (red), CESM-LR (blue) and CMIP5 (orange). For CMIP5, the yellow shading is added to represent one standard deviation calculated based on the 20 models (Supplementary Table 1). f. The box-and-whisker plot of MHW frequency grouped by continent for the LMEs, with the minimum and maximum (line end points), 25th and 75th percentile (boxes), medians (horizontal lines), and average (black points). Note that due to data availability, the OISST and GMPE data used in this study spans from 1982 to 2011, slightly different from the historical period of 1975–2004 used in climate modeling. The overlapping period of 1982–2011 that encompasses the model simulations and OISST or GMPE was also used in model evaluation, yielding similar results. Results show that CESM-HR is more realistic in reproducing the frequency of MHWs compared to the CESM-LR and CMIP5.

Fig. 2 Observed and simulated frequency of MHWs during the historical period

CESM-LR[41,42], and the better simulated mixed layer depth by CESM-HR[28].

The spatial (Fig. 3a–d; Supplementary Fig. 1b) and zonal (Fig. 3e) mean distributions of MHW mean intensity in CESM-HR are much closer to those of OISST and GMPE compared to CESM-LR and CMIP5, consistent with the comparison over the LMEs (Fig. 3f). It is noteworthy that large

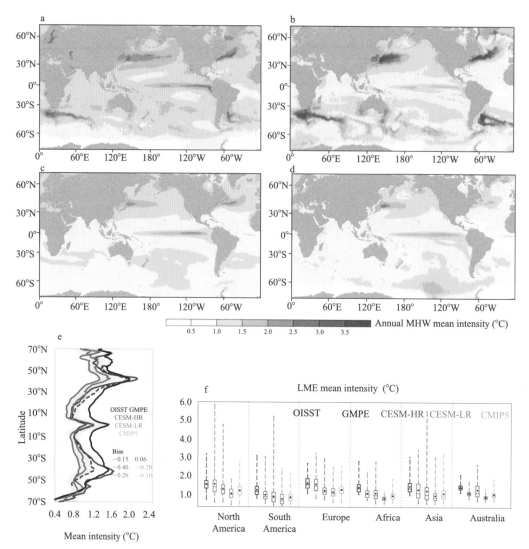

a–d. Spatial distribution of annual MHW mean intensity based on OISST data (a), as well as the simulation outputs in CESM-HR (b), CESM-LR (c) and CMIP5 (d). e. Zonal mean MHW mean intensity for OISST (solid black), GMPE (dashed gray), CESM-HR (red), CESM-LR (blue) and CMIP5 (orange). For CMIP5, the yellow shading is added to represent one standard deviation calculated based on the 20 models (Supplementary Table 1). f. The box-and-whisker plot of MHW mean intensity grouped by continent for the LMEs, with the minimum and maximum (line end points), 25th and 75th percentile (boxes), medians (horizontal lines), and average (black points). Note that due to data availability, the OISST and GMPE data used in this study spans from 1982 to 2011, slightly different from the historical period of 1975–2004 used in climate modeling. The overlapping period of 1982–2011 that encompasses the model simulations and OISST or GMPE was also used in model evaluation, yielding similar results. Results show that simulated MHW intensity by CESM-HR is in general closer to that in OISST and GMPE than simulated by CESM-LR and CMIP5.

Fig. 3 Observed and simulated mean intensity of MHWs during the historical period

differences exist in the intensity over the tropics and subtropics between OISST and GMPE, with smaller model biases when benchmarked against GMPE (Fig. 3e). The mean intensity of MHWs also exhibits large spatial variations, with higher intensity occurring in the western boundary current (WBC) regions, the eastern and central equatorial Pacific boundary current regions and the eastern boundary current regions (Fig. 3a; Supplementary Fig. 1b). The intensity over the WBC regions in CESM-LR and CMIP5 is underestimated, while in CESM-HR it is overestimated (Fig. 3b–d), leading to higher peak of the intensity in LMEs (that is, over South America and Asia in Fig. 3f) compared to OISST and GMPE. Nevertheless, the comparable maximum intensity exhibited in GMPE and CESM-HR over the LMEs in North America strongly support improvements in CESM-HR relative to CESM-LR and CMIP5.

The apparently high MHW intensity in the eddy-rich WBC regions, that is, the Scotian Shelf and Patagonian Shelf close to the North and South America, respectively, has been reported previously for a 0.1° high resolution ocean only model[43], or coupled ocean and sea ice model[24], possibly linked to the stronger internal variability of SST exhibited in high resolution models[43]. However, the relatively coarser resolution (0.25°) of the satellite SST data (see "Satellite-based observational dataset" in "Methods") might lead to underestimation of SST variability at mesoscale eddy-resolving resolution[21] (0.1°).

In general, both in the LME regions and on a global scale, CESM-HR is more skillful in reproducing the frequency and mean intensity of MHWs compared with the coarse resolution models (CESM-LR and CMIP5), lending some confidence for the following analysis.

Changes in MHWs under a Warming Climate

We define future MHWs using the "mean warming-inclusive threshold" and "future threshold" determined from simulations of the future (See "Definition of MHW" in "Methods") to isolate the effect of mean background warming and higher statistical moments of SST on MHWs. The number of annual MHW days is more highly correlated with changes of important basic biological groups in the world than the mean or maximum SST[6]. On the other hand, the intensity or the temperature anomaly during a MHW event can represent the level of acute heat stress for marine ecosystems and is closely linked to mortality of organisms such as intertidal barnacles[17].

On the basis of the mean warming-inclusive threshold, under RCP8.5, CESM-HR projects strong increases in annual MHW days (Fig. 4c) and average MHW intensity (Fig. 4d). Compared to the historical period, the mean annual MHW days between 70°N/S are projected to increase by 287.2 during 2071–2100 (Fig. 4c), and a permanent MHW state will be reached in many areas of the equatorial and subtropical regions. Many marine species in the equatorial region live near their high temperature ceiling and are highly sensitive to MHWs[44,45]. By contrast, the increase in annual MHW days in the North Atlantic and WBC region is much smaller. The mean intensity shows a mean increase of 1.2 °C based on CESM-HR (2071–2100; Fig. 4d), Moreover, the increase of intensity is much larger over the northern hemisphere with faster mean SST warming[46] (Supplementary Fig. 2a).

The mean warming dominates the changes in MHWs and explains 94% or more of the simulated changes, as inferred by the similarity between the changes directly estimated from the simulations (Fig. 4c, d) and the changes calculated based on the pseudo scenario with mean warming alone (Supplementary Fig. 3). The pseudo scenario is designed by adding a perturbation calculated based on the 30-year mean SST differences between future (2071–2100) and historical period (1975–2004) to the historical daily SST. The dominant role of mean warming in enhancing the changes in MHWs is consistent with previous results showing that the changes in the mean SST was the primary driver of the changes in MHW globally[11] or in coastal regions[12] during the historical period.

By contrast, by utilizing the future threshold, much smaller increases in annual MHW days and average MHW intensity are projected (Fig. 4a, b) by CESM-HR. The mean annual MHW days over 70°N/S are projected to increase by only 2.8 days during 2071–2100 compared to 1975–2004 (Fig. 4c), which is much lower than the result obtained

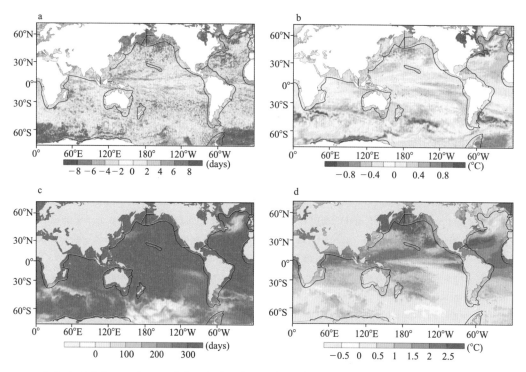

a–d. Projected changes in annual days (a,c) and mean intensity (b,d) of MHW in 2071–2100, based on future threshold (a,c) and mean warming-inclusive threshold (b,d). The areas surrounded by the black solid line and coastline represent the LMEs. Note that the color bar range is different in each plot. Results show mild increases based on future threshold in the annual MHW days and mean MHW intensity, while the changes are much stronger based on mean warming-inclusive threshold.

Fig. 4 Projected changes in annual MHW days and mean MHW intensity

using a mean warming-inclusive threshold. Besides annual days of MHWs, the mean intensity shows consistently mild increases over the oceans worldwide, with a mean increase of 0.2 °C (2071–2100) (Fig. 4b). Likewise, the application of similar methods yields comparably small changes in the annual MHW days over northeast Pacific Blob and MHW intensity over North Atlantic Ocean, respectively by the end of this century in RCP8.5 using coarse resolution simulations[13,22]. Moreover, the dipole feature of higher increase over northern hemisphere and lower increase over southern hemisphere exhibited from analysis using the mean warming-inclusive threshold disappears, resulting in more uniform increases in MHW intensity.

The analysis using CESM-LR and the CMIP5 multi-model ensemble in general supports the findings discussed above, further illustrating the dependence of MHW characteristics relative to the baseline climate (Supplementary Figs. 4, 5). However, CMIP5 and CESM-LR do not show WBC regions having distinct behavior relative to other mid-latitude regions (Supplementary Figs. 4b, 5b), while CESM-HR projects much larger changes over the major WBC areas including the Kuroshio Extension, the Gulf Stream, the Zapiola Anticyclone, the Agulhas Return Current, the East Australian Current and the South Pacific storm track (Fig. 4b). Moreover, almost all these major WBC regions except the East Australian Current delineate a distinctive meridional dipole intensity changes with increase over the poleward flank and decrease over the equatorward flank. The dipole feature can be explained by changes in the detrended SST variance (Supplementary Fig. 6), consistent with a previous study that demonstrated the relationship between SST variance and MHW intensity[13], with changes in SST variance and MHWs intensity likely attributable to the shifts in the frontal position in WBC regions[47]. Detrending SST before calculation of SST variance excludes the influence of greenhouse-induced long-term trends on SST variability[27]. The 30-year SST trends in the historical and future climate over LMEs are listed in Supplementary Table 2.

Future Changes of MHWs in LMEs with CESM-HR

Fisheries catch varies by two orders of magnitude among the LME regions, so it is useful to present the MHW days and intensity for the present and future for the different LME regions separately (Fig. 5). The mean and standard deviation of MHWs are shown in Supplementary Table 2. Historically, over the LME regions between 70°N/S, annual MHW days range from 27.4 to 40.6, with an average of 33.2 (x axis in Fig. 5a, c). With the mean warming-inclusive threshold, the annual MHW days over LMEs soar to 351.4 (y axis in Fig. 5c). The results using the future threshold largely suppresses the dominant effect of mean SST changes on the future MHW days. A total of 98% (except one) of LMEs show more MHWs days, with a mean annual increase of 2.8 days by the end of this

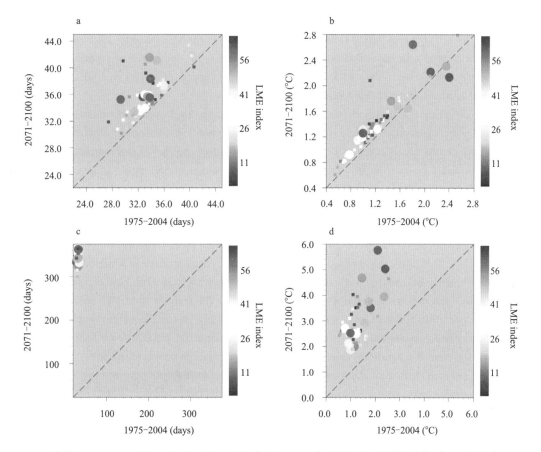

a–d. The mean annual days (a,c) and intensity (b,d) among the LMEs for MHWs defined on the basis of threshold for each period (1975–2004 and 2071–2100, respectively)(a,b), and mean warming-inclusive threshold (c,d). The smaller squares and larger circles represent the small and big fishery catch, respectively, with color indicating the index of the LMEs. Results show an enhanced MHW intensity and annual days over most of the LMEs in future even with a future threshold above mean warming.

Fig. 5 Comparison of MHW days and intensity between future and the historical period

century compared to the historical period (Fig. 5a). The increase in the mean annual MHW days is contributed by the increase in the persistence of MHWs, as indicated by the increase in the autocorrelation of SST despite of decreases in the frequency of MHWs[13].

Consistently, using mean warming-inclusive thresholds, we find that the mean intensity in LMEs increases by more than 100%, from 1.2 °C during the historical period to 2.9 °C by the end of the century. As expected, the mean intensity over the LMEs based on future thresholds yields a small increase of 0.2 °C. Despite this, 93% of LMEs display an increase in intensity (Fig. 5b). The increase is primarily contributed by the changes in the SST variance, reflected by a statistically significant correlation ($P<0.05$) between MHW intensity and SST variance in both historical and future periods (Supplementary Fig. 6). Given the vast diversity of geographical locations of the LMEs, forcing of the increased SST variance is equally diverse. Under greenhouse warming, dominant modes of climate variability, which strongly influence MHWs across the global ocean, are generally projected to increase in their variance. As such, increased SST variance, hence an increased intensity of MHWs over majority of the LMEs, at least in part attributable to strengthened ENSO variability[48], enhanced SST variability over north tropical Atlantic[49], increased frequency of stronger positive Indian Ocean Dipole[50] and stronger nonlinear relationship between evaporation and SST over the North Pacific[51] under greenhouse warming.

Importantly, there is a significant correlation of 0.9 ($P<0.05$) between the future and historical mean intensity over all LMEs, compared to otherwise 0.6 based on mean warming-inclusive threshold (Fig. 5b, d), emphasizing comparable severity of MHWs at present and future for the majority of LMEs. In other words, LMEs that are under stress now will continue to be so in the future but in addition to the stress due to the mean warming and the increased intensity. The change from a more scattered distribution (Fig. 5d) to the alignment almost in a straight line (Fig. 5b) is to a large extent because of the built-in increase in MHW intensity due to the mean warming in Fig. 5d that dominates the response when the mean warming-inclusive threshold is used.

Our result based on the future threshold show that marine species in most of the LMEs would still experience an increase in the threat of MHWs, if they were able to adapt to the slowly increasing mean warming. To highlight this point, we compare result for LMEs of the category I, which are the 15 LMEs with the largest fishing capacity (Supplementary Fig. 7) and for all other LMEs defined as category II (Fig. 1A in ref[52]). There is a generally larger changes in the mean annual MHW days and intensity in category I, compared to that of category II, indicative of a potentially more intensified impact of MHWs on LMEs with higher catches.

Organisms might adapt to climate change to a certain extent[44], but the rate of adaptation can vary widely among species[53,54], and spatial heterogeneity in the changes of SST might lead to differences in the extent to which marine species must adapt. Changes in MHWs defined using the future threshold are relevant if species can adapt fully to the future mean warming, which might not be possible[13] due to the rate at which SST is changing relative to what ecosystems have experienced in the past[55,56]. The increase in MHWs under the future threshold can be considered the "most optimistic" scenario for establishing the lower bound of climate change impact on marine ecosystems.

Conclusions

We find an increased intensity and annual days of MHWs over the majority of the LMEs in the future climate by applying a future-threshold definition of MHWs. Our result of a widespread increase of MHWs over LMEs implies that even if we assume that organisms in LMEs were able to adapt fully to the impact of the long-term mean warming, the LMEs would still face serious threats under global warming. Our result is based on a high fidelity simulation using a high-resolution model that provides improved simulation of MHWs in the LME regions. As computational power continues to improve, we expect that a multi-model ensemble of high-resolution model simulations will soon be possible to project future MHW changes under multiple climate forcing scenarios, to assess the associated uncertainty, and to provide early warning of the likely changes. Importantly, our initial result indicates that even under the most optimistic assumption, risks to LMEs are substantial. The result therefore has far-reaching ecological, social, and economic implications and calls for a response strategy from the impacted communities and policy makers.

Methods

▶ *Model descriptions*

Here, we use a high-resolution Earth System Model simulation, spanning 250 years from 1850 to 2100[28] that uses CMIP5 historical forcings until 2005 and the RCP8.5[16] (high-emission scenario) thereafter. The models in the high-resolution configuration of the CESM1.3 were used for the simulation. The atmospheric and land models have a nominal horizontal resolution of 0.25°, while a nominal horizontal resolution of 0.1° is used for the ocean and sea ice components. This high-resolution configuration of CESM1.3 is referred to as CESM-HR. At the resolutions of the individual components, the model allows for mesoscale eddies in the ocean to better delineate the interactions between the mesoscale phenomenon and large-scale circulation[28]. For comparison, simulations with CESM at a coarser spatial resolution of 1° in both atmosphere and ocean, referred to as CESM-LR, as well as the multi-model ensemble of 20 models participating in CMIP5[16] are also used in this study (Supplementary Table 1). All simulations during the historical period (1975–2004) for CMIP5, CESM-LR and CESM-HR, as well as the OISST/GMPE (See "Satellite-based observational data set" in "Methods") for 1982–2011 are interpolated to 1°. Note that the start time of OISST and GMPE is September 1981, thus January 1982 is selected as the start and December 2011 as the end for the 30-year comparison. Comparison of the future and historical MHW based on CESM-HR is performed at the spatial resolution of 0.25°.

▶ *LMEs*

The LMEs refer mainly to the coastal areas and the outer edge of coastal currents, including river basins and estuaries up to the seaward boundary of the continental shelfs or well-defined systems of currents without continental shelfs[25]. An LME usually include an area of 200,000 km^2 or more[25]. The LMEs are rich in biodiversity, including 95% of the global fish catch, though they cover only 22% of the total ocean area[52], providing goods and services to billions of people worth more than US$12.6 trillion annually[57]. In this study, the LMEs within 70°N/S are divided into 6 groups according to the adjoined continents (Fig. 1b): North America, South America, Europe, Africa, Asia and Australia. Some LMEs might be located between two continents. For example, the Mediterranean Sea lies between Europe and Africa, but to simplify our analysis, it is considered part of Europe.

▶ *Definition of MHW*

An MHW is a prolonged, discrete, anomalously warm water event[58]. Specifically, for each grid cell, a threshold for each day of a year is first determined based on the 90th percentile using daily-mean SSTs in the 11-day moving window centered on the specific day over a long, 30-year segment to ensure a sufficient sample size. This is followed by a 31-day moving average of the daily threshold. Five consecutive days or more with SST above the threshold is identified as a MHW event, and two events separated by an interval of two or fewer days are considered as one event. Note that a MHW defined above might not only occur in the warmer months, MHWs in colder months are also fatal for some creatures[60-62]. The number of days per event, denoted as duration, the number of annual

MHW events (frequency) and the average intensity representing the mean deviation of SST from the climatological mean within the event are calculated first, and then the total number of annual MHW days, as well as the mean intensity are derived. Note that the threshold over the high latitudes is affected by the melting rate of ice and snow which is not taken into account; therefore, the analysis in this study primarily focuses on regions within 70°N/S which are less likely affected by ice and snow cover.

Two thresholds are used in this study to calculate the future (2071–2100) MHWs: (1) the 90th percentile over the historical period[24,43] (1975–2004), called mean warming-inclusive threshold, as defined above; (2) the future (2071–2100) 90th percentile[22], called future threshold, provides a delineation of the extent to which MHW changes are associated with the mean warming or non-seasonal temperature changes[19].

▶ *Satellite-based observational dataset*

To compare the MHW index calculated from the simulations, the satellite-based NOAA OISST v2.1, referred to as OISST (https://www.ncdc.noaa.gov/oisst/)[63,64], is used to calculate the MHW index globally in the study. The OISST product has been widely used in MHW studies[5,15,59]. This data set was derived from remotely sensed SSTs by the Advanced Very High-resolution Radiometer (AVHRR) infrared satellite data and in-situ measurements. With a spatial resolution of 0.25° on daily scale globally, this product represents the water temperature in the top 0.5 m of the ocean. In addition, another global daily data set named Group for High Resolution Sea Surface temperature (GHRSST) multi-product ensemble (GMPE)[38,39] v2.0[65], including multiple SST data such as MyOcean OSTIA reanalysis, CMC 0.2 degree, AVHRR ONLY Daily 1/4 degree OISST, and MGDSST[40], with the spatial resolution at 0.25° is also used.

Date Availability

The raw CESM model output data are available from the iHESP data portal (https://ihesp.tamu.edu/products/ihesp-products/data-release/DataRelease_Phase2.html) and the QNLM data portal (http://ihesp.qnlm.ac). The CMIP5 data are available at https://esgf-node.llnl.gov/projects/cmip5/.

Code Availability

The CESM code used for the simulations is available at Zenodo via https://doi.org/10.5281/zenodo.3637771[66]. The code used to detect MHWs is available at https://github.com/ecjoliver/marineHeatWaves. All the other codes used in the data process, including the simulations and satellite data and visualization are available upon request to the corresponding authors.

Acknowledgements

Xiuwen Guo and Yang Gao are supported by the National Key Research and Development Program of China (2017YFC1404101), National Natural Science Foundation of China (42122039) and Fundamental Research Funds for the Central Universities (202072001). Shaoqing Zhang is supported by the National Key Research and Development Program of China (2017YFC1404104). The analysis was performed using the computing resources of the Center for High Performance Computing and System Simulation, Pilot National Laboratory for Marine Science and Technology (Qingdao). This research is completed through and supported by the International Laboratory for High Resolution Earth System Prediction (iHESP). Lixin Wu is supported by the National Key Research and Development Program of China (2019YFC1509100). Ping Chang acknowledges the support of the NSF Convergence Accelerator Program grant no. 2137684. Wenju Cai is supported by CSHOR, which is a joint research Centre for Southern Hemisphere Oceans Research between QNLM and CSIRO. Jakob Zscheischler acknowledges the Swiss National Science Foundation (Ambizione grant 179876) and the Helmholtz Initiative and Networking Fund (Young Investigator Group COMPOUNDX, grant agreement VH-NG-1537). L. Ruby Leung is supported by the Office of Science of the US Department of Energy

Biological and Environmental Research Regional and Global Model Analysis programme area. Pacific Northwest National Laboratory is operated for the US Department of Energy by Battelle Memorial Institute under contract DE-AC05-76RL01830. The National Center for Atmospheric Research (NCAR) is a major facility sponsored by the US National Science Foundation under cooperative agreement no. 1852977. Luanne Thompson's contribution to this material is based on work supported by the National Science Foundation under grant no. 2022874. Huiwang Gao is supported by the National Natural Science Foundation of China–Shandong Joint Fund (U1906215). We acknowledge the World Climate Research Programme, which coordinated and promoted CMIP5 through its Working Group on Coupled Modelling, and we thank the climate modelling groups for producing and making available their model outputs.

Author Contributions

Yang Gao and Shaoqing Zhang conceived the project; Xiuwen Guo performed the analysis and drafted the manuscript; Lixin Wu, Ping Chang, Wenju Cai, L. Ruby Leung and Luanne Thompson helped on the figure design and analysis; Jakob Zscheischler, Justin Small, Gokhan Danabasoglu and Huiwang Gao helped on the method and analysis. All authors contributed to the writing of the manuscript.

Competing Interests

The authors declare no competing interests.

References

1. Cheng L. How fast are the oceans warming?[J]. Science, 2019, 363: 1294-1294.
2. Mills K E, Pershing A J, Brown C J, et al. Fisheries management in a changing climate lessons from the 2012 ocean heat wave in the Northwest Atlantic[J]. Oceanography, 2013, 26: 191-195.
3. Oliver E C J, Donat M G, Burrows M T, et al. Longer and more frequent marine heatwaves over the past century[J]. Nat. Commun., 2018, 9: 1324.
4. Amaya D J, Miller A J, Xie S P, et al. Physical drivers of the summer 2019 North Pacific marine heatwave[J]. Nat. Commun., 2020, 11: 1903.
5. Oliver E C J, Benthuysen J A, Bindoff N L, et al. The unprecedented 2015/16 Tasman Sea marine heatwave[J]. Nat. Commun., 2017, 8: 16101.
6. Smale D A, Wernberg T, Oliver E C J, et al. Marine heatwaves threaten global biodiversity and the provision of ecosystem services[J]. Nat. Clim. Change, 2019, 9: 306-312.
7. Arafeh-Dalmau N, Schoeman D S, Montaño-Moctezuma G, et al. Marine heat waves threaten kelp forests[J]. Science, 2020, 367: 635-635.
8. Cavole L M, Demko A M, Diner R E, et al. Biological impacts of the 2013-2015 warm-water anomaly in the Northeast Pacific[J]. Oceanography, 2016, 29: 273-285.
9. Laufkotter C, Zscheischler J, Frolicher T L. High-impact marine heatwaves attributable to human-induced global warming[J]. Science, 2020, 369: 1621-1625.
10. Stillman J H. Heat waves, the new normal: Summertime temperature extremes will impact animals, ecosystems, and human communities[J]. Physiology, 2019, 34: 86-100.
11. Oliver E C J. Mean warming not variability drives marine heatwave trends[J]. Clim. Dyn, 2019, 53: 1653-1659.
12. Marin M, Feng M, Phillips H E, et al. A global, multiproduct analysis of coastal marine heatwaves: Distribution, characteristics, and long-term trends[J]. J. Geophys. Res. Oceans, 2021, 126: 1-17.
13. Oliver E C J, Benthuysen J A, Darmaraki S, et al. Marine heatwaves[J]. Annu. Rev. Mar. Sci., 2021, 13: 313-342.
14. Darmaraki S, Somot S, Sevault F, et al. Future evolution of marine heatwaves in the Mediterranean Sea[J]. Clim. Dyn., 2019, 53: 1371-1392.
15. Frölicher T L, Fischer E M, Gruber N. Marine heatwaves under global warming[J]. Nature, 2018, 560: 360-364.
16. Taylor K E, Stouffer R J, Meehl G A. An overview of CMIP5 and the experiment design[J]. Bull. Am. Meteorol. Soc., 2012, 93: 485-498.
17. Oliver E C J, Burrows M T, Donat M G, et al. Projected marine heatwaves in the 21st century and the potential for ecological impact[J]. Front. Mar. Sci., 2019, 6: 734.
18. Jacox M G, Alexander M A, Bograd S J, et al. Thermal displacement by marine heatwaves[J]. Nature, 2020, 584: 82-86.
19. Jacox M G. Marine heatwaves in a changing climate[J]. Nature, 2019, 571: 485-487.
20. Di Lorenzo E, Mantua N. Multi-year persistence of the 2014/15 North Pacific marine heatwave[J]. Nat. Clim.

Change, 2016, 6: 1042-1047.
21. Holbrook N J, Scannell H A, Gupta A S, et al. A global assessment of marine heatwaves and their drivers[J]. Nat. Commun., 2019, 10: 2624.
22. Plecha S M, Soares P M M, Silva-Fernandes S M, et al. On the uncertainty of future projections of marine heatwave events in the North Atlantic Ocean[J]. Clim. Dyn., 2021, 56: 2027-2056.
23. Benthuysen J, Feng M, Zhong L. Spatial patterns of warming off Western Australia during the 2011 Ningaloo Nino: Quantifying impacts of remote and local forcing[J]. Cont. Shelf Res., 2014, 91:232-246.
24. Pilo G S, Holbrook N J, Kiss A E, et al. Sensitivity of marine heatwave metrics to ocean model resolution[J]. Geophys. Res. Lett., 2019, 46: 14604-14612.
25. Sherman K. Adaptive management institutions at the regional level: The case of large marine ecosystems[J]. Ocean Coast Manag., 2014, 90: 38-49.
26. Belkin I M. Rapid warming of large marine ecosystems[J]. Prog. Oceanogr., 2009, 81: 207-213.
27. Alexander M A, Scott J D, Friedland K D, et al. Projected sea surface temperatures over the 21st century: Changes in the mean, variability and extremes for large marine ecosystem regions of Northern Oceans[J]. Elementa Sci. Anthrop., 2018, 6: 9.
28. Chang P, Zhang S, Danabasoglu G, et al. An unprecedented set of high-resolution earth system simulations for understanding multiscale interactions in climate variability and change[J]. J. Adv. Model. Earth Syst., 2020, 12: 1-52.
29. Hvas M, Folkedal O, Imsland A, et al. The effect of thermal acclimation on aerobic scope and critical swimming speed in Atlantic salmon, *Salmo salar*[J]. J. Exp. Biol., 2017, 220: 2757-2764.
30. Hittle K A, Kwon E S, Coughlin D J. Climate change and anadromous fish: How does thermal acclimation affect the mechanics of the myotomal muscle of the Atlantic salmon, *Salmo salar*[J]? J. Exp. Zool. Part A-Ecol. Integr. Physiol., 2021, 335: 311-318.
31. Forseth T, Barlaup B T, Finstad B, et al. The major threats to Atlantic salmon in Norway[J]. ICES J. Mar. Sci., 2017, 74: 1496-1513.
32. Norin T, Canada P, Bailey J A, et al. Thermal biology and swimming performance of Atlantic cod (*Gadus morhua*) and haddock (*Melanogrammus aeglefinus*) [J]. PeerJ, 2019, 7: e7784.
33. Hughes T P, Kerry J T, Álvarez-Noriega M, et al. Global warming and recurrent mass bleaching of corals[J]. Nature, 2017, 543: 373-377.
34. Liu G, Heron S F, Eakin C M, et al. Reef-scale thermal stress monitoring of coral ecosystems: New 5-km global products from NOAA Coral Reef Watch[J]. Remote Sens., 2014, 6: 11579-11606.
35. Aoki L R, McGlathery K J, Wiberg P L, et al. Seagrass recovery following marine heat wave influences sediment carbon stocks[J]. Front. Mar. Sci., 2021, 7: 576784.
36. Laurel B J, Rogers L A. Loss of spawning habitat and prerecruits of Pacific cod during a Gulf of Alaska heatwave[J]. Can. J. Fish. Aquat., 2020, 77: 644-650.
37. Perry A L, Low P J, Ellis J R, et al. Climate change and distribution shifts in marine fishes[J]. Science, 2005, 308: 1912-1915.
38. Dash P, Ignatov A, Martin M, et al. Group for High Resolution Sea Surface Temperature (GHRSST) analysis fields inter-comparisons—Part 2: Near real time web-based level 4 SST Quality Monitor (L4-SQUAM) [J]. Deep-Sea Res. II: Top. Stud. Oceanogr., 2012, 77: 31.
39. Martin M, Dash P, Ignatov A, et al. Group for High Resolution Sea Surface temperature (GHRSST) analysis fields inter-comparisons—Part 1: A GHRSST multi-product ensemble (GMPE) [J]. Deep-Sea Res. II: Top. Stud. Oceanogr., 2012, 77-80: 21-30.
40. Fiedler E K, McLaren A, Banzon V, et al. Intercomparison of long-term sea surface temperature analyses using the GHRSST Multi-Product Ensemble (GMPE) system[J]. Remote Sens. Environ., 2019, 222: 18-33.
41. Gent P R, McWilliams J C. Isopycna mixing in ocean circulation models[J]. J. Phys. Oceanogr., 1990, 20: 150-155.
42. Fox-Kemper B, Ferrari R, Hallberg R. Parameterization of mixed layer eddies. Part I: Theory and diagnosis[J]. J. Phys. Oceanogr., 2008, 38: 1145-1165.
43. Hayashida H, Matear R J, Strutton P G, et al. Insights into projected changes in marine heatwaves from a high-resolution ocean circulation model[J]. Nat. Commun., 2020, 11: 4352.
44. Vinagre C, Mendonça V, Cereja R, et al. Ecological traps in shallow coastal waters-Potential effect of heat-waves in tropical and temperate organisms[J]. Plos One, 2018, 13: e0192700.
45. Comte L, Olden J D. Climatic vulnerability of the world's freshwater and marine fishes[J]. Nat. Clim. Change, 2017, 7: 718-722.
46. Armour K C, Marshall J, Scott J R, et al. Southern Ocean warming delayed by circumpolar upwelling and equatorward transport[J]. Nat. Geosci., 2016, 9: 549-554.
47. Wu L, Cai W, Zhang L, et al. Enhanced warming over the global subtropical western boundary currents[J].

Nat. Clim. Change, 2012, 2: 161-166.
48. Cai W, et al. Santoso A, Collins M, Changing El Nino-Southern Oscillation in a warming climate[J]. Nat. Rev. Earth Environ., 2021, 2: 628-644.
49. Yang Y, Wu L, Guo Y, et al. Greenhouse warming intensifies north tropical Atlantic climate variability[J]. Sci. Adv., 2021, 7: eabg9690.
50. Cai W, Yang K, Wu L, et al. Opposite response of strong and moderate positive Indian Ocean Dipole to global warming[J]. Nat. Clim. Change, 2021, 11: 27-32.
51. Jia F, Cai W, Gan B, et al. Enhanced North Pacific impact on El Nino/Southern Oscillation under greenhouse warming[J]. Nat. Clim. Change, 2021, 11: 840-847.
52. Stock C A, John J G, Rykaczewski R R, et al. Reconciling fisheries catch and ocean productivity[J]. Proc. Natl. Acad. Sci. USA, 2017, 114: E1441-E1449.
53. Millien V, Lyons S K, Olson L, et al. Ecotypic variation in the context of global climate change: Revisiting the rules[J]. Ecol. Lett., 2006, 9: 853-869.
54. Tian L, Benton M J. Predicting biotic responses to future climate warming with classic ecogeographic rules[J]. Curr. Biol., 2020, 30: R744-R749.
55. Walther G R, Post E, Convey P, et al. Ecological responses to recent climate change[J]. Nature, 2002, 416: 389-395.
56. Sheridan J A, Bickford D. Shrinking body size as an ecological response to climate change[J]. Nat. Clim. Change, 2011, 1: 401-406.
57. Zhang S, Fu H, Wu L, et al. Optimizing high-resolution Community Earth System Model on a heterogeneous many-core supercomputing platform[J]. Geosci. Model Dev., 2020, 13: 4809-4829.
58. Costanza R, d'Arge R, de Groot, R, et al. The value of the world's ecosystem services and natural capital[J]. Nature, 1997, 387: 253-260.
59. Hobday A J, Alexander L V, Perkins S E, et al. A hierarchical approach to defining marine heatwaves[J]. Prog. Oceanogr., 2016, 141: 227-238.
60. Santelices B. Patterns of reproduction, dispersal and recruitment in seaweeds[J]. Oceanogr. Mar. Biol., 1990, 28: 177-276.
61. Lotze H K, Worm B, Sommer U. Strong bottom-up and top-down control of early life stages of macroalgae[J]. Limnol. Oceanogr., 2001, 46: 749-757.
62. Andrews S, Bennett S, Wernberg T. Reproductive seasonality and early life temperature sensitivity reflect vulnerability of a seaweed undergoing range reduction[J]. Mar. Ecol. Prog. Ser., 2014, 495: 119-129.
63. Huang B, Liu C, Banzon V, et al. Improvements of the Daily Optimum Interpolation Sea Surface Temperature (DOISST) Version 2.1[J]. J. Clim., 2021, 34: 2923-2939.
64. Reynolds R W, Smith T M, Liu C, et al. Daily high-resolution-blended analyses for sea surface temperature[J]. J. Clim., 2007, 20: 5473-5496.
65. Good S A. ESA sea surface temperature climate change initiative (SST_cci): GHRSST multi-product ensemble (GMPE), Centre for Environmental Data Analysis, v2.0[Z/OL]. 2020. https://doi.org/10.5281/zenodo.3637771.
66. RUO. lgan/cesm_sw_1.0.1: some efforts on refactoring and optimizing the Community Earth System Model (CESM1.3.1) on the Sunway TaihuLight supercomputer (Version cesm_sw_1.0.1) [Z/OL]. 2020. Zenodo https://doi.org/10.5281/zenodo.3637771.

Differential mobilization and sequestration of sedimentary black carbon in the East China Sea[①]

Jingyu Liu[1,2], Nan Wang[3], Cuimei Xia[1,2], Weifeng Wu[1,2], Yang Zhang[3], Guangxue Li[3], Yang Zhou[4], Guangcai Zhong[5], Gan Zhang[5], Rui Bao[1,2*]

1 Frontiers Science Center for Deep Ocean Multispheres and Earth System, Key Laboratory of Marine Chemistry Theory and Technology, Ministry of Education, Ocean University of China, Qingdao 266100, China
2 Laboratory for Marine Ecology and Environmental Science, Qingdao National Laboratory for Marine Science and Technology, Qingdao 266237, China
3 Frontiers Science Center for Deep Ocean Multispheres and Earth System, Key Laboratory of Submarine Geosciences and Prospecting Techniques, MOE and College of Marine Geosciences, Ocean University of China, Qingdao 266100, China
4 Guangzhou Marine Geologic Survey, Guangzhou 510760, China
5 State Key Laboratory of Organic Geochemistry, Guangzhou Institute of Geochemistry, Chinese Academy of Sciences, Guangzhou 510640, China
* Corresponding author: Rui Bao (baorui@ouc.edu.cn)

Abstract

Black carbon (BC) derived from incomplete combustion of biomass and fossil fuels on land can be mobilized and transported to the ocean. Burial of BC in the ocean sequesters atmospheric CO_2 into a long-term carbon sink, likely exerting a positive influence on mitigating global warming. However, the abundances, sources, and burial of sedimentary BC in marine sediments remain poorly constrained, hindering us from accurately understanding the mobilization and sequestration of BC and its roles in the ocean carbon cycle. Here, we investigate concentrations and isotopes (^{13}C and ^{14}C) of BC among grain size-fractionated surface sediments along a across-shelf transect from the Yangtze River prodelta to the Okinawa Trough to decipher the fate of BC in the East China Sea (ECS). Our results show that the bulk BC concentrations decrease firstly from the Yangtze River prodelta to the outer shelf and then increase to the Okinawa Trough. Grain size-fractionated BC concentrations vary obviously along the transect, which we mainly attribute to the differential mobilization of BC driven by hydrodynamic processes. We argue that BC is aged during the mobilization, which results in an older ^{14}C ages of BC found seaward. After considering biomass- and fossil-derived BC apportionments based on ^{14}C balance calculation, we think that BC aging may be verified by more fossil-derived BC burial in the Okinawa Trough. We estimate that BC may account for about 15% of sedimentary organic carbon (SOC), and up to about 30% of terrestrial SOC buried in the ECS. BC burial fluxes decrease along the transect, and are heterogeneous in different size fractions, indicating differential sequestration of BC in the shelf and trough. We further estimate that 685 Gg/a of BC is sequestered in the ECS, and 491 Gg/a in the prodelta area, with about 30% being continental biomass-derived BC. We suggest that increasing biomass-derived BC production on land and burying it in the ocean may serve as a powerful means for sequestrating

① 本文于2022年9月发表在 *Earth and Planetary Science Letters*，http://doi.org/10.1016/j.epsl.2022.117739。

atmospheric CO_2, potentially contributing to carbon neutrality.

Keywords: black carbon; radiocarbon; East China Sea; grain size; hydrodynamic processes

Introduction

Black carbon (BC) is ubiquitously formed during the incomplete combustion of biomass and fossil fuels on land[1-3]. It has a complex fate in the ocean after transportation either by fluvial systems or atmospheric depositions[4-8]. BC can reside over long-term timescales, its burial in the ocean acts as a carbon "sink", operating a slow but important function in global carbon cycles[4,9-13]. In the past few decades, the importance of BC as a vital component of total organic carbon in the land-ocean carbon cycles has been extensively recognized[4,10,13,14]. According to recent studies, a large amount of particulate BC (about 17–37 Tg/a[14]) and dissolved BC (about 27–66 Tg/a[15]) are exported to the marginal seas by rivers. Atmospheric depositions could also deliver about 1–10 Tg/a particulate BC[16,17] and about 1–2 Tg/a dissolved BC[17] into the ocean. Most of this BC is ultimately buried in marine sediments[4,13,18,19], which acts as a huge long-term carbon sink.

Previous studies found that about 0.63–1.39 Tg/a of sedimentary BC is buried in the East China Sea (ECS)[12,20], accounting for about 1%–3% of the global BC input flux to the ocean. It was estimated that about 10%–30% of the sedimentary organic carbon (SOC) in the ECS is BC[12,20], which is comparable with other global marginal sea systems (about 3%–35%[4,9]). In this context, the abundances, sources, and burial of sedimentary BC in the ECS should be focused due to their importance in understanding the ultimate fate of BC in the marginal seas. Nevertheless, to date, most studies investigated the spatial distributions of BC abundances in the ECS[12,20,21], and the mobilization and sequestration of BC in the ocean receive far less attention, leaving a large knowledge gap about the fate of sedimentary BC on the continental shelf.

The transect from the Yangtze River prodelta to the Okinawa Trough (Fig. 1) provides an appropriate area to study the fate of sedimentary BC from a source-to-sink perspective. Although the fate of SOC along this across-shelf transect in the ECS has been well investigated[22-25], the mobilization and sequestration of BC on the shelf have not been thoroughly studied. Pilot studies suggested that the transportation and burial of SOC in the ECS are closely associated with hydrodynamic processes, such as resuspension and bedload processes[24-28]. Huang et al.[12] noticed that the concentration of sedimentary BC in the ECS is highly correlated with the sediment grain size, showing that the clay and silt are more prone to BC accumulation than the sand. Fang et al.[29] suggested the distributions of BC abundances in the shelf sediments are influenced by the regional hydrodynamic processes. These investigations inferred that the mobilization and sequestration of sedimentary BC are probably also associated with hydrodynamic processes. The sites along the transect are influenced by different hydrodynamic processes[22,23], providing a location to deconvolve the fate of BC in the ECS. Furthermore, the Okinawa Trough could act as a sink for seaward transport of SOC across the shelf[23]. Therefore, investigating the characteristics of sedimentary BC along the transect could reveal the mobilization and sequestration of BC.

The influence of hydrodynamic processes on the mobilization and sequestration of BC could be deciphered by examining the abundances and isotopes of grain size-fractionated BC in surface sediments. Previous studies performed on the transport and burial of SOC in different grain-size surface sediments of the ECS and East Siberia shelf seas[27,28,30]. The sediment grain size is closely related to the hydrodynamic processes and organic matter-mineral interactions, which influence the mobilization and effective sequestration of SOC[27,28,30-32]. Although the characteristics of bulk BC in surface sediments of the ECS were investigated[12,21], few studies focus on the abundances and isotopes of grain size-fractionated

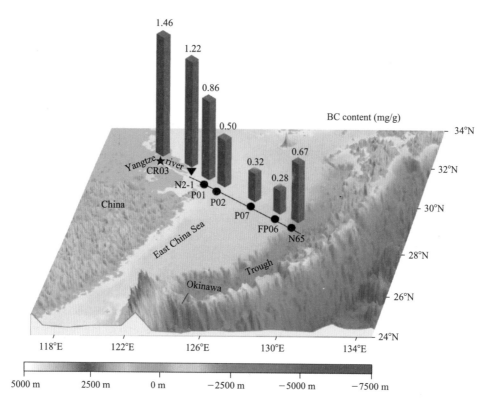

Particulate BC data in the Yangtze River and estuary (star CR03, and triangle N2-1) are from Wang et al.[21,43]

Fig. 1 Site locations and BC concentrations (mg/g) along a transect from the Yangtze River to the Okinawa Trough

BC[33,34], hindering our ability to shed light on the mobilization and sequestration of sedimentary BC.

More importantly, previous studies considered SOC as a whole when studying the oceanic fate of SOC[26-28], not dividing it into BC and non-BC organic carbon (abbreviated as OC here and calculated as OC = SOC − BC)[9,34]. In fact, the BC and OC have different cycling fates[7,35]. BC cycles in a slow loop due to its refractory nature, while OC cycles in a rapid loop[4,6,36]. Therefore, considering BC and OC separately, and studying their distinct cycling processes, could help us understand their different fates in the ocean.

Fossil- and biomass-derived BC play different roles in sequestering atmospheric CO_2[6]. The formation of biomass-derived BC converts labile biogenic carbon to refractory carbon, isolating the atmospheric CO_2 from the rapid carbon cycle between atmosphere-biosphere reservoirs and sequestering it into a slow carbon cycle[4,9]. In contrast, the formation of fossil-derived BC just converts one form of geological carbon to another[6]. The former is of great significance for reducing human CO_2 emissions and contributing to carbon neutrality under the ongoing global warming[36,37]. Apportioning the sources of BC by radiocarbon (^{14}C) analysis and quantifying their burial fluxes in the ocean are crucial for us to accurately estimate the potential role of BC in global climate change.

The objective of this study is to reveal the mobilization and sequestration of BC by in-depth investigations of the abundances and isotope compositions of grain size-fractionated BC along a transect from the Yangtze River prodelta to the Okinawa Trough. In particular, we conduct a comprehensive ^{14}C analysis of grain size-fractionated BC and SOC to decipher the distinct cycling processes between BC and OC in the ocean. Furthermore, we make source apportionments, estimate fluxes of BC in the ECS, and discuss the implications of BC sequestration as a long-term carbon sink for carbon neutrality.

Sites and Materials

Surface sediments (0–2 cm, Site P01, P02, P07, FP06, and N65) were collected along a well-characterized transect in the ECS from the terrestrial-influenced Yangtze River prodelta to the marine-dominated Okinawa Trough in 2013 and 2015 (Fig. 1). Water depth and distance from the estuary increase seaward along the transect (Table S1). These sites represent areas from the inner shelf (29 m water depth) to the deep ocean (966 m), offering an opportunity to investigate the accumulation, residence, and sequestration of BC in the ECS.

In the study area, input of terrestrial materials is mainly from the Yangtze River[22,23] and is influenced by hydrodynamics processes[38]. Approximately half of the sediments discharged by the Yangtze River are deposited on the inner shelf, while a part of the remainder is carried southward by the strong Zhejiang-Fujian coastal current, and the other part is transported across the shelf[23,38,39]. Previous studies documented that the sortable sediments (20–63 μm), as well as the associated SOC, in the Yangtze River prodelta could be influenced by intensive resuspension, and partly transported to the open ocean by the across-shelf hydrodynamic processes[22,24,25]. In contrast, the middle-outer shelf with coarse grain-size sediments is mainly influenced by tidal and bottom currents[38]. The Okinawa Trough could act as a sediment sink, because sediments from the shelf in the ECS may accumulate in the trough after complicated hydrodynamic processes[23,39,40].

Methods

▶ *Sieving surface sediments*

In this study, samples of Site P01 and N65 were dry-sieved to <20, 20–32, 32–63, 63–125, 125–250 μm by Sonic Sifter Separator (Model L3P-25, Advantech Manufacturing, INC, USA). These grain size divisions are related to different transport behaviors and hydrodynamic processes[41], likely influencing the mobilization and sequestration of associated sedimentary organic matter[27,28]. The BC characteristics of these grain size fractions can be used for comparing with our available SOC data[27,28]. Based on these data of BC, OC, and SOC, we will discuss their different fates across the transect. A Horizontal Pulse Accessory (Advantech L3-N8) was used for < 63 μm sediments to avoid electrostatic charges and disperse aggregation. The recovery rate of the Sonic Sifter Separator could be up to 97% when dealing with about 10 g sediments, and the major loss occurs on the adsorption by fine sieves. The sieved samples were stored in pre-combusted glass vials (450 °C, 5 h) and put in a desiccator until the next steps. Different grain size samples of P02, P07, and FP06 were from wet-sieved residues in our previous study[28]. The samples of 20–32 μm and 32–63 μm in P07 were combined to 20–63 μm due to insufficient mass of the sediments.

▶ *Organic geochemistry analyses*

The organic geochemistry analysis methods of BC, OC, and SOC are shown here. SOC is equivalent to the TOC of previous studies[25,27,28], which is composed of BC and OC. Here, we define the non-BC SOC fraction as OC by the difference between SOC and BC[9,34].

Concentration analyses of BC, OC, and SOC

The pretreatment methodologies of SOC used in this study follow those described in Bao et al. (2016)[27]. Briefly, samples were acid-fumigated (37% HCl) at 60 °C for 72 h and then NaOH-fumigated for another 72 h to remove inorganic carbon. The Chemo-Thermal Oxidation (so-called CTO-375 method[2]) was used to isolate BC in this study. Sieved samples were ground, weighed into a 5 mm × 8 mm pre-combusted silver capsule (450 °C, 5 h), and then combusted in a tube furnace at 375 °C for 24 h under continuous airflow of 200 mL/min[3].

The ramped temperature rate was set to 2.5 °C/min to avoid potential artificial charring[11]. Then a in situ acid-pretreatment process was conducted (1 M HCl) for 12 h to remove carbonates[2,3]. After this, the samples were dried at 60 °C for 48 h and stored in a desiccator before analysis.

The concentrations of BC and SOC were measured using an EA-IsoLink elemental analyzer with a MAS 200R autosampler coupled to a MAT253 plus isotope ratio mass spectrometer via a ConFlo Ⅳ universal interface (Thermo Fisher Scientific, Germany) at the Key Lab of Submarine Geosciences and Prospecting Techniques, Ministry of Education, Ocean University of China. The analytical uncertainty of the BC concentrations is conservatively estimated to be < 10% through duplicate analysis of selected samples (n = 12). The harbor sediment standard reference by the National Institute of Standards and Technology (NIST), SRM 1941b, was used for quality control on our CTO-375 pretreatment procedures of BC. Our results of BC in SRM 1941b, (5.9 ± 0.4) mg/g (n = 8), could be compared to other results using the same standard in the previous literature ((5.8 ± 0.5) mg/g[3]; (6.0 ± 0.2) mg/g[11]).

Isotope analyses of BC, OC, and SOC

The stable isotope analyses of SOC and BC were conducted with the concentration measurements. The isotopic results are reported relative to the Vienna Pee Dee Belemnite (VPDB) with an external analytical precision of ± 0.006‰ (1σ, n = 42). For ^{14}C analysis of SOC and BC, the pretreated samples (n = 14) were sent to National Ocean Sciences Accelerator Mass Spectrometry Facility (NOSAMS) at the Woods Hole Oceanographic Institution (WHOI) and Ocean University of China radiocarbon Accelerator Mass Spectrometry Center (OUC-CAMS). The SOC, $\delta^{13}C$, and $\Delta^{14}C$ values of P01, P02, P07, and FP06 in different grain size sediments are from Bao et al.[25,27,28] except 125–250 μm of FP06. The $\Delta^{14}C$ value of SOC in the >63 μm fraction of N65 is calculated by the mass balance of bulk and other fractional sediments. The $\delta^{13}C$ and $\Delta^{14}C$ values of OC are presented in Fig. 3, Fig. 4, and Table S2, and calculated by mixing model of SOC and BC according to mass percentage. Source apportionments of BC, fossil BC, and SOC were conducted based on carbon isotope values of BC and SOC (see methods in Supplementary material).

▶ *Burial fluxes of sedimentary BC*

According to the distribution characteristics of BC concentrations among different grain sizes, we calculate the sedimentary BC fluxes of different fractions in the ECS by the following equation:

$$F = C \times \omega \qquad (1)$$

where F is the BC flux (Gg/a, $1G = 10^9$) of each size fraction, C is the BC concentration in different grain size fractions, and ω is the sediment mass flux (Gg/a) of each grain size fraction according to mass proportion. In the Okinawa Trough, we estimate a bulk sediment flux of about 68,600 Gg/a based on the sediment accumulation rate of 0.049 $g \cdot cm^{-2} \cdot a^{-1}$ [23] and about 140,000 km^2 area of the Okinawa Trough[42]. We cite the sediment fluxes in the prodelta and on the middle-outer shelf from Deng et al.[40]. The detailed data and sources of sediment mass fluxes are summarized in Table S9.

Results and Discussion

▶ *Differential mobilization of sedimentary BC*

Variability of bulk BC concentrations across the Yangtze River-Okinawa Trough transect

Bulk BC concentrations decrease from 1.46 mg/g in the water column of the Yangtze River (particulate BC[43]) to 0.28 mg/g in sediments on the outer shelf, and then increase to 0.67 mg/g in sediments in the Okinawa Trough (Fig. 1). The BC concentrations of bulk surface sediments in the ECS, varying from 0.28 mg/g to 0.86 mg/g (Fig. 1), could be comparable to previous studies in this region (0.30–1.70 mg/g[12,21]). The spatial distribution of bulk BC concentrations along the transect is related to mean grain size of sediments (Fig. S1), which are fine in the prodelta and Okinawa Trough with high BC concentrations but very coarse on the outer shelf with low BC concentrations. The relationship between bulk BC concentrations and sediment grain size is also observed in a previous study[12]. Sediment grain size is closely related to the hydrodynamic processes[31], such as resuspension and bedload processes[27]. The correlation between bulk BC concentration and mean grain size in our study sites (Fig. S1) indicates that hydrodynamic processes may have a significant influence on BC concentrations in sediments.

Variability of grain size-fractionated BC concentrations across the Yangtze River-Okinawa Trough transect

Our results show that BC concentrations among different grain size fractions range from 0.23 mg/g to 1.69 mg/g along the transect (Fig. 2b). The highest BC concentrations at all the sites are in the <20 μm fraction, and the lowest BC concentrations are in the 63–125 μm fraction except at site P01 (Fig. 2b). These distribution patterns of BC concentrations among different grain size sediments may be related to the BC-mineral interactions. Fine-sized minerals with high surface areas such as clay are thought to adsorb and protect organic matters[31,32]. BC in fine-sized sediments may interact closely with minerals[33], making it difficult for desorption during transportation, likely explaining the high BC concentrations of the <20 μm fraction. In contrast, minerals in coarse sediments weakly interact with organic matter[31], helping explain the low BC concentrations in coarse-grained sediments.

Differential BC mobilization triggered by hydrodynamic processes in the ECS may also be responsible for the heterogeneous distribution of BC concentrations among different grain size sediments. Sediments can be differentially transported by diverse hydrodynamic processes in the ECS[24,28], which is reflected by the different

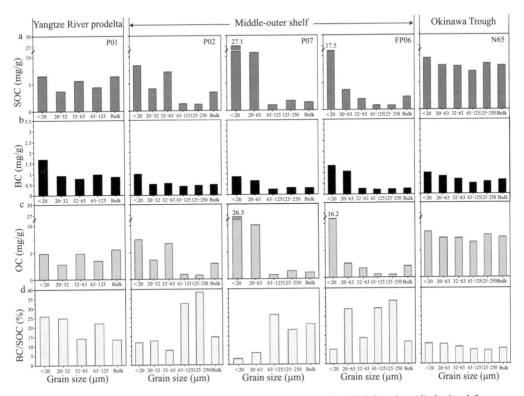

Fig. 2 SOC, BC, OC (SOC-BC) concentrations (a–c), and BC/SOC ratios (d) derived from different grain size fractions and bulk sediments along the across-shelf transect

grain size distribution patterns[27]. In the prodelta area with fine-grained sediments, resuspension processes preferentially transport particles of sortable silt fraction (10–63 μm, see Bao et al. (2016)[27] and references therein). On the middle-outer shelf, bedload processes mobilize the coarse-grained sediments under the stronger flow conditions (see Bao et al. (2019)[28] and references therein). In the Okinawa Trough, the low bottom current velocity (<0.03 m/s) indicates a relatively weak hydrodynamic process[44], which is not conducive to marked movement of sediments. This differential mobilization of different grain size particles will lead to differential mobilization of SOC adsorbed on these particles[27,28]. BC can also be absorbed on the particles[7,34], and sorted with these grain size sediments under hydrodynamic processes[4,34]. A part of sediments in the prodelta area is transported across-shelf to the Okinawa Trough by hydrodynamic processes[23,38,39]. Hence, hydrodynamic processes may cause the differential mobilization of BC with different grain size sediments in the ECS, which contributes to the heterogeneous distribution of BC concentrations among different grain size sediments.

The BC concentrations among different grain size sediments exhibit similar distribution pattern, but less obvious features to SOC or OC concentrations (Fig. 2). There is a positive correlation between grain size-fractionated BC and SOC concentrations (Fig. S2), implying that BC and SOC may undergo similar mobilization in the ECS. However, the changes of SOC or OC concentrations among grain size-fractionated sediments are more significant than BC, particularly with higher SOC or OC concentrations in fine-size fractions than coarse-size fractions (Fig. 2). Since BC is more refractory than OC[2,3,45], the effect of mineral adsorption and protection on fine-grained OC from degradation may be stronger than that of BC. This adsorption and protection effect cause more OC accumulated in fine-grained sediments and less OC in coarse-grained sediments under strong resuspension and bedload processes[31], which is reflected in more obvious varying distribution patterns of OC concentrations among different grain size in the prodelta and on the middle-outer shelf. The preferential mobilization of BC, SOC, and OC in fine-grained sediments from the prodelta and coarse sediments from the middle-outer shelf to the Okinawa Trough may lead to the uniform distributions of BC, SOC, and OC concentrations in the Okinawa Trough (Fig. 2). The weak hydrodynamic processes in the Okinawa Trough also contribute to the uniform distributions of BC, SOC, and OC concentrations among different grain size sediments.

▶ *Ages and sources of sedimentary BC*

The influence of hydrodynamic processes on ages of sedimentary BC

In general, the $\Delta^{14}C_{BC}$ values are from –62.83‰ to –65.04‰ among different grain size sediments at P01, and –69.07‰ to –71.67‰ at N65 (Fig. 3). The age offsets of BC between P01 and N65 are from 1145 ± 330 to 1700 ± 335 years (average 1550 ± 332 years) in corresponding size fractions (Fig. S3). Obviously, the BC ages in the Okinawa Trough are older than the Yangtze River prodelta. BC from aerosols can be aged in the water column and incorporated or adsorbed into the sinking particulate OC to the sediments[7,17]. Some researchers have proposed that the fossil SOC from Taiwan rivers can be transported to the Okinawa Trough[39]. While these additional sourced BC may be responsible for the older BC ages in the Okinawa Trough, the BC from aerosols and Taiwan rivers input may show obvious distinct ages and concentrations in different size fractions, yet this potential is incompatible with the homogenous ages and abundances of grain size-fractionated BC in the Okinawa Trough observed in this study (Fig. 3). For example, the aged BC from aerosols is submicron size level[5], which cannot significantly influence the BC ages and abundances of coarse grain size fraction. In this scenario, the additional input of aged BC from Taiwan rivers input or atmospheric depositions, may be not a main reason for older BC ages in the Okinawa Trough.

The increasing ^{14}C ages of grain size-fractionated

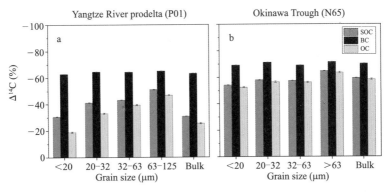

Fig. 3 $\Delta^{14}C$ values of BC, OC, and SOC derived from different grain size and bulk sediments of the Yangtze River prodelta (a, P01) and the Okinawa Trough (b, N65)

sedimentary BC seaward may imply BC aging during the lateral transport. Sedimentary BC absorbed on the particles could be resuspended and dispersed repeatedly during the mobilization on the seafloor, like SOC[27,28]. Under this process, some pre-aged sedimentary BC could be subject to prolonged transport in which they absorb and re-absorb within the particles, leading to the BC ^{14}C aging. In addition, given that the sedimentary BC, particularly for soot and graphitic BC, is refractory[3,45] during the transport processes[12], BC selective degradation are unlikely to account for the older BC ages in the Okinawa Trough. Together, we suggest that the BC aging during the mobilization under the hydrodynamic processes may be the main reason for the older ages of sedimentary BC in the Okinawa Trough.

Different ages between BC and SOC or OC

In this study, we obtain new $^{14}C_{SOC}$ data of the Okinawa Trough, with ranges from −54.36‰ to −64.85‰ (Fig. 3b). The distribution of $\Delta^{14}C_{SOC}$ values in Okinawa Trough among different grain size sediments is consistent with the grain size-fractionated $\Delta^{14}C_{SOC}$ distributions in the ECS[28], which have high $\Delta^{14}C_{SOC}$ values in fine grain size and low $\Delta^{14}C_{SOC}$ values in coarse grained sediments, demonstrating that old SOC prevails in coarse sediments and young SOC dominates in fine sediments. The distribution of $\Delta^{14}C_{SOC}$ values varying from −30.29‰ to −50.99‰ among different grain size sediments in P01 is more heterogeneous than in N65 (Fig. 3). The $\Delta^{14}C_{OC}$ values ranging from −18.90‰ to −47.00‰ in P01 and −52.59‰ to −63.70‰ in N65 also show similar distributions (Fig. 3). This indicates the ages of SOC or OC become older and more homogenous among different size fractions after mobilization, which could be attributed to the selective degradation of SOC or OC under the intensive hydrodynamic processes. The preferential degradation of fresh organic matter absorbed on fine-size particles and the movement aging of organic matter during mobilization lead to older ages of SOC or OC in the <20 μm fraction. Meanwhile, there may be only movement aging for organic matter on coarse particles due to poor preservation of fresh organic matter on >125 μm fraction[27,28]. This selective degradation of SOC or OC may be partly responsible for the diminished difference of $\Delta^{14}C_{SOC}$ between different size fractions after mobilization.

The smaller variations of $\Delta^{14}C_{BC}$ in different size fractions compared with $\Delta^{14}C_{OC}$ and $\Delta^{14}C_{SOC}$ (Fig. 3) indicate that sedimentary BC is more refractory than SOC or OC. The decreasing age offsets between BC and SOC or OC from the prodelta to the trough (Fig. S4) highlight that BC is more stable than SOC or OC during the mobilization. Therefore, BC may have a different cycling fate from SOC or OC in the ocean, which is reflected in the fact that the turnover time of sedimentary BC is much longer than SOC or OC[4,6].

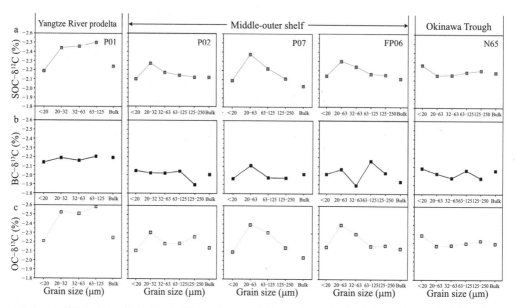

Fig. 4　$\delta^{13}C$ values of SOC (a), BC (b), and OC (SOC-BC) (c) derived from different grain size fractions and bulk sediments along the cross-shelf transect

Source apportionments of grain size-fractionated sedimentary BC

The dual carbon discriminant diagram for source apportionments of BC shows that our dataset falls within the end-member range of C3 and C4 biomass, fossil fuel, and graphitic BC (Fig. 5a)[9,18,21,46,47]. Our $\Delta^{14}C$ values of sedimentary BC (−69.07‰ to −71.67‰) in the Okinawa Trough are lower than that in the ECS shelf (−21.68‰ ± 12.3‰ to −58.55‰ ± 9.5‰[21]), implying more ^{14}C-depleted BC burial in the Okinawa Trough. Besides, the $\delta^{13}C_{BC}$ values of −1.89‰ to −2.16‰ in the middle-outer shelf and −1.97‰ to −2.08‰ in the Okinawa Trough are generally higher than −2.14‰ to −2.21‰ in the Yangtze River prodelta (Fig. 4b). Dickens et al.[18] found that the ^{14}C-dead petrogenic graphitic BC buried in the coastal areas and open ocean has marine source-like values of $\delta^{13}C_{BC}$ (about (2.03 ± 0.10)‰). The $\delta^{13}C$ values of fossil BC from fuel vary from −2.30‰ to −2.80‰[46,21]. Based on the $\delta^{13}C_{BC}$ values, we calculated the percentages of graphite- and fossil fuel-sourced BC by two end-member model (Table S6). There is obviously more graphitic BC burial in the Okinawa Trough (Fig. S5). Wu et al.[26] and Huang et al.[12] also suggested that there is much graphitic BC buried in the outer shelf. Taken together, we argue that the burial of graphitic BC could enhance towards offshore in the ECS.

BC aging during the mobilization could affect the burial of BC from different sources. The results of $\Delta^{14}C_{BC}$ indicate about 32% ± 2% of biomass-sourced BC ($BC_{biomass}$) and about 68% ± 2% of fossil-sourced BC (BC_{fossil}) in the Yangtze River prodelta, and about 26% ± 2% of $BC_{biomass}$ and about 74% ± 2% of BC_{fossil} in the Okinawa Trough (Fig. 5b). The proportions of $BC_{biomass}$ and BC_{fossil} among different grain size sediments in the prodelta and the trough are relative homogenous (Fig. 5b), revealing their similar sources of grain size-fractionated BC and reinforcing the interpretation of BC mobilization. Along the transect, the increase of BC_{fossil} proportion seaward may be related to the BC aging during the mobilization. Indeed, a higher graphitic BC percentage is found in the 20–63 μm, a sortable silt fraction easy to be entrained during the resuspension-deposition loops (see Bao et al. (2016)[27] and references therein), in the Okinawa Trough (Fig. S5), which indicates that the proportion of graphitic

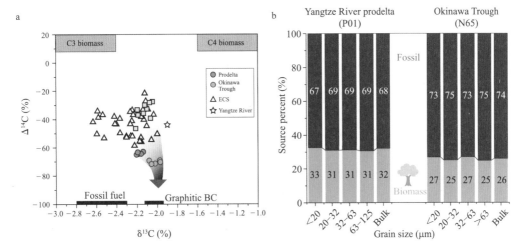

The end-member values of C3 and C4 biomass, fossil fuel, and graphitic BC are shown[9,21,46,47]. Data of sedimentary BC in the ECS (open triangle), particulate BC in the Yangtze River (star) and Yellow River (square) are cited from Wang et al.[21,43]. The gray shadow area indicates a trend towards graphitic BC end-member from the ECS shelf to the Okinawa Trough. The >63 μm sediment is combined by the 63–125 μm and 125–250 μm size fractional sediments.

Fig. 5 Plot of $\Delta^{14}C$ (%) versus $\delta^{13}C$ (%) values for sedimentary BC in the ECS (a) and source percentages of biomass and fossil fuel among different grain size and bulk sediments of the Yangtze River prodelta (P01) and the Okinawa Trough (N65) (b)

BC in sedimentary BC in the 20–63 μm fraction may increase under hydrodynamic sorting processes.

Marine settings may play an important role in modifying the proportion of different sourced BC which bury on the seafloor. Our results of source apportionments indicate about 32% ± 2% $BC_{biomass}$ in the Yangtze River prodelta and about 26% ± 2% in the Okinawa Trough, which is similar to the $BC_{biomass}$ relative contribution to total BC emissions in China (about 26%[48]) (Fig. 6a). ^{14}C analysis of the Yangtze River particulate BC showed that the about 49% particulate BC is from biomass combustion[43]. These estimations may suggest that particulate BC is getting old during transportation from the Yangtze River to the ECS. Alternatively, after entering the ocean, biomass-sourced particulate BC may be leached to dissolved BC by chemical oxidation or biodegradation, then undergoes degradation[6,8,43]. These processes may also cause the $BC_{biomass}$ proportion to decrease from the Yangtze River to the ECS. The above evidences together indicate that marine settings may regulate the relative proportion of $BC_{biomass}$ and BC_{fossil} in the ocean.

▶ *Differential sequestration of sedimentary BC*

The burial of SOC in the ocean could exert influence on climate change by regulating the atmospheric CO_2 concentrations[35]. More importantly, BC, as the inert component of SOC, may play a much more vital role in carbon sequestration as it can reside in the ocean for millennial or even geological timescales[9,10,49]. Our results show that BC accounts for 3%–38% of the SOC in the ECS (Fig. S2, Table S2), which is comparable to the values of about 15%–30% in the global ocean[4,9,13]. The BC/SOC ratios vary from 0.14 to 0.26 among different grain size fractions in the Yangtze River prodelta, while ranging from 0.07 to 0.11 in the Okinawa Trough (Fig. 7a). These spatial distributions of BC/SOC indicate that sedimentary BC accounting for SOC in the ECS is heterogeneous, which may imply differential sequestration of BC in the ocean.

The relationship between BC concentrations and terrestrial SOC% (Terr-SOC%) is investigated

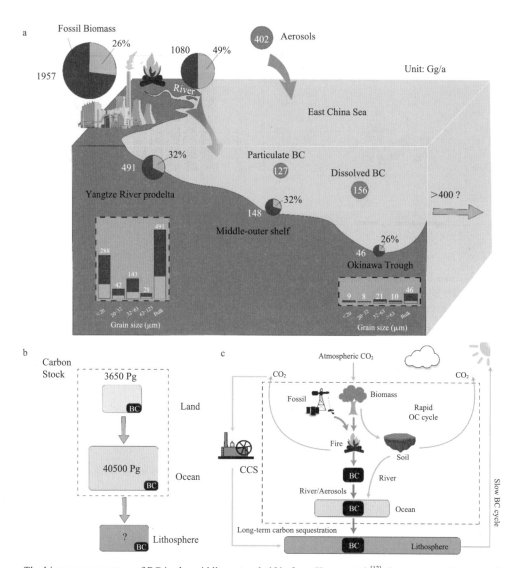

The biomass percentage of BC in the middle-outer shelf is from Huang et al.[12]. Source apportionment of river particulate BC is calculated after Wang et al.[43]. See Table S9 for detailed data and references. Carbon stock in land, ocean, and lithosphere is revised after Jones et al.[49]. The question mark indicates uncertain geological carbon stock. The CCS is referred to the technique of carbon capture and storage[51].

Fig. 6 BC input and burial fluxes in the ECS (a), carbon stock in land, ocean, and lithosphere (b), and a concept model of BC sequestering atmospheric CO_2 into the ocean (c)

to evaluate the effect of BC on the burial of Terr-SOC. Terr-SOC is a key component of SOC in marine sediments, which has more complex sources and variable fates[30]. BC shows a high correlation with Terr-SOC in bulk samples while no obvious correlation in different grain size fractions (Fig. S6). The distribution of grain size-fractionated BC concentrations and Terr-SOC% may be influenced by the hydrodynamic processes, leading to poor correlations between grain-size fractionated BC and Terr-SOC.

BC accounts for 34% of Terr-SOC in the Yangtze River prodelta and 27% in the Okinawa Trough (Fig. 7b). Particularly in the <20 μm fraction of the Yangtze River prodelta, BC accounts for 82% of Terr-SOC (Fig. 7b), suggesting that BC in

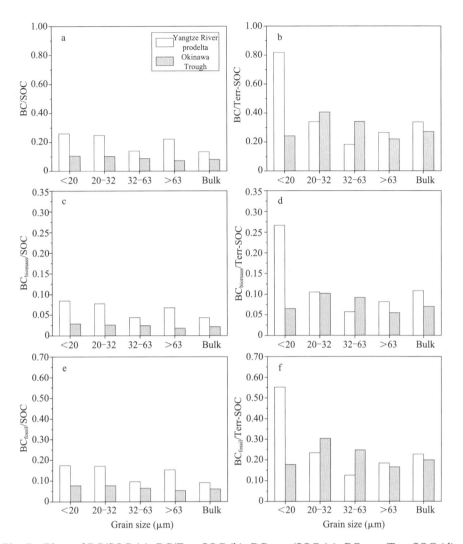

Fig. 7　Plots of BC/SOC (a), BC/Terr-SOC (b), $BC_{biomass}$/SOC (c), $BC_{biomass}$/Terr-SOC (d), BC_{fossil}/SOC (e), and BC_{fossil}/Terr-SOC (f) versus sediment grain size in the Yangtze River prodelta (P01) and the Okinawa Trough (N65)

fine-grained sediments plays a critical role in the sequestration of Terr-SOC. The ratios of biomass- and fossil-sourced BC to Terr-SOC ($BC_{biomass}$/Terr-SOC and BC_{fossil}/Terr-SOC) in the 20–63 μm fraction of the Okinawa Trough is higher than that of the Yangtze River prodelta (Fig.7b, d, f), which may be related to the hydrodynamic sorting of these grain-sized particles. The BC adsorbed on these particles could be preferentially transported to the open ocean relative to Terr-SOC during the sorting processes[18], resulting in a higher proportion of BC in Terr-SOC sequestration.

BC burial fluxes in different size fractions along the transect are estimated to determine the differential sequestration of BC in the ECS (Fig. 6a). Most of the sedimentary BC is buried in the prodelta area of the ECS (491 Gg/a), which is consistent with the study of Huang et al.[12]. The BC burial fluxes in the middle-outer shelf (148 Gg/a) and the Okinawa Trough (46 Gg/a) are lower than that in the Yangtze River prodelta (Fig. 6a). Taken together, we estimate that 685 Gg/a BC is buried in the ECS, similar to that of (630 ± 728) Gg/a from Fang et al.[20], but lower than that of 1390 Gg/a from Huang et al.[12]. In addition, about 42% of sedimentary BC in the ECS is buried in the <20 μm fraction sediments of the prodelta area (288 Gg/a), suggesting that the fine-grained mud areas in the inner shelf exert the largest potential to sequester BC.

▶ *BC sequestration in the ocean and its implications for carbon neutrality*

Human activities have greatly accelerated the processes of geological carbon entering the atmosphere by burning fossil fuels[37], resulting in the unbalanced release of a large amount of CO_2 into the atmosphere, which has caused profound negative effects on the climate. It is urgent to reduce carbon emissions and increase carbon sinks to offset the effects of excessive anthropogenic CO_2 release. One of the promising solutions is using natural processes such as increasing the marine carbon sink to remove anthropogenic CO_2 emissions. Atmospheric CO_2 is converted by primary producers to organic matter and ultimately buried in marine sediments, serving as an important pathway for the marine carbon sink[50]. These processes could be helpful for "carbon neutrality", a "net zero" emissions by anthropogenic CO_2 removals[37].

BC, as a crucial carbon sink in terrestrial and marine carbon cycles, exhibits a considerable carbon stock on land, and in the ocean and lithosphere (Fig. 6b)[49]. It transfers atmospheric CO_2 into the ocean and eventually into the lithosphere to achieve long-term carbon sequestration, playing a similar role to artificial capture of CO_2 from industrial activities and injection into carbon storage[37,51]. The carbon sequestration by the BC pathway could save a considerable amount of money in neutralizing equivalent CO_2 relative to that the carbon capture and storage way nowadays[51]. So far, most studies focused on the function of biochar in soils for carbon neutrality[36,51], few studies pay attention to the contribution of BC burial in the ocean. According to the pilot studies, China emitted about 1957 Gg BC per year[48], and about 1080 Gg and about 402 Gg is transported to the ECS by rivers and aerosols per year, respectively (Fig. 6a)[12], accounting for about 80% of BC emissions from China. If the BC sink can be fully utilized[36,37,51], for instance, increasing BC burial flux in the ocean, it would be an economic pathway to remove CO_2 from the atmosphere and may contribute to carbon neutrality.

It is worth noting that $BC_{biomass}$ and BC_{fossil} play different roles in the sequestration of atmospheric CO_2[6]. The formation of $BC_{biomass}$, followed by its burial, transfers atmospheric CO_2 into long-term carbon cycling pools. However, the formation of BC_{fossil} just converts the fossil-fuel carbon to BC, another inert carbon, without net sequestration of atmospheric CO_2. $BC_{biomass}$ cycles slowly and is stored and transported in soil, river, and finally buried in the ocean as a real carbon sink, which exert positive impacts on long-term climate change mitigation[6,36,51]. According to the percentage of $BC_{biomass}$ combined with total BC flux in this study, we estimate that about 216 Gg/a of $BC_{biomass}$ buried in the ECS could sequester a considerable amount of atmospheric CO_2 into sediments. We believe that increasing $BC_{biomass}$ production by smoldering biomass fuel or raising the amount of biomass fuel pyrolysis and burying it in the ocean could contribute to carbon neutrality[37,51]. However, the CO_2 emission during the biomass combustion process should be considered, because we have to make sure that the climate mitigation effect of $BC_{biomass}$ overwhelms the warming effect of CO_2 emission.

Conclusions and Implications

This study presents a comprehensive investigation of concentrations and isotopes of BC among different fractional sediments along a cross-shelf transect in the ECS, shedding light on the differential mobilization and sequestration of sedimentary BC on the shelf and trough. We find that bulk BC concentrations are closely associated with the mean grain size of sediments. BC concentrations among different size fractions are heterogeneous, which may be due to the differential mobilization of different fractionated BC under hydrodynamic processes in the ECS and BC-mineral interactions. The isotopic characteristics of BC suggest that BC may be aged during the mobilization from the Yangtze River prodelta to the Okinawa Trough. BC aging could exert an influence on the relative proportion of biomass- and fossil-sourced BC in the ocean. We also find that most sedimentary BC is buried in the prodelta area with fine grain size sediments, indicating differential sequestration of BC in the ECS. There are about 685 Gg/a BC and about 216 Gg/a $BC_{biomass}$ burial in the ECS, indicating BC, particularly $BC_{biomass}$, has a huge potential to neutralize excessive CO_2 emitted by human beings.

Our findings suggest that BC aging during mobilization could be a main reason for the older BC ages in the trough, leading to fossil-derived BC (calculated by ^{14}C) increases in the Okinawa Trough. Given the CTO-375 method used in this study could involve the ^{14}C-dead graphitic BC, the possibility of additional graphitic BC input to the Okinawa Trough cannot be excluded. It is necessary to further determine the provenance of graphitic BC in the Okinawa Trough in the future. Furthermore, we find that BC has a different cycling fate than OC or SOC in the ocean, due to their distinct degradation nature. In this scenario, special attention should be paid to the applications of organic geochemical proxies which are closely associated with OC or SOC, because their properties may change under the effects of resuspension and bedload processes. In contrast, the small variability of $\Delta^{14}C_{BC}$ under hydrodynamic processes implies that the geochemical characteristics of sedimentary BC remain relatively stable during the mobilization, suggesting that BC-related proxies may be more reliable.

Acknowledgements

This study is supported by the Special Project for Introduced Talents Team of Southern Marine Science and Engineering Guangdong Laboratory (Guangzhou) (Grant No. GML2019ZD0209), National Natural Science Foundation of China (Grants: 92058207 and 42076037), and the Fundamental Research Funds for the Central Universities (Grant: 2020042010). This paper is also granted by the State Key Laboratory of Organic Geochemistry, GIGCAS (Grant: SKLOG202020), and Taishan Young Scholars (Grant: tsqn202103030), and Shandong Natural Science Foundation (Grant: ZR2021JQ12).

Author Contributions

Jingyu Liu: Conceptualization, formal analysis, investigation, methodology, writing (original draft). Nan Wang: Investigation, methodology, writing (review and editing). Cuimei Xia: Investigation, writing (review and editing). Weifeng Wu: Methodology, writing (review and editing). Yang Zhang, Guangxue Li, Yang Zhou, Guangcai Zhong and Gan Zhang: Writing (review and editing). Rui Bao: Conceptualization, funding acquisition, project administration, supervision, writing (review and editing).

Competing Interests

The authors declare that they have no known competing financial interests or personal relationships that could have appeared to influence the work reported in this paper.

References

1. Goldberg E D. Black Carbon in the Environment[M]. New York: John Wiley, 1985.
2. Gustafsson Ö, Haghseta F, Chan C, et al. Quantification of the dilute sedimentary soot phase: Implications for PAH speciation and bioavailability[J]. Environ. Sci. Technol., 1997, 31: 203-209.
3. Gustafsson Ö, Bucheli T D, Kukulska Z, et al. Evaluation of a protocol for the quantification of black carbon in sediments[J]. Glob. Biogeochem. Cycles, 2001, 15: 881-890.
4. Middelburg J J, Nieuwenhuize J, van Breugel P. Black carbon in marine sediments[J]. Mar. Chem., 1999, 65: 245-252.
5. Jurado E, Dachs J, Duarte C M, et al. Atmospheric deposition of organic and black carbon to the global oceans[J]. Atmospheric Environ., 2008, 42: 7931-7939.

6. Mitra S, Zimmerman A R, Hunsinger G, et al. Black carbon in coastal and large river systems[M]//Bianchi T, Allison M, Cai W J. Biogeochemical Dynamics at Major River-Coastal Interfaces. New York: Cambridge University Press, 2013.
7. Coppola A I, Ziolkowski L A, Masiello C A, et al. Aged black carbon in marine sediments and sinking particles[J]. Geophys. Res. Lett., 2014, 41: 2427-2433.
8. Qi Y, Fu W, Tian J, et al. Dissolved black carbon is not likely a significant refractory organic carbon pool in rivers and oceans[J]. Nat. Commun., 2020, 11: 5051.
9. Lohmann R, Bollinger K, Cantwell M, et al. Fluxes of soot black carbon to South Atlantic sediments[J]. Glob. Biogeochem. Cycles, 2009, 23: GB1015.
10. Masiello C A, Louchouarn P. Fire in the ocean[J]. Science, 2013, 340: 287-288.
11. Yang W, Guo L. Abundance, distribution, and isotopic composition of particulate black carbon in the northern Gulf of Mexico: Yang et al., black carbon in seawater[J]. Geophys. Res. Lett., 2014, 41: 7619-7625.
12. Huang L, Zhang J, Wu Y, et al. Distribution and preservation of black carbon in the East China Sea sediments: Perspectives on carbon cycling at continental margins[J]. Deep Sea Res. Part II Top. Stud. Oceanogr., 2016, 124: 43-52.
13. Zhang X, Xu Y, Xiao W, et al. The hadal zone is an important and heterogeneous sink of black carbon in the ocean[J]. Commun. Earth Environ., 2022, 3: 25.
14. Coppola A I, Wiedemeier D B, Galy V, et al. Global-scale evidence for the refractory nature of riverine black carbon[J]. Nat. Geosci., 2018, 11: 584-588.
15. Fang Y, Chen Y, Huang G, et al. Particulate and dissolved black carbon in coastal China seas: Spatiotemporal variations, dynamics, and potential implications[J]. Environ. Sci. Technol., 2021, 55: 788-796.
16. Bird M I, Wynn J G, Saiz G, et al. The pyrogenic carbon cycle[J]. Annu. Rev. Earth Planet. Sci., 2015, 43: 273-298.
17. Bao H, Niggemann J, Luo L, et al. Aerosols as a source of dissolved black carbon to the ocean[J]. Nat. Commun., 2017, 8: 510.
18. Dickens A F, Gélinas Y, Masiello C A, et al. Reburial of fossil organic carbon in marine sediments[J]. Nature, 2004a, 427: 336-339.
19. Jaffé R, Ding Y, Niggemann J, et al. Global charcoal mobilization from soils via dissolution and riverine transport to the oceans[J]. Science, 2013, 340: 345-347.
20. Fang Y, Chen Y, Hu L, et al. Large-river dominated black carbon flux and budget: A case study of the estuarine-inner shelf of East China Sea, China[J]. Sci. Total Environ., 2019, 651: 2489-2496.
21. Wang Y, Li T, Zhang R, et al. Fingerprinting characterization of sedimentary PAHs and black carbon in the East China Sea using carbon and hydrogen isotopes[J]. Environ. Pollut., 2020, 267: 115415.
22. Iseki K, Okamura K, Kiyomoto Y. Seasonality and composition of downward particulate fluxes at the continental shelf and Okinawa Trough in the East China Sea[J]. Deep Sea Res. Part II Top. Stud. Oceanogr., 2003, 50: 457-473.
23. Oguri K, Matsumoto E, Yamada M, et al. Sediment accumulation rates and budgets of depositing particles of the East China Sea[J]. Deep Sea Res. Part II Top. Stud. Oceanogr., 2003, 50: 513-528.
24. Zhu Z Y, Zhang J, Wu Y, et al. Bulk particulate organic carbon in the East China Sea: Tidal influence and bottom transport[J]. Prog. Oceanogr., 2006, 69: 37-60.
25. Bao R, Zhao M, McNichol A, et al. On the origin of aged sedimentary organic matter along a river-shelf-deep ocean transect[J]. J. Geophys. Res. Biogeosci., 2019a, 124: 2582-2594.
26. Wu Y, Eglinton T, Yang L, et al. Spatial variability in the abundance, composition, and age of organic matter in surficial sediments of the East China Sea[J]. J. Geophys. Res. Biogeosci., 2013, 118: 1495-1507.
27. Bao R, McIntyre C, Zhao M, et al. Widespread dispersal and aging of organic carbon in shallow marginal seas[J]. Geology, 2016, 44: 791-794.
28. Bao R, Blattmann T M, McIntyre C, et al. Relationships between grain size and organic carbon ^{14}C heterogeneity in continental margin sediments[J]. Earth Planet. Sci. Lett., 2019b, 505: 76-85.
29. Fang Y, Chen Y, Tian C, et al. Flux and budget of BC in the continental shelf seas adjacent to Chinese high BC emission source regions[J]. Glob. Biogeochem. Cycles, 2015, 29: 957-972.
30. Tesi T, Semiletov I, Dudarev O, et al. Matrix association effects on hydrodynamic sorting and degradation of terrestrial organic matter during cross-shelf transport in the Laptev and East Siberian shelf seas[J]. J. Geophys. Res. Biogeosci., 2016, 121: 731-752.
31. Coppola L, Gustafsson Ö, Andersson P, et al. The importance of ultrafine particles as a control on the distribution of organic carbon in Washington Margin and Cascadia Basin sediments[J]. Chem. Geol., 2007, 243: 142-156.
32. Blattmann T M, Liu Z, Zhang Y, et al. Mineralogical control on the fate of continentally derived organic matter in the ocean[J]. Science, 2019, 366: 742-745.
33. Dickens A F, Gélinas Y, Hedges J I. Physical separation of combustion and rock sources of graphitic black

carbon in sediments[J]. Mar. Chem., 2004b, 92: 215-223.

34. Oen A M P, Cornelissen G, Breedveld G D. Relation between PAH and black carbon contents in size fractions of Norwegian harbor sediments[J]. Environ. Pollut., 2006, 141: 370-380.

35. Masiello C A. New directions in black carbon organic geochemistry[J]. Mar. Chem., 2004, 92: 201-213.

36. Lehmann J, Cowie A, Masiello C A, et al. Biochar in climate change mitigation[J]. Nat. Geosci., 2021, 14: 883-892.

37. IPCC Scientists. The Intergovernmental Panel on Climate Change (IPCC) Special Report on Global Warming of 1.5°C[R]. IPCC, 2018.

38. Liu J P, Li A C, Xu K H, et al. Sedimentary features of the Yangtze River-derived along-shelf clinoform deposit in the East China Sea[J]. Cont. Shelf Res., 2006, 26: 2141-2156.

39. Kao S J, Dai M H, Wei K Y, et al. Enhanced supply of fossil organic carbon to the Okinawa Trough since the last deglaciation[J]. Paleoceanography, 2008, 23: PA2207.

40. Deng B, Zhang J, Wu Y. Recent sediment accumulation and carbon burial in the East China Sea[J]. Glob. Biogeochem. Cycles, 2006, 20: GB3014.

41. Blair T C, McPherson J G. Grain-size and textural classification of coarse sedimentary particles[J]. J. Sediment. Res., 1999, 69: 6-19.

42. Meng X, Zhang X, Han B, et al. Features of geophysical field and crustal structure of Okinawa trough and its adjacent regions[J]. Mar. Geol. Front., 2015, 31: 1-6. (In Chinese with English abstract)

43. Wang X, Xu C, Druffel E M, et al. Two black carbon pools transported by the Changjiang and Huanghe Rivers in China[J]. Glob. Biogeochem. Cycles, 2016, 30: 1778-1790.

44. Yuan P, Wang H, Bi N, et al. Temporal and spatial variations of oceanic fronts and their impact on transportation and deposition of fine-grained sediments in the East China Shelf Seas[J]. Mar. Geol. & Quat. Geol., 2019, 40: 25-42. (In Chinese with English abstract)

45. Hammes K, Schmidt M W I, Smernik R J, et al. Comparison of quantification methods to measure fire-derived (black/elemental) carbon in soils and sediments using reference materials from soil, water, sediment and the atmosphere[J]. Glob. Biogeochem. Cycles, 2007, 21: GB3016.

46. Gustafsson Ö, Kruså M, Zencak Z, et al. Brown clouds over South Asia: Biomass or fossil fuel combustion?[J]. Science, 2009, 323: 495-498.

47. Bird M I, Ascough P L. Isotopes in pyrogenic carbon: a review[J]. Org. Geochem., 2012, 42: 1529-1539.

48. Wang R, Tao S, Wang W, et al. Black carbon emissions in China from 1949 to 2050[J]. Environ. Sci. Technol., 2012, 46: 7595-7603.

49. Jones M W, Santín C, van der Werf G R, et al. Global fire emissions buffered by the production of pyrogenic carbon[J]. Nat. Geosci., 2019, 12: 742-747.

50. Zhang Y, Zhao M, Cui Q, et al. Processes of coastal ecosystem carbon sequestration and approaches for increasing carbon sink[J]. Sci. China Earth Sci., 2017, 60: 809-820.

51. Hepburn C, Adlen E, Beddington J, et al. The technological and economic prospects for CO_2 utilization and removal[J]. Nature, 2019, 575: 87-97.

Persistently high efficiencies of terrestrial organic carbon burial in Chinese marginal sea sediments over the last 200 years[①]

Meng Yu[1,2,3], Timothy I. Eglinton[3*], Negar Haghipour[3,5], Nathalie Dubois[3,4], Lukas Wacker[5], Hailong Zhang[1,2], Gui'e Jin[1,2], Meixun Zhao[1,2*]

1 Frontiers Science Center for Deep Ocean Multispheres and Earth System, and Key Laboratory of Marine Chemistry Theory and Technology, Ministry of Education, Ocean University of China, Qingdao 266100, China
2 Laboratory for Marine Ecology and Environmental Science, Qingdao National Laboratory for Marine Science and Technology, Qingdao 266237, China
3 Geological Institute, Department of Earth Sciences, ETH Zürich, 8092 Zürich, Switzerland
4 Swiss Federal Institute of Aquatic Science and Technology, Eawag, Department of Surface Waters - Research and Management, 8600 Dübendorf, Switzerland
5 Laboratory for Ion Beam Physics, Department of Physics, ETH Zürich, 8093 Zürich, Switzerland
* Corresponding authors: Timothy I. Eglinton (timothy.eglinton@erdw.ethz.ch), Meixun Zhao (maxzhao@ouc.edu.cn)

Abstract

Constraining the origins, transport history, and burial efficiency of terrestrial organic carbon (OC_{terr}) accumulating in marine sediments is of fundamental importance for understanding the carbon cycle on a range of spatial and temporal scales. While there is abundant evidence that OC composition and age influences the sequestration of OC_{terr} in surface sediments, little is known about longer-term controls on OC_{terr} sources and burial efficiencies in response to natural and anthropogenic processes that influence marginal sea sediments. Here, we use bulk and molecular-level carbon isotopic ($\delta^{13}C$ and $\Delta^{14}C$) measurements to examine depth-related variations in the sources, ages and burial efficiency of OC_{terr} in a sediment core that captures deposition over the last 200 years in the central Yellow Sea mud area, the largest mud deposit in eastern Chinese marginal seas. The similar ^{14}C ages of terrestrial higher plant long-chain fatty acid biomarkers (1830–2700 a) to those of sedimentary OC (1890–3360 a) suggest the continuous and dominant supply of pre-aged OC ($OC_{pre-aged}$). Two carbon isotopic mixing models are used to apportion contributions from different terrestrially-derived OC pools. A dual carbon mixing model based on bulk OC and molecular $\delta^{13}C$ and $\Delta^{14}C$ values showed $OC_{pre-aged}$ and fossil OC (OC_{fossil}) accounted for 52% ± 3% and 13% ± 2% of sedimentary OC, respectively; while a binary mixing model based on bulk $\delta^{13}C$ values showed OC_{terr} accounted for 45% ± 3% of sedimentary OC. Estimates of high burial efficiency for OC (about 60%) and especially for the different terrestrial OC pools (>80%) over the last 200 years highlight that refractory millennial-aged terrestrial OC inputs from the Yellow River combined with stable depositional conditions promote OC preservation in the Yellow Sea. Notably, $OC_{pre-aged}$ and OC_{fossil} exhibited contrasting fates, with higher burial efficiency for

---
[①] 本文于2022年9月发表在 *Chemical Geology* 第606卷，http://doi.org/10.1016/j.chemgeo.2022.120999。

the former. These observations are attributed to differences in mineral associations, transport pathways as well as changes in sediment provenance and hydrodynamic regime. Furthermore, our results suggest that initial sequestration of OC_{terr} in surface sediments presages long-term burial in deeper sediments despite the existence of temporal variability. Although not directly influenced by riverine discharge, persistent and efficient sequestration of millennial-aged terrestrial OC renders the distal marginal sea mud area a long-term carbon sink on (at least) centennial to millennial timescales.

Keywords: terrestrial organic carbon; high burial efficiency; molecular-level isotopes (^{13}C and ^{14}C); marginal seas

Introduction

Organic carbon (OC) burial in marine sediments is an important sink in the global carbon cycle, with up to 90% of the burial occurring on continental margins[1]. Marginal seas are highly dynamic and variable in terms of OC sources and sedimentary processes. Rivers export large fluxes of terrestrial OC (OC_{terr}) from the continents to the ocean, influencing the atmospheric CO_2 inventory on a range of timescales through the balance between oxidation and sedimentary burial of OC_{terr} depending on its type and origin[2-4]. Hydrodynamic processes during sediment transport as well as post-depositional alteration further influence the chemical compositions, reactivities and ages of OC_{terr}[5-7]. Constraining the burial efficiency of OC_{terr} accumulating in marine sediments is of fundamental importance for understanding the dynamics of the carbon cycle, yet the complex origins, composition and processes acting upon OC_{terr} delivered to marginal seas render it challenging to determine its sedimentary fate[5]. Multiple approaches are required to constrain the distribution and preservation of different components of OC_{terr} (e.g., soil derived pre-aged OC, plant OC, and fossil OC from rock erosion and/or anthropogenic inputs), and to distinguish contributions from marine OC. Characteristics that have been exploited include those based on bulk (e.g., C/N ratio, $\delta^{13}C$ and $\Delta^{14}C$) and molecular (e.g., phytoplankton, higher plant and soil bacterial biomarkers) parameters[8-10]. Heterogeneous OC sources can be further constrained by contents and carbon isotopic compositions ($\delta^{13}C$ and $\Delta^{14}C$) of specific biomarkers that retain the isotopic signals of their corresponding carbon sources and obviate interferences from other carbon pools in order to aid in interpretation of bulk OC isotopic signatures[11-13]. Compound-specific radiocarbon analysis has also been used to investigate OC_{terr} storage in intermediate reservoirs, aging and transport in fluvial systems, estuarine systems and coastal seas[12,13], shedding further light on the fate of OC_{terr} buried in marginal seas.

The eastern Chinese marginal seas (including the Bohai Sea, Yellow Sea and East China Sea) comprise a globally important continental shelf sea system. It receives large amounts of sediment and OC_{terr} discharged by the Yellow River and Yangtze River[14], and accounts for about 10% of OC burial on continental margins globally[15,16]. Several mud deposits exist as a consequence of the interaction of coastal and ocean currents and act as the OC sink. The central Yellow Sea mud area (Fig. 1) is the largest of these deposits and accounts for 21% of fine-grained sediment within the eastern Chinese marginal seas[16,17]. Recent assessments show large spatial variability of OC_{terr} burial but reveal generally higher burial efficiency of OC_{terr} in the central Yellow Sea than in other Chinese marginal seas, and also higher than the global average for river-dominated margins[15,18-20]. However, these and other assessments of "burial efficiency" have generally been based on OC_{terr} residing in the upper few centimeters of marine sediments (typically in the surface layer mixed by physical and biological processes), and thus does not constitute long-term burial in the strictest sense[8,21,22]. Although surface sediments represent a critical interface through

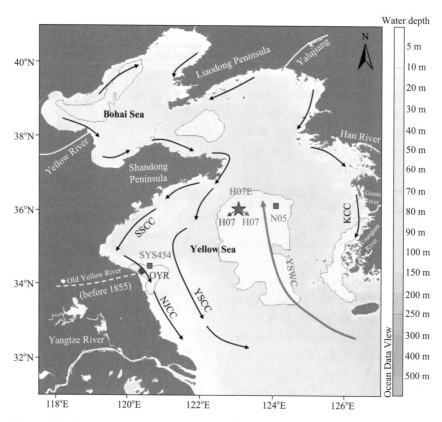

H07, N05 and SYS454 (pink square) are from Tao et al.[27]; H07 and OYR (blue dot) are from Yu et al.[20]. Mud areas are indicated by light grey shade[17]. YSWC: Yellow Sea Warm Current; YSCC: Yellow Sea Coastal Current; SSCC: South Shandong Coastal Current; NJCC: North Jiangsu Coastal Current; KCC: Korea Coastal Current. The map was generated by Ocean Data View software[28].

Fig. 1 Locations of core H07E (red star) in the central Yellow Sea mud area, as well as previously reported surface sediment samples

which OC_{terr} must pass, a full assessment of burial potential requires evaluation of OC_{terr} fate over longer timescales[20].

One of the confounding factors that complicate assessments of OC burial is the increasing levels of human perturbation of marginal seas. In particular, anthropogenic activity has have markedly altered the fluvial supply of sediment and OC_{terr}[6,23]. Owing to the sharp decline of sediment loads from the Yellow River and Yangtze River in recent decades, the fluxes of fluvial OC_{terr} have also decreased accordingly, and surface sediment mass balances indicate a shift in the importance of fluvial inputs versus coastal erosion (e.g., sediment supply from the abandoned Yellow River delta[19]). Despite the growing awareness of the impact of human activities on the marginal sea carbon cycle, most studies of the fate of OC_{terr} thus far have focused on surface-most sediments. Critically, information is lacking on variations in the content and composition of OC_{terr} in deeper sediments that have accumulated over longer timescales, as well as how this influences assessment of the burial potential of OC_{terr}.

This study focused on decadal- to centennial-scale variations in OC_{terr} burial in Chinese marginal sea sediments through down-core measurements of the contents and carbon isotopic compositions ($\delta^{13}C$ and $\Delta^{14}C$) of total OC (TOC) and n-fatty acid (n-FA) biomarkers, mineral-specific surface area (SA), and grain size of sediment samples from a box core

(H07E) collected in the central Yellow Sea mud area (Fig. 1). The main objectives were (1) to assess down-core/temporal variability of specific OC_{terr} pools constrained by both bulk OC and molecular-level characteristics; (2) to assess the main factors influencing the sequestration of the different OC_{terr} pools; and (3) to evaluate whether "proto-burial" efficiencies of OC_{terr} manifested in surface sediments presage longer-term burial in deeper sediments subjected to protracted diagenetic processes. To our knowledge, this study provided the first constraints on OC_{terr} burial efficiency in a marginal sea depocenter based on high-resolution records of bulk and compound-specific ^{14}C data.

Materials and Methods

▶ *Study area and sampling*

The Yellow Sea is a semi-enclosed, shallow marginal sea in the northwestern Pacific, covering about 4×10^5 km^2 with an average depth of 44 m. Surface-ocean circulation consists of northward Yellow Sea Warm Current (YSWC, a branch of the Kuroshio Current) and southward coastal currents, including the Yellow Sea Coastal Current (YSCC) (Fig. 1) that form a cyclonic gyre in the western Yellow Sea. The Yellow Sea receives sediments from surrounding rivers, including the Yellow River, Yangtze River, and several Korean rivers (Han River, Geum River and Youngsan River). Under the influence of winter basin-wide cyclonic gyre and the Yellow Sea cold water mass, fine-grained sediments are transported, focused and deposited in the relatively quiescent sedimentary environment comprising the central mud area[17,24,25]. This area represents the primary depocenter for both fine-grained sediment and OC burial[16]. Provenance analysis of sediment sources and budget show that these sediments are mainly derived from the modern and Old Yellow River[24,26].

Box core H07E (35°59.456′N, 122°59.956′E; water depth 71 m; total core length 43 cm) from the central Yellow Sea mud area (Fig. 1) was retrieved during the cruise of the National Natural Science Foundation of China in August, 2015. Sediments were sliced onboard at 1 cm intervals and were stored at −20 °C and subsequently freeze-dried before laboratory analyses.

▶ *Stratigraphy*

The age model of core H07E was established using the down-core activities of excess ^{210}Pb ($^{210}Pb_{ex}$). ^{137}Cs activities were measured simultaneously. Freeze-dried and homogenized sediment samples (about 7 g) were placed in sealed tubes and left to equilibrate in order to measure ^{222}Rn and its daughters for 20 d. Radionuclide activities of ^{210}Pb, ^{226}Ra and ^{137}Cs were determined by γ-spectrometry in a High-purity Germanium (HPGe) Well Detector (gamma spectrometer; Canberra Industries) at Eawag (Dubendorf, Switzerland).

▶ *Bulk sediment parameter analyses*

About 1 g of unground sediment sample was heated at 350 °C for 12 h in order to remove organic matter[21]. Mineral-specific surface area (SA) was measured using a 5-point Brunauer-Emmet-Teller (BET) method on a NOVA 4000 surface area analyzer (Quantachrome Instrument). Grain size distribution was measured using a Malvern Mastersizer 2000. The following grain size fractions were defined and quantified: < 4 μm for clay, 4–63 μm for silt and > 63 μm for sand.

About 1 g of homogenized bulk sediment sample was decarbonated using 6 N HCl at room temperature and then rinsed with Milli-Q water several times before being oven-dried at 55 °C. Total OC content (TOC%) and total nitrogen content (TN%) were determined using a Thermo Flash 2000 Elemental Analyzer (EA) with a standard deviation of 0.02% (mass fraction, $n = 6$) and 0.002% (mass fraction, $n = 6$), respectively. Stable carbon isotope ($δ^{13}C$) values were determined using an EA coupled to a Thermo Delta V mass spectrometer, with a standard deviation of less than 0.02% ($n = 6$). Results were reported relative to Vienna Pee Dee Belemnite (VPDB).

For radiocarbon analysis, a portion of pre-weighed sediment samples was fumigated with 12

N HCl (60 °C, 72 h)[29] and subsequently neutralized over NaOH pellets (60 °C, 72 h). The radiocarbon composition was determined by a gas-ion source Mini Carbon Dating System (MICADAS) coupled with an elemental analyzer at the Laboratory for Ion Beam Physics, ETH Zürich[30]. All the radiocarbon data were reported as $\Delta^{14}C$ (‰) values and corresponding conventional ^{14}C ages (years before 1950 CE[70]). For down-core sediments, all $\Delta^{14}C$ values have been decay-corrected for ^{14}C loss since time of deposition (Eq. 1).

$$\Delta^{14}C = (Fm \times e^{\lambda(1950-y)} - 1) \times 1000 \quad (1)$$

where Fm is the fraction modern, λ is the radiocarbon decay constant (1/8267) and y is the year of deposition derived from the ^{210}Pb stratigraphy.

▶ *Biomarker extraction and compound-specific carbon ($\delta^{13}C$ and $\Delta^{14}C$) analyses*

The extraction, purification and isolation of target n-FAs for compound-specific ^{13}C and ^{14}C analyses were carried out at the Biogeoscience Group at ETH Zürich[20]. Compound-specific ^{13}C analysis of purified saturated fatty acid methyl esters (FAMEs) was performed in duplicate by GC-isotope ratio mass spectrometry (GC-IRMS) and $\Delta^{14}C$ values were measured using gas-ion source MICADAS system at ETH Zürich[31]. Target n-FAs for $\Delta^{14}C$ analysis are C_{16} FA, C_{18} FA, C_{24} FA and $C_{26+28+30}$ FAs (combined for sufficient carbon mass) in this study. After correction for procedural blanks and derivative carbon from methanol in the FAMEs, all $\Delta^{14}C$ values were decay-corrected (Eq. 1).

▶ *Carbon isotope mixing models*

Binary mixing model

Marine sedimentary OC is classified as either OC_{terr} or marine OC (OC_{marine}) based on a conventional $\delta^{13}C_{TOC}$ approach using a binary mixing model (Eqs. 2, 3), which has been widely applied to marginal sea sediments[9,10,15].

$$\delta^{13}C_{TOC} = f_{terr} \times \delta^{13}C_{terr} + f_{marine} \times \delta^{13}C_{marine} \quad (2)$$
$$1 = f_{terr} + f_{marine} \quad (3)$$

where $\delta^{13}C_{TOC}$ is the measured value of the sample.

In the Yellow Sea sediments, the terrestrial endmember $\delta^{13}C$ value ($\delta^{13}C_{terr}$) was obtained from measured $\delta^{13}C$ values of Yellow River particulate OC (POC), which is –2.39‰ ± 0.06‰ ($n = 34$[20]), and –2.00‰ ± 0.10‰ is assigned as the marine OC endmember ($\delta^{13}C_{marine}$) to be consistent with previously published studies[20,32]. Although the sharply contrasting $\delta^{13}C$ and $\Delta^{14}C$ values of POC from the Geum River imply a limited contribution to central Yellow Sea sedimentary OC[20,33], the average $\delta^{13}C$ value (–2.40‰) of western Korean Rivers (Han River, Geum River and Youngsan River[32]), was close to that of the Yellow River, and thus the estimated OC_{terr} is taken to represent total fluvially derived OC_{terr}.

Dual carbon isotope mixing model

To quantify the relative fractional contribution of different aged sources, i.e., modern/contemporary OC (OC_{modern}), pre-aged (soil) OC ($OC_{pre-aged}$), and fossil OC (OC_{fossil}), a dual carbon isotope mixing model based on bulk and molecular $\delta^{13}C$ and $\Delta^{14}C$ values[34] is applied to the down-core samples (Eqs. 4–6), which has been previously applied to Bohai Sea and Yellow Sea surface sediments[20,27].

$$\Delta^{14}C_{TOC} = f_{modern} \times \Delta^{14}C_{modern} + f_{pre-aged} \times \Delta^{14}C_{pre-aged} + f_{fossil} \times \Delta^{14}C_{fossil} \quad (4)$$
$$\delta^{13}C_{TOC} = f_{modern} \times \delta^{13}C_{modern} + f_{pre-aged} \times \delta^{13}C_{pre-aged} + f_{fossil} \times \delta^{13}C_{fossil} \quad (5)$$
$$1 = f_{modern} + f_{pre-aged} + f_{fossil} \quad (6)$$

where $\delta^{13}C_{TOC}$ and $\Delta^{14}C_{TOC}$ are the measured values of sediment samples. The $\delta^{13}C$ and $\Delta^{14}C$ values of OC_{modern}, $OC_{pre-aged}$ and OC_{fossil} are constrained by carbon isotopic compositions of specific biomarkers characteristic of different end-members. Although the selected biomarkers generally comprise only a small portion of bulk OC, they can retain the isotopic signals of the corresponding carbon sources and avoid interference from other carbon pools[11]. For down-core sediments, all $\Delta^{14}C$ values have been decay corrected since the deposition time (derived from the ^{210}Pb dating results) prior to the above quantitative source apportionment. $\Delta^{14}C$ values of n-C_{16} FA and n-$C_{26+28+30}$ FAs are assigned as the

$\Delta^{14}C$ endmember values of OC_{modern} and $OC_{pre-aged}$ respectively (See Section "OC characteristics"). According to the down-core variation of bulk OC characteristic, four sub-sets of $\Delta^{14}C$ FAs were assigned as endmember values in four intervals (prior to 1880s, 1880s to 1920s, 1920s to 1970s and 1970s to present), respectively. Because $\delta^{13}C$ values of biomarkers in core sediments could be altered by microbial activity and degradation[35,36], especially more obvious for short-chain FAs (See Section "OC characteristics"), the $\delta^{13}C$ values of OC_{modern}, $OC_{pre-aged}$ and OC_{fossil} are assigned based on surface sediments. The $\delta^{13}C$ values of n-C_{16} FA and n-$C_{26+28+30}$ FAs are based on the corresponding values of H07 surface sediment, which was collected from the same location during the cruise (Fig. 1[20]). Previously published $\delta^{13}C$ and $\Delta^{14}C$ values of n-C_{16+18} Alkanes in the Bohai Sea and Yellow Sea surface sediments are assigned as OCfossil endmembers. An offset of 0.4‰–0.7‰ to biomarker $\delta^{13}C$ values was then applied to account for the $\delta^{13}C_{bulk} - \delta^{13}C_{lipid}$ fractionation resulting from biosynthetic fractionation and related effects[37,38]. Here we assumed $\delta^{13}C$ values for OC_{modern} (mainly for OC_{marine}) endmembers would not change substantially during the past centuries because the decrease in $\delta^{13}C$ values of dissolved inorganic carbon (DIC) in Pacific Ocean from 1960s caused by Suess effect is only about 0.02‰[39]. A Bayesian Markov chain Monte Carlo (MCMC) simulation was applied to minimize errors from arbitrary assignments of endmember values[20,40].

Results

▶ *Bulk sediment characteristics*

The down-core profile of the excess ^{210}Pb in core H07E showed no obvious mixed layer and a mean sedimentation rate was estimated to be 0.22 cm/a based on the constant flux constant sedimentation rate model[41], which is consistent with previously reported ranges for the central mud area (0.2–0.3 cm/a, Fig. S1 and Table S1). Processes such as bioturbation by benthic infauna or physical disturbance could influence the stratigraphy in the upper horizons. However, the absence of a clear mixed layer and well preserved primary sedimentary structure indicated by radionuclide profiles in this study area suggest a relatively stable hydrodynamic environment with a limited magnitude of benthic disturbance. We therefore conclude that core H07E provides a history of sediment accumulation spanning approximately the past 200 years (Fig. 2), and we have used the resulting age model assuming a linear sedimentation rate in order to correct down-core ^{14}C data for radioactive decay since deposition.

Bulk sediment and OC properties are listed in Table S2. The sediment was uniformly fine-grained, being composed of 42%–50% (ave. 46%) clay, 50%–57% (ave. 53%) silt, and only 0.1%–0.9% (ave. 0.2%) sand, exhibiting minor temporal variations (Fig. 2a). Mean grain size ranged from 6.1 μm to 7.8 μm (ave. 6.8 μm ± 0.4 μm), showing some variations but without a temporal trend (Fig. 2a). Mineral SA ranged from 24.2 to 35.9 m^2/g (ave. 31.1 m^2/g ± 2.0 m^2/g), and exhibited a slight increase above 10 cm (Fig. 2b). TOC/SA ratios (i.e., TOC loadings) ranged from 0.30 mg/m^2 to 0.45 mg/m^2 (ave. 0.35 mg/m^2 ± 0.03 mg/m^2) which showed relatively stable values, with the exception of one sample at 12.5 cm more than 0.40 mg/m^2 that was due to an anomalously low SA value (Fig. 2b).

TOC% varied from 1.0% to 1.3% (ave. 1.1% ± 0.1%), with a slightly increasing trend above 10 cm (Fig. 2c). TOC% and TN% displayed a significant correlation with a positive TN intercept (indicating the presence of inorganic N). Thus, C/N molar ratio (after correction for inorganic N) ranged from 7.9 to 8.7, with a mean value of 8.4 ± 0.1. The $\delta^{13}C_{TOC}$ values varied within a small range, from −2.21‰ to −2.15‰ (ave., −2.17‰ ± 0.01‰), exhibiting a slightly decreasing trend that begins above 20 cm and is most pronounced above 10 cm (Fig. 2d). The $\Delta^{14}C_{TOC}$ values (corrected for decay since deposition) showed an overall increasing trend up-core from −33.2‰ to −21.5‰ (ave. −26.3‰ ± 3.0‰, $R^2 = 0.86$, $P < 0.01$, Fig. 2d), with corresponding

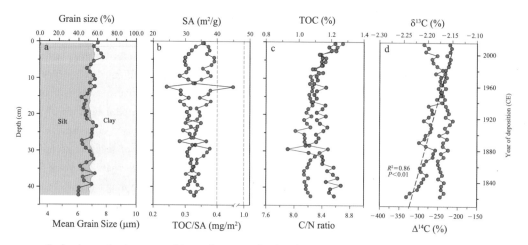

a. Grain size grain size composition and mean grain size (blue). b. SA (red) and TOC/SA ratio (blue). Dashed blue lines represent the TOC/SA ratios between 0.4 mg/m² and 1.0 mg/m²[5]. c. TOC (red) and C/N molar ratio (blue). d. $\delta^{13}C$ (red) and $\Delta^{14}C$ (blue, decay-corrected). The age scale on the right side is approximate due to the uncertainty of the age model; the same applies to Fig. 3, Fig. 6, Fig. 7.

Fig. 2　Down-core profiles of bulk geochemical parameters in the core H07E

conventional ^{14}C ages ranging from 3360 a to 1890 a. As shown in Fig. S2, radioactive decay since the time of deposition of the lowermost sediment interval (42–43 cm) amounts to a $\Delta^{14}C$ shift of 1.5‰, which is similar to the measurement uncertainty of 0.6‰ to 1.0‰, indicating that the decay correction in this study is minor and does not impact the overall interpretation.

▶ *Carbon isotopic compositions of n-FAs*

Four sediment horizons (0–2 cm, 14–17 cm, 24–27 cm and 38–41 cm) were selected for compound-specific carbon isotope analysis (Fig. S2). To obtain sufficient quantities of target compounds for radiocarbon measurement, two or three continuous 1 cm intervals were combined, resulting in a time integration of approximately 5 to 10 years per sample. The $\delta^{13}C$ and $\Delta^{14}C$ values of *n*-FAs are shown in Fig. 3 and Table S3. Overall, $\delta^{13}C$ FAs ranged from −2.88‰ to −2.58‰ (Fig. 3a), with $\delta^{13}C_{18}$ FA values exhibiting the most depleted values (−2.88‰ to −2.74‰) among FAs in different layers. The $\delta^{13}C_{16}$ FA values (−2.78‰ to −2.58‰) were higher than those of longer-chain homologues in the upper- and lower-most depth intervals, but similar or lower at intermediate depths. The $\delta^{13}C_{24}$ FA values (−2.73‰ to −2.69‰) and $\delta^{13}C_{26+28+30}$ FAs values (−2.77‰ to −2.71‰) showed less variability across the depth profiles.

The decay-corrected $\Delta^{14}C$ FAs values for each depth interval generally exhibited a decreasing trend with increasing chain length (Fig. 3b). The $\Delta^{14}C_{16}$ FA and $\Delta^{14}C_{18}$ FA values were similar and varied within a relatively narrow range, from −3.8‰ ± 1.4‰ to +0.9‰ ± 0.8‰ (corresponding ^{14}C ages from 416a to modern), and from −3.4‰ ± 0.9‰ to +0.5‰ ± 0.9‰ (corresponding ^{14}C ages from 284 a to modern), respectively. The highest values for C_{16} FA and C_{18} FA were both observed in core-top sediments. The $\Delta^{14}C_{24}$ FA and $\Delta^{14}C_{26+28+30}$ FAs values displayed a larger range, varying from −13.6‰ ± 1.1‰ to −23.7‰ ± 0.8‰ (corresponding ^{14}C ages from 1184 a to 2289 a) and from −20.3‰ ± 0.8‰ to −27.6‰ ± 0.8‰ (corresponding ^{14}C ages from 1830 a to 2700 a), respectively.

Discussion

▶ *Characteristics and changes of sedimentary OC sources over the last 200 years*

OC characteristics

Core H07E is characterized by the absence

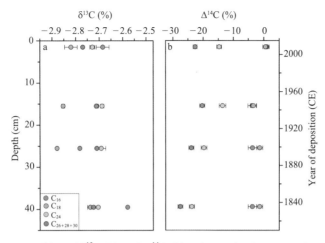

Fig. 3 Down-core profiles of $\delta^{13}C$ (a) and $\Delta^{14}C$ (b) values of n-C_{16} FA, n-C_{18} FA, n-C_{24} FA and abundance-weighted average n-$C_{26+28+30}$ FAs, which have been decay-corrected from the time of deposition

of an obvious mixed layer, consistent with previously published data (Fig. S1 and Table S1), and uniformly high percentages of clay and silt, reflecting a relatively stable sedimentary environment[16,17]. The oxygen penetration depth was shallow (<1 cm) in this mud area and limited oxygen availability combined with low bottom-water temperatures (generally < 10 °C) likely inhibit diagenetic processes[42]. The down-core profile of TOC loadings showed no strong trend, suggesting minimal post-depositional remineralization of OC, and the uniformly low values relative to typical continental shelf sediments (Fig. 2) implies that the remaining OC in this fine-grained sediment deposit must be closely associated with (and stabilized by) available mineral surfaces. Thus, temporal variations in geochemical properties, especially carbon isotope compositions, are more strongly influenced by historical variations in OC sources rather than by diagenetic processes.

Down-core $\delta^{13}C_{TOC}$ values (−2.17‰ ± 0.01‰) exhibited less variability than decay-corrected $\Delta^{14}C_{TOC}$ values (from −21.5‰ to −33.2‰), and both fell within reported ranges for surface sediments in this area[20]. TOC-^{14}C ages (1890–3360 a) are much older than the sediment deposition age (< 200 a), implying the supply of significant proportions of pre-aged and/or fossil OC together with recently fixed, modern OC (Fig. 4a). The relatively constant $\delta^{13}C_{TOC}$ values suggest no marked temporal variations in the proportions of marine versus terrestrial OC source endmembers, but the more variable $\Delta^{14}C_{TOC}$ values imply time-varying contributions from different aged OC. Compound-specific radiocarbon analysis of biomarkers provide further constraints on the OC sources and exclude interferences from other carbon pools[11,13,20]. Short-chain (C_{16} and C_{18}) n-FAs, which consistently have higher $\Delta^{14}C$ values than the longer-chain homologues (C_{24} FA and $C_{26+28+30}$ FAs) in the down-core profile, seem to primarily trace the surface ocean DIC pool that is influenced by exchange with atmospheric CO_2 (Fig. S3), and reflect recent carbon fixation by contemporary OC sources. $\delta^{13}C$ values of short-chain n-FAs showed larger temporal variation likely because of alteration by microbial activity and degradation[35,36]. In contrast, the vascular plant wax-derived $C_{26+28+30}$ FAs have lower $\Delta^{14}C$ values (ave. −23.6‰ ± 3‰, n = 4). The corresponding ^{14}C ages (1830–2700 a) imply millennial-scale storage of terrestrial higher plant OC in intermediate reservoirs (e.g., soil) prior to their delivery to marginal sea sediments, as previously reported in riverine and marine surface sediments[12,13]. The similarity between $\Delta^{14}C_{26+28+30}$

FAs values and bulk $\Delta^{14}C_{TOC}$ values throughout the core, as well as the corresponding $\delta^{13}C$ values of these long-chain FAs (−2.73‰ ± 0.03‰), implies a substantial and continuous supply of pre-aged OC, predominantly of C3 plant origin, over the past 200 years (Fig. 4b and Fig. S3).

The gradual decrease in decay-corrected $\Delta^{14}C_{TOC}$ values downcore may be a consequence of selective removal of younger and more labile marine OC in deeper sediments, but it is not the case for $\delta^{13}C_{TOC}$ values and C/N ratios (Fig. 2). The slightly decreasing $\delta^{13}C_{TOC}$ values over the uppermost 10 cm of the core, corresponding to deposition since the mid-20th Century might be partly influenced by the Suess effect (the decrease in $\delta^{13}C$ values of surface ocean DIC in Pacific Ocean since the 1960s is about 0.02‰[39]), but also likely indicate increasing proportions of OC_{terr}. The increasing $\Delta^{14}C_{TOC}$ values towards the present may at least partly reflect biosynthetic incorporation of ^{14}C-enriched "bomb-derived" carbon that penetrated the surface ocean and influenced mixed layer DIC $\Delta^{14}C$ values since the 1960s[43-45], however there is no clear increase of $\Delta^{14}C_{TOC}$ near or since the 1960s (Fig. S3). Up-core increasing $\Delta^{14}C$ values in sediments deposited prior to the 1960s (i.e., at sediment depths greater than ca. 12 cm) must reflect the changes of the contributions of different OC sources and ages. An ^{14}C-enrichment of long-chain FAs from post-bomb horizons over those from pre-bomb horizons was observed in sediment records from the Cariaco Basin, Saanich Inlet and Mackenzie Delta[45], as well as the Bay of Bengal[46], suggesting the incorporation of bomb radiocarbon. In contrast, $\Delta^{14}C$ data of both short- and long-chain FAs isolated from a single depth horizon from the latter half of the 20th Century in core H07E does not show evidence of a bomb effect (Fig. S3). Moreover, the down-core variations in $\Delta^{14}C_{26+28+30}$

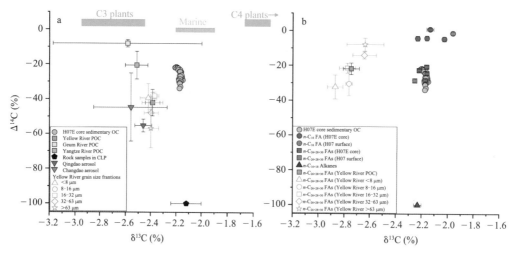

a. Possible sources of OC including typical C3 and C4 plants from Chinese Loess Plateau[48], marine phytoplankton[49], Yellow River POC and grain size fractions[20,47,50], Geum River POC[33], Yangtze River POC[50,51], and rock samples from the western Chinese Loess Plateau[52], as well as aerosols from coastal city (Qingdao and Changdao) in Shandong peninsula[53]. $\Delta^{14}C$ values of modern sources and ancient rocks are assigned as 0±5‰ and −100‰ respectively[8]. b. The δ13C and $\Delta^{14}C$ values of source-specific biomarkers (n-C_{16} FA, n-$C_{26+28+30}$ FAs and n-C_{16+18} Alkanes), following correction for $\delta^{13}C_{bulk}$−$\delta^{13}C_{lipid}$ fractionation (by 0.4‰–0.7‰ in $\delta^{13}C$ values). The $\delta^{13}C$ FAs and $\Delta^{14}C$ FAs values of surface sediment of H07 site are from Yu et al.[20]; the $\delta^{13}C$ and $\Delta^{14}C$ values of n-C_{16+18} Alkanes are from Tao et al.[27]; the $\delta^{13}C$ and $\Delta^{14}C$ values of n-$C_{26+28+30}$ FAs of Yellow River POC and grain size fractions are from Yu et al.[47,54]. All $\Delta^{14}C$ values of core sediments have been decay-corrected from the time of deposition.

Fig. 4 Scatter plots of $\delta^{13}C$ versus $\Delta^{14}C$ values of H07E core-sediment OC (blue dots) and plausible endmembers

FAs fall within the range of spatial variation in central Yellow Sea (−21.7‰ to −28.5‰[20,27]), and among finer grain size fractions of Yellow River suspended particles (Fig. 4b[47]). We therefore attribute the down-core variations of $\Delta^{14}C_{TOC}$ values to time-varying contributions from different pools of aged OC over the past 200 years, which are estimated in the next section.

Quantification of OC sources

For sedimentary OC source apportionment, two different carbon isotopic mixing models are applied (Section "Carbon isotope mixing models") and corresponding results are shown in Table S4 and Fig. S4. Fractional contributions (%) of OC_{marine} to TOC ranged from 46% to 61% (ave., 55% ± 3%), and OC_{terr}% ranged from 39% to 54% (45% ± 3%) with a slightly increasing trend in the proportion of the latter since the mid-20th Century (Fig. S4a). Fractional contributions of different components of aged OC to TOC showed different temporal trends, and reveal a predominance of $OC_{pre-aged}$ (Fig. S4b). OC_{modern}%, $OC_{pre-aged}$% and OC_{fossil}% ranged from 32% to 37% (ave. 35% ± 1%), from 47% to 59% (52% ± 3%), and from 9% to 18% (13% ± 2%), respectively. OC_{modern}% showed a positive relationship with OC_{marine}%, suggesting OC_{modern} is dominated by OC_{marine} (Fig. 5a). However, OC_{marine}% is higher than OC_{modern}% because of the limitation of the binary mixing model in accounting for the widespread contributions of pre-aged OC to Bohai Sea-Yellow Sea sediments. There is also a positive relationship between $OC_{non-modern}$% (i.e., $OC_{pre-aged}$ + OC_{fossil})

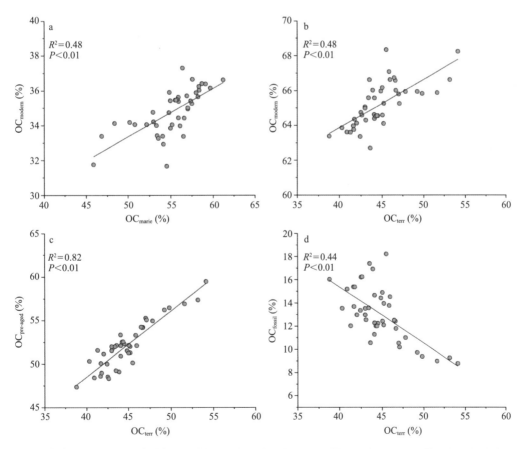

Fig. 5 Plots of fractional contribution (%) of OC_{marine} and OC_{modern} (a), OC_{terr} and $OC_{non-modern}$ (b), OC_{terr} and $OC_{pre-aged}$ (c), and OC_{terr} and OC_{fossil} (d)

and $OC_{terr}\%$ (Fig. 5b) as well as parallel temporal (down-core) variation (Fig. S4c), suggesting a common continental (fluvial) origin. $OC_{pre\text{-}aged}\%$ was strongly correlated with $OC_{terr}\%$, suggesting the predominance of $OC_{pre\text{-}aged}$ in terrestrial OC pools (Fig. 5c). Although both $OC_{pre\text{-}aged}$ and OC_{fossil} are of terrestrial origin and are sub-pools of terrestrial OC, they showed different depth variations. The overall decreasing trend of $OC_{fossil}\%$ and an increasing trend of $OC_{pre\text{-}aged}\%$ towards the present (Fig. S4), which would result in the negative relationship between $OC_{pre\text{-}aged}\%$ and $OC_{fossil}\%$ ($R^2 = 0.77$, $P < 0.01$). As a result, $OC_{fossil}\%$ exhibits a negative relationship with $OC_{terr}\%$ (Fig. 5d). The different temporal variations between $OC_{pre\text{-}aged}\%$ and $OC_{fossil}\%$ may reflect contrasting hydrodynamic sorting processes during transport given that OC_{fossil} is mostly associated with coarser particles, limiting long-range transport to offshore areas[20,47]. Both bulk and molecular based isotope approaches reveal the high proportion of terrestrial OC that has accumulated in the central Yellow Sea mud area over the last 200 years.

The absolute contents of different OC pools (mass fraction, %) showed similar temporal variations to those of fractional contributions (Fig. 6). OC_{terr} contents increased markedly from the mid-20th Century onwards, while OC_{marine} contents remained relatively stable. $OC_{pre\text{-}aged}$ contents increased over a similar time frame, while OC_{modern} contents remained invariant, but OC_{fossil} contents showed a subtle but systematic decreasing trend towards the present. Because $OC_{pre\text{-}aged}$ is the dominant component of non-modern OC, $OC_{non\text{-}modern}$ contents closely follow OC_{terr} contents. Increasing accumulation of terrestrial OC is also supported by the increasing contents of higher plant alkane biomarkers in sediment cores from the same area[55]. It is well established that sediment load from the Yellow River and Yangtze River has declined significantly since the 1950s under the impact of both natural and anthropogenic forcing[56]. Thus, riverine export of terrestrial OC must also have decreased accordingly. However, supply of terrigenous sediment and OC to the central Yellow Sea is modulated via the coastal currents and circulation within the marginal sea system[25], and the latter has maintained sedimentation rates despite diminished riverine inputs. In addition, anthropogenic activities, such as urbanization and energy consumption, have partially contributed to the increasing inputs of OC as reflected by export of aliphatic hydrocarbons and black carbon to the Yellow Sea[57,58]. Qi et al.[19] proposed that decreasing riverine inputs would result in increased OC contributions (from 44% to 77% after 1990s) to the central Yellow Sea from erosion of Old Yellow River delta sediments. Because the $\Delta^{14}C$ values of OC in the Old Yellow River delta surface sediments (−43.2‰ ± 0.6‰, $n = 2$) are similar to modern Yellow River POC (−42‰ ± 7.6‰, $n = 34$), and both sediments have a similar provenance (the Chinese Loess Plateau), it is difficult to distinguish these two sources based on bulk carbon isotopes. The $\Delta^{14}C_{26+28+30}$ FAs values from Old Yellow River delta sediments (−39.2‰ ± 1.3‰, $n = 2$) are older than those of Yellow River POC (−21.2‰ ± 2.6‰, $n = 18$) and surface sediments near the modern Yellow River delta (−22.6‰[20]). Although aging processes during lateral transport[59,69] render it difficult to resolve inputs from Old Yellow River delta erosion, erosional supply from the modern Yellow River delta would likely have proportionally increased with decreasing riverine inputs. Increasing OC_{terr} contents and decreasing OC_{modern} contents with decreasing Yangtze River POC load have also been observed in a 70-year sedimentary record from the East China Sea mud area, likely as a consequence of the increasing erosion from the lower reaches of the Yangtze River and subaqueous delta[60]. The decreasing OC_{fossil} contents are likely caused by decreasing riverine inputs, coupled with hydrodynamic sorting effects that limit long-range transport of coarser particles derived from rock erosio[47].

▶ *Temporal evolution of OC component loadings and burial*

Mineral protection plays an important role in OC

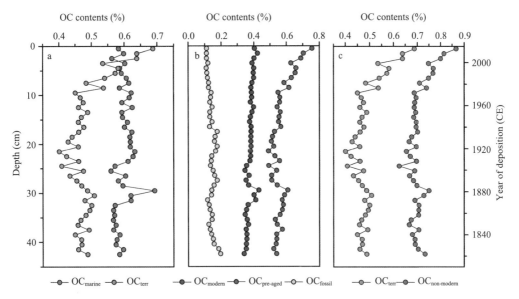

Fig. 6   Temporal variations of contents (mass fraction, %) of OC_{marine} and OC_{terr} (a), OC_{modern}, $OC_{pre-aged}$, and OC_{fossil} (b), and OC_{terr} and $OC_{non-modern}$ (c)

preservation in marine sediments[61-63]. Normalizing OC content to mineral SA (i.e., OC loadings) provides constraints on degradation and preservation of OC, taking into account potential effects of both physical sorting and mineral association[5]. Clay mineral composition does not show significant variability in the central Yellow Sea mud area over the past century[64] and SA does not vary significantly with depth in core H07E (Fig. 2b), allowing assessment in terms of temporal variations of loadings of TOC and specific components. Downcore TOC loadings (ave. (0.35 ± 0.03) mg·m^{-2}) are generally lower than those of typical riverine and shelf sediments (0.4–1.0 mg·m^{-2}, Fig. 2b[5]), but significantly higher than those of lower Yellow River suspended particulate matter (SPM, (0.16 ± 0.04) mg·m^{-2}, $n = 18$[54]) and Old Yellow River delta surface sediments ((0.20 ± 0.04) mg·m^{-2}[19]). Higher TOC loadings, combined with higher $\delta^{13}C$ values of marine sediments than those of riverine suspended particles suggest the addition of marine OC.

Similar to TOC loadings, normalizing specific OC component content to SA helps evaluate temporal changes of OC accumulation during last 200 years. Due to the positive relationships between the fractional abundance of OC_{modern} and OC_{marine}, as well as between $OC_{non-modern}$ and OC_{terr}, we find that corresponding loadings co-vary with each other (Fig. 7). If the OC composition did not vary significantly, OC component loadings would co-vary with TOC loadings and thus with each other. OC_{modern} and OC_{marine} loadings were relatively stable, exhibiting a slightly decreasing trend from the late 20th Century onwards (Fig. 7). This potentially reflects a weaker association between OC derived from marine productivity and minerals that enhances exposure to degradation[5]. OC_{terr} and $OC_{non-modern}$ loadings both exhibited an increasing trend since the early 20th Century (Fig. 7). $OC_{pre-aged}$ and OC_{fossil} loadings revealed contrasting temporal variations, with a general increase for $OC_{pre-aged}$ and a decrease for OC_{fossil} since the early 20th Century (Fig. 7). Based on these trends, we attribute the increased accumulation of terrestrial OC since the last century to increasing proportions of pre-aged OC, as discussed in Section "Characteristics and changes of sedimentary OC sources over the last 200 years". Contrasting fates of sedimentary $OC_{pre-aged}$ and OC_{fossil} have previously been inferred from analysis of a suite of Bohai Sea-Yellow Sea surface

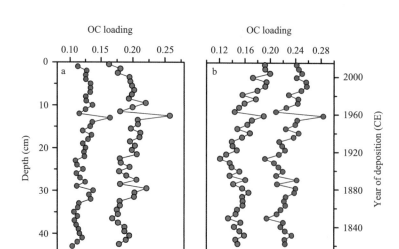

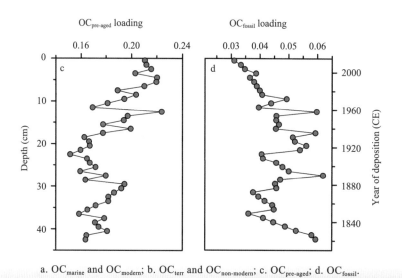

a. OC_{marine} and OC_{modern}; b. OC_{terr} and $OC_{non-modern}$; c. $OC_{pre-aged}$; d. OC_{fossil}.

Fig. 7 Temporal variations of different OC component loadings (mg·m^{-2})

sediments, implying either preferential degradation of OC_{fossil} or addition of $OC_{pre-aged}$ following fluvial delivery to marginal sea systems[20].

To quantify the burial of different terrestrial OC pools in the central Yellow Sea mud area over the past 200 years, we compare sedimentary loadings of different terrestrial OC pools with their corresponding loadings in riverine samples[15,20,21]. Clay mineralogy of Yellow Sea sediments deposited over the past century indicated the modern Yellow River and Old Yellow River delta remained the primary sediment sources[64]. A sediment budget constructed on centennial timescales indicated that the central Yellow Sea sediment primarily originates from the Yellow River, with a lesser contribution from erosion from the Old Yellow River delta[17]. Loadings of OC_{terr}, $OC_{pre-aged}$ and OC_{fossil} from Yellow River suspended samples ((0.16 ± 0.04) mg·m^{-2}, (0.07 ± 0.02) mg·m^{-2} and (0.05 ± 0.01) mg·m^{-2}) are similar to those from Old Yellow River delta samples ((0.14 ± 0.05) mg·m^{-2}, (0.07 ± 0.04) mg·m^{-2} and 0.05 mg·m^{-2}[20,27]), suggesting a common origin (i.e., the Chinese Loess Plateau). The linear regression-derived average OC loadings of Old Yellow River delta samples are also consistent with Yellow River suspended samples (Fig. S5). These lines of evidence indicate that OC loadings of suspended samples exported by the Yellow River have likely remained similar despite the large changes in sediment flux since the 1950s and the relocation of the river mouth from the Old Yellow River since 1855. Taking into account of hydrodynamic sorting effects during transport, the OC loadings from finer grain size fractions of Yellow River (0.16–0.19 mg·m^{-2} for < 32 μm fractions, Fig. S6) are consistent with total suspended particles. Thus, it is reasonable to compare the central Yellow Sea mud area core results with Yellow River samples to further evaluate OC burial efficiency on centennial timescales. The average down-core sedimentary loading of $OC_{pre-aged}$ in core H07E is higher than that of the Yellow River, while that of OC_{terr} and OC_{fossil} are similar in magnitude. These results imply that efficient preservation of terrestrial OC delivered by the Yellow River has been a persistent phenomenon over the past 200 years.

▶ *Implications for the OC burial in distal mud area*

The dramatical decrease in the riverine sediment fluxes is a consequence of anthropogenic activities and climate change since the 1950s[56], thus the use of mass-balance budgets based on surface sediments to evaluate the burial efficiencies of different OC pools is prone to large uncertainty[21]. We adopt the concept of burial efficiency of OC_{terr} constrained by changes of the loadings of OC_{terr} from riverine inputs to marine sediments (Eq. 7[15,21]), consistent with previous studies[20,22,65,66]. The assumption is that mineral particles remain largely unaffected by fluvial export, and OC_{terr} is associated with and protected by mineral surfaces during export[5,15,61,63]. Recent studies using this approach have mostly focused on surface sediment samples[18-20], here we extend the approach to different aged terrestrial OC pools, constrained by molecular carbon isotope data (Eqs. 8, 9), in order to constrain historical changes in OC_{terr}, $OC_{pre-aged}$ and OC_{fossil} loadings and assess longer-term burial of terrestrial OC pools in the central Yellow Sea mud area.

OC_{terr} burial efficiency = OC_{terr} loading (sediment)/ OC_{terr} loading (riverine SPM) × 100%
(7)

$OC_{pre-aged}$ burial efficiency = $OC_{pre-aged}$ loading (sediment)/ $OC_{pre-aged}$ loading (riverine SPM) × 100%
(8)

OC_{fossil} burial efficiency = OC_{fossil} loading (sediment)/ OC_{fossil} loading (riverine SPM) × 100%
(9)

The burial efficiency for these three pools of OC was very high, 97% ± 12% for OC_{terr}, 90% ± 15% for OC_{fossil}, and greater than 100% for $OC_{pre-aged}$ (Fig. S7). Thus, down-core patterns are broadly consistent with those determined using a similar approach for surface sediments[20,27]. The increasing efficiency of OC_{terr} burial since the mid-20th Century is mainly caused by a concomitant increase in $OC_{pre-aged}$ burial efficiency, while that of OC_{fossil} decreased from about 90% to 60% over the same time period. In the context of the decreasing riverine sediment and OC flux during recent decades, these inputs may have been supplanted by erosion of lower reach or delta sediments as well as current-driven resuspension processes that promoted the long-range transport of $OC_{pre-aged}$[20]. Even comparing sedimentary OC_{terr} loadings to those from the Yellow River finer grain size fractions (0.16–0.19 mg·m^{-2} for < 32 μm fractions) in order to account for potential hydrodynamic sorting effects, average burial efficiency of OC_{terr} is > 80%. Burial efficiencies of $OC_{pre-aged}$ in excess of 100% could reflect uncertainties associated with the molecular isotope-based source apportionment approach, mineral associations, temporal variability in riverine loadings, as well as the existence of additional sources. The calculated sedimentary $OC_{pre-aged}$ could reflect the supply of fluvial $OC_{pre-aged}$ as well as aging during lateral transport and sorting effects[59,69]. The latter is widespread in marginal seas but difficult to quantify because transport aging decreases $\Delta^{14}C$ values but may not change δ¹³C values[67]. If we assume Yellow River POC is totally composed of $OC_{pre-aged}$, we still obtain a burial efficiency of about 110%. Using biomarkers to represent a certain OC pool may thus introduce uncertainties. Such uncertainties could be reduced by combining this approach with others that constrain pools based on other properties, such as using ramped pyrolysis/oxidation analysis[63]. Considering a scenario of OC supply from the Yangtze River to H07E using isotopic signatures (Fig. 4a) and riverine $OC_{pre-aged}$ and OC_{fossil} in SPM loadings of 0.21 mg·m^{-2} and 0.08 mg·m^{-2}, respectively[51,66], corresponding burial efficiencies are about 90% and 60%, respectively.

Irrespective of potential additional sources, the calculated burial efficiency for $OC_{pre-aged}$ is higher than OC_{fossil}, which could be caused by degradation of OC_{fossil} during long-range transport, the degree and mode of mineral associations, and hydrodynamic sorting effects[20,62,63]. The down-core results suggest that initial sequestration of OC_{terr} in Yellow Sea surface sediments presages long-term burial in deeper sediments, despite the existence of temporal variability in supply. Together, the new down-core data and previously published surface sediment results[20,27] strongly imply sustained and efficient burial of Yellow River-derived $OC_{pre-aged}$ in the central Yellow Sea over the past 200 years.

Our biomarker isotope-based estimate of OC_{terr} burial efficiency can be compared with those for TOC and marine OC using other approaches. Based on the measured sediment accumulation rate of 0.14 g·m^{-2}·a^{-1}, TOC accumulation rates in H07E core ranged from 14.1 g·m^{-2}·a^{-1} to 18.3 g·m^{-2}·a^{-1} (ave. (15.6 ± 0.8) g·m^{-2}·a^{-1}). This is within the estimated values for the Bohai Sea-Yellow Sea surface sediments[16], but much higher than that of average global shelf sediments (4.15 g·m^{-2}·a^{-1}[68]). Accounting for the TOC remineralization rate of (12.0 ± 1.7) g·m^{-2}·a^{-1} in the central Yellow Sea mud area[42], the average TOC deposition rate is (27.6 ± 1.9) g·m^{-2}·a^{-1}. This implies an overall OC burial efficiency of about 60%, which is largely dominated by the sequestration of pre-aged (i.e., refractory) terrestrial OC. The area of the central Yellow Sea mud depocenter is 99.3 × 10³ km²[17] and the average TOC burial rate is estimated about (1.5 ± 0.1) Tg·a^{-1} over the last 200 years, which corresponds to 42% of TOC burial in the entire Yellow Sea system (3.6 Tg·a^{-1}[16]).

Despite being only indirectly influenced by the riverine inputs, the efficient burial of pre-aged terrestrial OC and contemporary marine OC renders the central Yellow Sea mud area a carbon sink on (at least) centennial to millennial timescales. Environmental factors contributing to the efficient burial of OC in this location include cyclonic cold eddy-enhanced focusing and deposition of fine-grained sediments, the predominance of refractory and pre-aged OC_{terr} inputs closely associated with detrital mineral phases, coupled with low oxygen contents and temperatures of bottom and interstitial waters[25,42]. Furthermore, our findings suggest that the sequestration of OC_{terr} in surface sediments presages longer-term burial in deeper sediments in the central Yellow Sea. Although our findings are based on a single sediment core, the relatively homogeneous depositional characteristics and OC sources in this area[16,20] make the results of the single core representative of a broad area that accounts for a substantial fraction of OC burial in the Chinese marginal seas. The novel approach adopted in this study serves as an example to constrain the temporal evolution in burial efficiencies of specific OC_{terr} pools in marine sediments. Further investigations that take into account information from additional cores from this location, as well as focusing on different river-dominated margins and sedimentary environments (e.g., distal mud area vs. inner shelf mud area) are needed to better constrain the fates of terrestrial OC in marginal sea sediments on a range of temporal scales, as well as in response to natural and anthropogenic processes.

Conclusions

This study investigated variations in the sources and burial of OC in the central Yellow Sea mud area sediments deposited over the past 200 years, using bulk and molecular-level carbon isotopic measurements as well as sedimentological data. The $\Delta^{14}C$ values of TOC (−21.5‰ to −33.2‰, corresponding to ^{14}C age of 1890–3360 a) and higher plant long-chain n-FA biomarkers (−20.3‰ ± 0.8‰ to −27.6‰ ± 0.8‰, 1830–2700 a), reveal that continuous supply and sequestration of pre-aged terrestrial OC accounts for much of the buried OC. Quantification of different terrestrial OC pools based on carbon isotopic mixing models showed terrestrial OC (OC_{terr}), pre-aged OC ($OC_{pre-aged}$) and fossil OC (OC_{fossil}) accounted for 45% ± 3%, 52% ± 3% and 13% ± 2% of sedimentary TOC, respectively. Corresponding estimates of burial efficiency of TOC (about 60%) and different terrestrial OC pools (>80%) are high, and reveal the importance of terrestrial OC burial in the central Yellow Sea mud area. These high OC burial efficiencies are attributed to the supply of refractory, millennial-aged terrestrial OC derived from the Yellow River, coupled with depositional conditions that are conducive to organic matter preservation.

Different temporal patterns of $OC_{pre-aged}$ and OC_{fossil}, and consistently higher burial efficiencies of the former reveal contrasting fates of $OC_{pre-aged}$ and OC_{fossil}. The different preservation pattern between $OC_{pre-aged}$ and OC_{fossil} may be linked to differences in the degree and mode association with detrital mineral phases. Both the contents and loadings of OC_{terr} and $OC_{pre-aged}$ showed increasing trend from past decades towards the present, while OC_{fossil} decreased, likely as a consequence of changes in sediment provenance and the hydrodynamic regime.

Down-core results suggest that initial sequestration of OC_{terr} in the Yellow Sea surface sediments presages longer-term burial in deeper sediments despite the existence of temporal variability. The distal Yellow Sea mud area acts as long-term carbon sink, particularly with respect to terrestrial OC pools. Further characterization and quantification of different pools of terrestrial OC in marine sediments is necessary to assess the role of shelf seas in carbon cycle on a range of timescales, as well as on-going anthropogenic perturbations to the heterogeneous, dynamic systems.

Acknowledgements

We thank the Editor and three anonymous reviewers for constructive suggestions. We would like to thank Weijie Hao and crews of *R/V Dongfanghong II* for sampling help, Daniel B. Montluçon, Li Li and Yang Ding for laboratory assistance. We also thank all members of Laboratory for Ion Beam Physics at ETH Zürich for ^{14}C measurements. This study was supported by the National Natural Science Foundation of China (Meng Yu, No. 41906032), the Marine S&T Fund of Shandong Province for the Pilot National Laboratory for Marine Science and Technology (Qingdao) (Meixun Zhao, No. 2022QNLM040003-2), the Swiss National Science Foundation (Timothy I. Eglinton, "CAPS-LOCK2" Grant #200020_163162, "CAPS-LOCK3" Grant #200020_184865), the Fundamental Research Funds for the Central Universities (Meng Yu, No. 202072011), the Postdoctoral Applied Research Project of Qingdao (Meng Yu, No. 2019196), the "111" Project (B13030), and the China Scholarship Council (Meng Yu, No. 201506330021). This is MCTL (Key Laboratory of Marine Chemistry Theory and Technology) contribution #276 and OUC-CAMS contribution #1.

Competing Interests

The authors declare that they have no known competing financial interests or personal relationships that could have appeared to influence the work reported in this paper.

References

1. Hedges J I, Keil R G. Sedimentary organic matter preservation: An assessment and speculative synthesis[J]. Mar. Chem., 1995, 49: 81-115.
2. Galy V, Peucker-Ehrenbrink B, Eglinton T. Global carbon export from the terrestrial biosphere controlled by erosion[J]. Nature, 2015, 521: 204-207.

3. Leithold E L, Blair N E, Wegmann K W. Source-to-sink sedimentary systems and global carbon burial: A river runs through it[J]. Earth-Sci. Rev., 2016, 153: 30-42.
4. Bianchi T S, Cui X, Blair N E, et al. Centers of organic carbon burial and oxidation at the land-ocean interface[J]. Org. Geochem., 2018, 115: 138-155.
5. Blair N E, Aller R C. The fate of terrestrial organic carbon in the marine environment[J]. Annu. Rev. Mar. Sci., 2012, 4: 401-423.
6. Bauer J E, Cai W J, Raymond P A, et al. The changing carbon cycle of the coastal ocean[J]. Nature, 2013, 504: 61-70.
7. Ausín B, Bruni E, Haghipour N, et al. Controls on the abundance, provenance and age of organic carbon buried in continental margin sediments[J]. Earth Planet. Sci. Lett., 2021, 558: 116759.
8. Goñi M A, Yunker M B, Macdonald R W, et al. The supply and preservation of ancient and modern components of organic carbon in the Canadian Beaufort Shelf of the Arctic Ocean[J]. Mar. Chem., 2005, 93: 53-73.
9. Lamb A L, Wilson G P, Leng M J. A review of coastal palaeoclimate and relative sea-level reconstructions using $\delta^{13}C$ and C/N ratios in organic material[J]. Earth-Sci. Rev., 2006, 75: 29-57.
10. Belicka L L, Harvey H R. The sequestration of terrestrial organic carbon in Arctic Ocean sediments: A comparison of methods and implications for regional carbon budgets[J]. Geochim. Cosmochim. Acta, 2009, 73: 6231-6248.
11. Eglinton T I, Benitez-Nelson B C, Pearson A, et al. Variability in radiocarbon ages of individual organic compounds from marine sediments[J]. Science, 1997, 277: 796-799.
12. Eglinton T I, Galy V V, Hemingway J D, et al. Climate control on terrestrial biospheric carbon turnover[J]. Proc. Natl. Acad. Sci. U.S.A., 2021, 118: e2011585118.
13. Kusch S, Mollenhauer G, Willmes C, et al. Controls on the age of plant waxes in marine sediments: A global synthesis[J]. Org. Geochem. 2021, 157: 104259.
14. Milliman J D, Meade R H. World-wide delivery of river sediment to the oceans[J]. J. Geol., 1983, 91: 1-21.
15. Burdige D J. Burial of terrestrial organic matter in marine sediments: A re-assessment[J]. Global Biogeochem. Cycles, 2005, 19: GB4011.
16. Hu L, Shi X, Bai Y, et al. Recent organic carbon sequestration in the shelf sediments of the Bohai Sea and Yellow Sea, China[J]. J. Mar. Syst., 2016, 155: 50-58.
17. Qiao S, Shi X, Wang G, et al. Sediment accumulation and budget in the Bohai Sea, Yellow Sea and East China Sea[J]. Mar. Geol., 2017, 390: 270-281.
18. Hou P, Yu M, Zhao M, et al. Terrestrial biomolecular burial efficiencies on continental margins[J]. J. Geophys. Res. Biogeosciences, 2020, 125: e2019JG005520.
19. Qi L, Wu Y, Chen S, et al. Evaluation of abandoned Huanghe Delta as an important carbon source for the Chinese marginal seas in recent decades[J]. Journal of Geophysical Research: Oceans, 2021, 126: e2020JC017125.
20. Yu M, Eglinton T I, Haghipour N, et al. Contrasting fates of terrestrial organic carbon pools in marginal sea sediments[J]. Geochim. Cosmochim. Acta, 2021, 309: 16-30.
21. Keil R G, Mayer L M, Quay P D, et al. Loss of organic matter from riverine particles in deltas[J]. Geochim. Cosmochim. Acta, 1997, 61: 1507-1511.
22. Vonk J E, Giosan L, Blusztajn J, et al. Spatial variations in geochemical characteristics of the modern Mackenzie Delta sedimentary system[J]. Geochim. Cosmochim. Acta, 2015, 171: 100-120.
23. Syvitski J P M, Vörösmarty C J, Kettner A J, et al. Impact of humans on the flux of terrestrial sediment to the global coastal ocean[J]. Science, 2005, 308: 376-380.
24. Alexander C R, DeMaster D J, Nittrouer C A. Sediment accumulation in a modern epicontinental-shelf setting: The Yellow Sea[J]. Mar. Geol., 1991, 98: 51-72.
25. Shi X, Shen S, Yi H I, et al. Modern sedimentary environments and dynamic depositional systems in the southern Yellow Sea[J]. Chin. Sci. Bull., 2003, 48: 1-7.
26. Yang S Y, Jung H S, Lim D, et al. A review on the provenance discrimination of sediments in the Yellow Sea[J]. Earth-Sci. Rev., 2003, 63: 93-120.
27. Tao S, Eglinton T I, Montluçon D B, et al. Diverse origins and pre-depositional histories of organic matter in contemporary Chinese marginal sea sediments[J]. Geochim. Cosmochim. Acta, 2016, 191: 70-88.
28. Schlitzer R. Ocean Data View. 2018. Available on https://odv.awi.de/.
29. Komada T, Anderson M R, Dorfmeier C L. Carbonate removal from coastal sediments for the determination of organic carbon and its isotopic signatures, $\delta^{13}C$ and $\Delta^{14}C$: Comparison of fumigation and direct acidification by hydrochloric acid[J]. Limnol. Oceanogr. Methods, 2008, 6: 254-262.
30. McIntyre C P, Wacker L, Haghipour N, et al. Online ^{13}C and ^{14}C Gas Measurements by EA-IRMS–AMS at ETH Zürich[J]. Radiocarbon, 2017, 59: 893-903.
31. Christl M, Vockenhuber C, Kubik P W, et al. The ETH Zurich AMS facilities: Performance parameters and

reference materials[J]. Nucl. Instrum. Methods Phys. Res., Sect. B, 2013, 294: 29-38.

32. Yoon S H, Kim J H, Yi H I, et al. Source, composition and reactivity of sedimentary organic carbon in the river-dominated marginal seas: A study of the eastern Yellow Sea (the northwestern Pacific)[J]. Cont. Shelf Res., 2016, 125: 114-126.

33. Kang S, Kim J H, Ryu J S, et al. Dual carbon isotope ($\delta^{13}C$ and $\Delta^{14}C$) characterization of particulate organic carbon in the Geum and Seomjin estuaries, South Korea[J]. Mar Pollut Bull, 2020, 150: 110719.

34. Drenzek N J, Montluçon D B, Yunker M B, et al. Constraints on the origin of sedimentary organic carbon in the Beaufort Sea from coupled molecular ^{13}C and ^{14}C measurements[J]. Mar. Chem., 2007, 103: 146-162.

35. Volkman J K, Barrett S M, Blackburn S I, et al. Microalgal biomarkers: a review of recent research development[J]. Org. Geochem., 1998, 29: 1163-1179.

36. Häggi C, Pätzold J, Bouillon S, et al. Impact of selective degradation on molecular isotope compositions in oxic and anoxic marine sediments[J]. Org. Geochem., 2021, 153: 104192.

37. Collister J W, Rieley G, Stern B, et al. Compound-specific $\delta^{13}C$ analyses of leaf lipids from plants with differing carbon dioxide metabolisms[J]. Org. Geochem., 1994, 21: 619-627.

38. Schouten S, Breteler W C M K, Blokker P, et al. Biosynthetic effects on the stable carbon isotopic compositions of algal lipids: Implications for deciphering the carbon isotopic biomarker record[J]. Geochim. Cosmochim. Acta, 1998, 62: 1397-1406.

39. Swart P K, Greer L, Rosenheim B E, et al. The ^{13}C Suess effect in scleractinian corals mirror changes in the anthropogenic CO_2 inventory of the surface oceans[J]. Geophys. Res. Lett., 2010, 37: L05604.

40. Andersson A, Deng J, Du K, et al. Regionally-varying combustion sources of the January 2013 severe haze events over eastern China[J]. Environ Sci Technol, 2015, 49: 2038-2043.

41. Sanchez-Cabeza J A, Ruiz-Fernández A C. ^{210}Pb sediment radiochronology: An integrated formulation and classification of dating models[J]. Geochim. Cosmochim. Acta, 2012, 82: 183-200.

42. Zhao B, Yao P, Bianchi T S, et al. The remineralization of sedimentary organic carbon in different sedimentary regimes of the Yellow and East China Seas[J]. Chem. Geol., 2018, 495: 104-117.

43. Levin I, Hesshaimer V. Radiocarbon: A unique tracer of global carbon cycle dynamics[J]. Radiocarbon, 2020, 42: 69-80.

44. Bröder L, Tesi T, Andersson A, et al. Historical records of organic matter supply and degradation status in the East Siberian Sea[J]. Org. Geochem., 2016, 91: 16-30.

45. Vonk J E, Drenzek N J, Hughen K A, et al. Temporal deconvolution of vascular plant-derived fatty acids exported from terrestrial watersheds[J]. Geochim. Cosmochim. Acta, 2019, 244: 502-521.

46. French K L, Hein C J, Haghipour N, et al. Millennial soil retention of terrestrial organic matter deposited in the Bengal Fan[J]. Scientific Reports, 2018, 8: 11997.

47. Yu M, Eglinton T I, Haghipour N, et al. Molecular isotopic insights into hydrodynamic controls on fluvial suspended particulate organic matter transport[J]. Geochim. Cosmochim. Acta, 2019b, 262: 78-91.

48. Liu W, An Z, Zhou W, et al. Carbon isotope and C/N ratios of suspended matter in rivers: An indicator of seasonal change in C4/C3 vegetation[J]. Appl. Geochem., 2003, 18: 1241-1249.

49. Fry B, Sherr E B. $\delta^{13}C$ measurements as indicators of carbon flow in marine and freshwater ecosystems[J]. Marine Science, 1984, 27: 13-47.

50. Wang X, Ma H, Li R, et al. Seasonal fluxes and source variation of organic carbon transported by two major Chinese Rivers: The Yellow River and Changjiang (Yangtze) River[J]. Global Biogeochem. Cycles, 2012, 26: GB2025.

51. Wu Y, Eglinton T I, Zhang J, et al. Spatio-temporal variation of the quality, origin and age of particulate organic matter transported by the Yangtze River (Changjiang)[J]. J. Geophys. Res. Biogeosciences, 2018, 123: 2908-2921.

52. Liu W, Yang H, Ning Y, et al. Contribution of inherent organic carbon to the bulk $\delta^{13}C$ signal in loess deposits from the arid western Chinese Loess Plateau[J]. Org. Geochem., 2007, 38: 1571-1579.

53. Yu M, Guo Z, Wang X, et al. Sources and radiocarbon ages of aerosol organic carbon along the east coast of China and implications for atmospheric fossil carbon contributions to China marginal seas[J]. Sci. Total Environ., 2018, 619-620: 957-965.

54. Yu M, Eglinton T I, Haghipour N, et al. Impacts of natural and human-induced hydrological variability on particulate organic carbon dynamics in the Yellow River[J]. Environmental Science & Technology, 2019a, 53: 1119-1129.

55. Xing L, Zhao M, Zhang H, et al. Biomarker records of phytoplankton community structure changes in the Yellow Sea over the last 200 years[J]. J. Ocean Univ. China, 2009, 39: 317-322. (In Chinese with English abstract)

56. Wang H, Saito Y, Zhang Y, et al. Recent changes of sediment flux to the western Pacific Ocean from major

rivers in East and Southeast Asia[J]. Earth-Sci. Rev., 2011, 108: 80-100.
57. Liu L Y, Wei G L, Wang J Z, et al. Anthropogenic activities have contributed moderately to increased inputs of organic materials in marginal seas off China[J]. Environ Sci Technol, 2013, 47: 11414-11422.
58. Fang Y, Chen Y, Lin T, et al. Spatio-temporal trends of elemental carbon and char/soot ratios in five sediment cores from Eastern China marginal seas: Indicators of anthropogenic activities and transport patterns[J]. Environ Sci Technol., 2018, 52: 9704-9712.
59. Bao R, Uchida M, Zhao M, et al. Organic carbon aging during across-shelf transport[J]. Geophys. Res. Lett., 2018, 45: 8425-8434.
60. Sun X, Fan D, Liu M, et al. The fate of organic carbon burial in the river-dominated East China Sea: Evidence from sediment geochemical records of the last 70 years[J]. Org. Geochem., 2020, 143: 103999.
61. Keil R G, Tsamakis E, Fuh C B, et al. Mineralogical and textural controls on the organic composition of coastal marine sediments: Hydrodynamic separation using SPLITT-fractionation[J]. Geochim. Cosmochim. Acta, 1994, 58: 879-893.
62. Blattmann T M, Liu Z, Zhang Y, et al. Mineralogical control on the fate of continentally derived organic matter in the ocean[J]. Science, 2019, 366: 742-745.
63. Hemingway J D, Rothman D H, Grant K E, et al. Mineral protection regulates long-term global preservation of natural organic carbon[J]. Nature, 2019, 570: 228-231.
64. Zhang Z, Chu Z. Modern variations in clay minerals in mud deposits of the Yellow and East China Seas and their geological significance[J]. The Holocene, 2017, 28: 386-395.
65. Galy V, France-Lanord C, Beyssac O, et al. Efficient organic carbon burial in the Bengal fan sustained by the Himalayan erosional system[J]. Nature, 2007, 450: 407-410.
66. Wu Y, Eglinton T, Yang L, et al. Spatial variability in the abundance, composition, and age of organic matter in surficial sediments of the East China Sea[J]. J. Geophys. Res. Biogeosciences, 2013, 118: 1495-1507.
67. Marwick T R, Tamooh F, Teodoru C R, et al. The age of river-transported carbon: A global perspective[J]. Global Biogeochem. Cycles, 2015, 29: 122-137.
68. Berner R A. Burial of organic carbon and pyrite sulfur in the modern ocean: Its geochemical and environmental significance[J]. Am. J. Sci., 1982, 282: 451-473.
69. Bröder L, Tesi T, Andersson A, et al. Bounding cross-shelf transport time and degradation in Siberian-Arctic land-ocean carbon transfer[J]. Nat. Commun., 2018, 9: 806.
70. Stuiver M, Polach H A. Discussion; reporting of ^{14}C data[J]. Radiocarbon, 1977, 19: 355-363.

第二篇
非碳能源替代方案与关键技术

Part II
Emission-free Energy Solutions and Key Technologies

Semitransparent polymer solar cell/triboelectric nanogenerator hybrid systems: Synergistic solar and raindrop energy conversion for window-integrated applications[①]

Tong Liu[1], Yang Zheng[1], Yunxiang Xu[1], Xianjie Liu[2], Chuanfei Wang[1*], Liangmin Yu[3,4], Mats Fahlman[2], Xiaoyi Li[1*], Petri Murto[5*], Junwu Chen[6], Xiaofeng Xu[1*]

1 College of Materials Science and Engineering, Ocean University of China, Qingdao 266100, China.
2 Laboratory of Organic Electronics, Department of Science and Technology (ITN), Linköping University, SE-60174 Norrköping, Sweden.
3 Key Laboratory of Marine Chemistry Theory and Technology, Ministry of Education, Ocean University of China, Qingdao 266100, China.
4 Open Studio for Marine Corrosion and Protection, Pilot National Laboratory for Marine Science and Technology, Qingdao 266237, China.
5 Yusuf Hamied Department of Chemistry, University of Cambridge, Cambridge, CB2 1EW, United Kingdom.
6 Institute of Polymer Optoelectronic Materials & Devices, State Key Laboratory of Luminescent Materials & Devices, South China University of Technology, Guangzhou 510640, China.
* Corresponding authors: Xiaofeng Xu (email: xuxiaofeng@ouc.edu.cn); Petri Murto (pm707@cam.ac.uk); Xiaoyi Li (email: lixiaoyi@ouc.edu.cn); Chuanfei Wang (wangchuanfei@ouc.edu.cn)
Tong Liu and Yang Zheng contribute equally to this work

Abstract

Development of photovoltaic (PV)-derived hybrid power systems can overcome the weather-dependent electricity production and increase the amount of dispatchable renewable energy generation. Herein, monolithic hybrid devices are developed *via* rational integration of high-performance semitransparent polymer solar cells (ST-PSCs) and liquid−solid triboelectric nanogenerators (TENGs). High-performance PSCs with efficiencies of 17.4% for rigid and 15.7% for flexible devices are achieved. Further electrode modifications and integration of transparent TENGs synergistically balance the above-bandgap photon harvesting and transparency in a broad wavelength range (380−1000 nm), yet significantly reduce the transmittance in the near-infrared wavelength range (1000−2500 nm) of hybrid devices. The hybrid devices simultaneously provide high visible light transparency, good color fidelity, efficient heat resistance and possibility to integrate on rigid and flexible substrates. The hybrid devices attain a high solar conversion efficiency of 10.1% under 1 sun, indicating efficient light-to-electricity conversion (a maximum electrical power output: 101 W·m^{-2}) on sunny days. The hybrid devices can also generate a maximum electrical power output of 2.62 W·m^{-2} through waterdrop energy conversion, implying complementary green electricity production on rainy days.

---

① 本文于2022年12月发表在*Nano Energy*第103卷，https://doi.org/10.1016/j.nanoen.2022.107776。

The controlled ambient temperature and specific transmittance windows provided by the hybrid devices sustain plant growth and highlight their great potential in agricultural applications. Gratifyingly, this work demonstrates the first example of ST-PSC/TENG hybrid systems for scaling up renewable power generation in different weather conditions, considering architectural and agricultural applications.

Keywords: hybrid energy conversion; green electricity production; organic photovoltaic cells; semitransparent polymer solar cells; triboelectric nanogenerator

Introduction

Access to affordable, clean and reliable energy derived from natural processes has become a cornerstone of the world's prosperity and sustainability. Sunlight is the largest energy source for our planet. The total amount of solar energy incident on Earth exceeds the world's current and anticipated energy requirements. The use of solar power has become increasingly attractive due to its inexhaustible supply, climate change mitigation and non-polluting character. The most common form of solar power is harnessed by photovoltaic (PV) cells, which convert light directly into electricity. Today, electricity production from solar power plants has become cost-competitive compared to that generated from fossil fuel power plants. Most current PV modules are powered by crystalline silicon-based PV cells, representing >95% of PV market sales worldwide. For residential and commercial settings, most PV cells are assembled into large and utility-scale systems and deploying on ground, rooftop and water bodies. Since urban areas currently house about 55% f the world's population lives and consume about 75% of the global primary energy supply, there is also a growing consensus that many aspects of our life can be powered by many distributed PV systems for on-site use. Introduction of PV elements as integrated parts in energy-saving buildings and as off-grid power for wearable, portable, Internet of Things and agricultural applications have gained wide attention[1,2].

The past decade has seen the rapid development of new PV materials and technologies, relying on the improvements in transparency, flexibility, integrability and appearance[3,4]. Organic photovoltaics (OPVs) have demonstrated appealing characteristics such as low-cost, light-weight, flexible, compatible for large-scale manufacturing and a short energy payback time[5-10]. The recent development of non-fullerene acceptors and ternary strategies have significantly enhanced the light-to-electricity conversion of OPVs, leading to outstanding power conversion efficiencies (PCEs) of >19% in 2022[11-18]. Organic semiconductors feature a wide choice of materials, strong but narrow absorption, thin active layers (about 100 nm), high absorption coefficients (about 10^5 cm^{-1}) and tunable colors, these competitive properties have promoted extensive research on semitransparent polymer solar cells (ST-PSCs), delivering rapid PCE growth of >12%, visible light transmission (VLT) of >20%, vivid colors and flexibility[19-32]. It is estimated that residential and commercial sectors have accounted for >40% of the total energy consumption[33]. Windows have become one of the largest contributors to the heat gain and loss in buildings. To achieve global climate objectives by 2050, the energy consumption in buildings should be reduced while new technologies to better integrate renewable energy sources are still being developed. Recent milestones of ST-PSCs have attracted growing attention in many window-integrated applications. Building-integrated ST-PSCs enable us to obtain both practical and aesthetic values for next-generation energy-saving and self-powered architecture, addressing the increasing challenges to urban energy and food security[34-37]. Compared to other color-tinted and rigid PV technologies, the development of ST-PSCs represents one of the most promising market niches for OPVs[38,39].

Despite the impressive advances in past decades, there are several inherent challenges for real-world applications of PV technologies, which can be rationalized as follows: (1) Solar energy is an intermittent energy source, which strongly relies on location, weather, time of the day and seasons of the year. Solar power inevitably drops in low light conditions (i.e., on cloudy, rainy, snowy and dirt days), exhibiting interrupted electricity production. (2) A maximum theoretical efficiency of single-junction PV cells has been determined by the Shockley-Queisser (SQ) limit. About 32% of solar power (1000 $W \cdot m^{-2}$) falling on an ideal silicon-based PV cell (bandgap: about 1.1 eV) can be converted into electricity[40]. Lower PCEs (15%–22%) are generally obtained from industrially produced solar modules.

In order to scale up renewable power generation, design of PV-derived hybrid systems for multi-source energy harvesting (i.e., thermal, hydrodynamic and mechanical energy) has drawn growing attention[1,41]. Like solar energy, mechanical energy from falling water is another renewable energy source ubiquitous in nature. Based on different liquid-solid interfacial interactions, triboelectric nanogenerators (TENGs) can efficiently convert water energy (i.e., raindrops and water waves) into electricity[42]. In this regard, TENGs can spontaneously compensate for the electric power generation of PV cells on rainy days[43-47]. As a result, the PV/TENG hybrid systems represent a great potential to complement vulnerable aspects of individual PV and TENG components, converting more renewable energy and boosting green electricity output under changing weather conditions and in different application scenarios[48,49]. The generated electrical energy can be stored in energy storage devices (i.e., capacitors and batteries) and released when the demand for electricity increases. Recent development of PV/TENG hybrid systems has mainly focused on exploring dual-energy conversion and efficient electricity production via material, interface and device engineering[50-56]. However, it is noted that current PV/TENG hybrid systems do not fully combine the advantages of next-generation PV technologies (i.e., transparency, color-tunability, flexibility and integrability). Although semitransparent PV cells are emerging as a promising energy solution for window-integrated photovoltaics, none of semitransparent PVs have been integrated with TENGs to date. Future development of semitransparent PV/TENG hybrid system would be highly desired, providing promising opportunities to ensure green electricity production, transparency, color perception and window insulation for versatile on-site use (i.e., window panes, skylights, building facades and greenhouses). To the best of our knowledge, such a proof-of-concept has not been reported yet.

In this contribution, ST-PSC/TENG hybrid systems were developed via a proper combination of high-performance ST-PSCs and TENGs based on a monolithic design (TENGs on top and ST-PSCs on bottom) (Fig. 1). In the first stage, a series of opaque PSCs were fabricated, taking the benefits of a ternary strategy. Rigid and flexible PSCs attained high PCEs of 17.4% and 15.7% under 1 sun, respectively. Based on electrode modulations, optimized ST-PSCs showed simultaneously high PCEs of 12.7% and VLT of >20%, leading to well-balanced solar power generation, transparency and color-neutrality. Transparent TENGs were integrated onto best-performing ST-PSCs via shared electrodes, delivering rigid and flexible ST-PSC/TENG hybrid devices. A maximum electric power output of 2.62 $W \cdot m^{-2}$ with a current of about 120 μA, transferred charges of about 60 nC and a voltage of about 100 V were achieved under water droplet stimuli. For the first time, our work endows ST-PSC with a new function of waterdrop power conversion. We proposed a feasibility study and new experimental approaches to explore the dual-energy conversion, optical and heat-insulating properties of the ST-PSC/TENG hybrid devices, all of which were well correlated to their device structures. Furthermore, plant cultivation experiments disclose that the hybrid systems can sustain plant growth due to the controlled transmission and ambient temperature. Our work shines a light on the facile fabrication of ST-PSC/TENG hybrid systems and highlights the great potential of developing ST-PSC/TENG hybrid systems for energy-wise window-integrated applications.

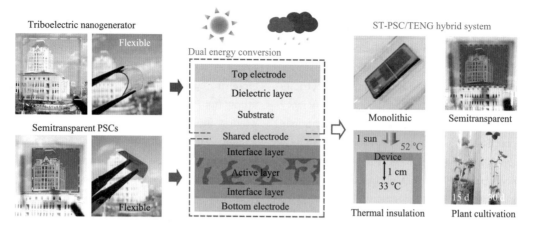

Fig. 1　Schematic illustration and digital photographs of the ST-PSC/TENG hybrid systems developed in this study

Results and Discussion

▶ *Fabrication and characterization of opaque PSCs*

Conjugated materials named PBDB-TF and Y6 are one of the best-performing combinations in the state-of-the-art PSCs[57]. Electron-donor PBDB-TF is an alternating copolymer[58]. Small molecule Y6 is one of the most representative non-fullerene electron-acceptors[59-62]. Recently, the incorporation of a third building block into the PBDB-TF backbone via copolymerization has overcome some drawbacks of PBDB-TF and further enhanced photovoltaic performance[63-68]. In this study, a terpolymer (named PBDB-TF-T10) was employed as the electron donor[69]. Compared to PBDB-TF, the merits of PBDB-TF-T10 are rationalized as follows: (1) The donor−acceptor 1− donor−acceptor 2 (D−A_1−D−A_2)-type backbone of PBDB-TF-T10 can broaden the absorption spectrum for sunlight harvesting, increase the solubility for solution-processing and optimize miscibility and film morphology. (2) Incorporating double-ester-substituted thiophenes as the third building block down-shifted the energy levels of PBDB-TF-T10, delivering about 0.05 V higher photovoltage of PSCs. (3) PBDB-TF-T10 can provide a broad molecular weight window and attain steady PCEs of >16% in PSCs. Apart from Y6, a fullerene electron-acceptor (named $PC_{71}BM$) was employed as the third component (the second acceptor), forming ternary PSCs. The chemical structures of PBDB-TF-T10, Y6 and $PC_{71}BM$ are depicted in Fig. 2a.

The absorbance spectrum of PBDB-TF-T10 in film ranged from 350 nm to 750 nm, exhibiting a maximum absorption coefficient of about $6.4×10^4$ cm^{-1} at 610 nm. The absorbance spectrum of Y6 in film was in a wavelength range of 425−1000 nm, showing the highest absorption coefficient of about $11.0×10^4$ cm^{-1} at 825 nm (Fig. 2b). The complementary absorption of PBDB-TF-T10 and Y6 afforded broad absorption (350−1000 nm) in the blend film, covering the solar spectrum in the whole visible (380−780 nm) and some near-infrared (NIR) (780−1000 nm) wavelength ranges. The absorbance spectra of the ternary blends in the short-wavelength range (350−620 nm) were further improved along with the increased content of $PC_{71}BM$ (Fig. S1), offering a prerequisite for improving the photocurrent of ternary PSCs. As depicted in the energy level diagram, the highest occupied molecular orbital (HOMO) and the lowest unoccupied molecular orbital (LUMO) levels of PBDB-TF-T10 were −5.60 eV and −3.72 eV, respectively (Fig. S2a). Y6 featured down-shifted HOMO (−5.67 eV) and LUMO (−4.12 eV)

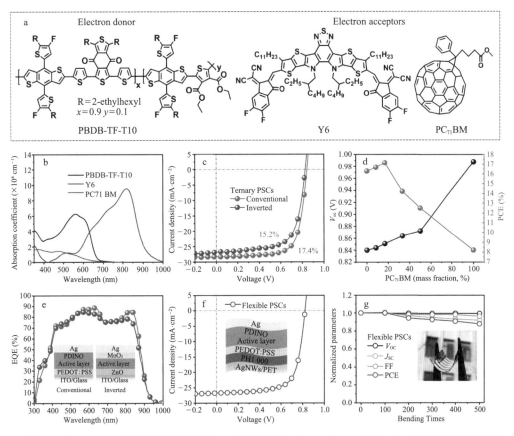

a. Chemical structures of PBDB-TF-T10, Y6 and PC$_{71}$BM; b. Absorption coefficients of PBDB-TF-T10, Y6 and PC$_{71}$BM in thin films; c. J–V curves of opaque PSCs; d. V_{oc} and PCEs of opaque PSCs with different PC$_{71}$BM content; e. EQE spectra of opaque PSCs (inset images show device structures of conventional and inverted PSCs); f. J–V curves of flexible PSCs (the inset image shows the device structure); g. Device parameter decay versus bending cycles of flexible PSCs (the inset image shows a digital photograph of flexible PSC).

Fig. 2 Fabrication and characterization of opaque PSCs

levels compared to PBDB-TF-T10. The HOMO and LUMO levels of Y6 were in between those of PC$_{71}$BM. This ternary system did not show a cascade energy level diagram (Fig. S2b). The charge/energy transfer pathways were correlated with the photovoltaic performance in the following sections.

Opaque PSCs were fabricated in both conventional and inverted configurations. All PSCs were characterized under 1 sun (AM 1.5 G and 1000 W·m^{-2}). The photovoltaic performance was optimized by varying the active layer thickness (90–180 nm) and adding solvent additives (1-chloronaphthalene, 0.3%–0.8% (volume fraction) v) (Table S1 and Table S2). Current density-voltage (J–V) curves of conventional and inverted PSCs are plotted in Fig. 2c, Fig. S3a and Fig. S4a. Photovoltaic parameters of best-performing PSCs are summarized in Table 1. Binary PSCs (PBDB-TF-T10:Y6 = 1:1.2) in conventional and inverted structures afforded high PCEs of 16.3% and 14.6%, respectively. For ternary PSCs, variations in open-circuit voltage (V_{oc}), short-circuit current density (J_{sc}), fill factor (FF) and PCEs as a function of the PC$_{71}$BM content are summarized in Fig. 2d and Table S3. A negligible increase in V_{oc} from 0.84 V to 0.85 V was found when the PC$_{71}$BM content was below 20%. Similar results have been demonstrated in previous reports[70-72]. Apart from charge transfer, energy transfer from PCBM to Y6 has been evidenced in several ternary systems[73,74]. Therefore, the small content of $PC_{71}BM$ (<20%) can harvest additional sunlight and contribute to the evident J_{sc} improvement, yet exhibited negligible increase in V_{oc}. When the $PC_{71}BM$ content was increased from 20% to 100%, a quasilinear increase in V_{oc} from 0.85 V to 0.99 V was recorded. The composition-dependent V_{oc} behavior and reduced V_{oc} loss indicate the good miscibility and molecular intermixing of the materials in the ternary system, leading to an alloy model mechanism of the exciton dissociation and charge transfer. However, both J_{sc} and FF showed a significant decrease. As a result, a maximum PCE of 17.4% was attained in conventional PSCs with an optimized ratio of PBDB-TF-T10:Y6:$PC_{71}BM$ = 1:1:0.2. PCEs of ternary PSCs were improved by about 6.7% compared to binary PSCs. A similar trend of photovoltaic parameters was observed in inverted PSCs, which resulted in a higher PCE of 15.2% compared to binary PSCs. The binary and ternary blend films presented fine microstructures without large phase separation and polymer domains.

The small root-mean-square (RMS) roughness of 0.86–1.09 nm indicates the low surface roughness of all blend films (Fig. S5). External quantum efficiency (EQE) spectra of all PSCs were measured (Fig. 2e, Fig. S3b and Fig. S4b). Broad EQE profiles were found in the wavelength range of 300–1000 nm, which agrees well with the absorbance spectra of the blend films. For each PSC, the photocurrent density calculated by integrating the EQE spectra with an AM 1.5G solar spectrum was comparable to J_{sc}, exhibiting a mismatch of <5%. The results confirm the accuracy of the J–V characterization of PSCs.

Exciton dissociation, charge carrier transport and recombination losses were characterized to understand the superior photovoltaic performance of ternary PSCs. The exciton dissociation probability (P_{diss}) was evaluated via the J–V characteristics, which refers to the probability that an exciton dissociates to free charges under a short-circuit condition[75]. The ternary PSCs in both conventional and inverted structures showed slightly higher Pdiss compared to the binary counterparts (Fig. S6). Hole and electron mobilities of blend films

Table 1 Photovoltaic parameters of optimized opaque PSCs and ST-PSCs

Device	Active layer	V_{oc} (V)	J_{sc} (mA·cm^{-2})	FF	PCE (%)
Opaque PSCs					
Conventional[a]	binary[d]	0.84	26.9 (25.6)[f]	0.72	16.3g (16.0 ± 0.3)[h]
	ternary[e]	0.85	27.6 (26.3)	0.74	17.4 (17.1 ± 0.3)
Inverted[b]	binary	0.83	26.3 (25.1)	0.67	14.6 (14.1 ± 0.5)
	ternary	0.83	26.9 (25.6)	0.68	15.2 (14.9 ± 0.3)
Flexible[c]	ternary	0.82	27.0 (25.7)	0.71	15.7 (15.3 ± 0.4)
ST-PSCs[i]					
Conventional	ternary	0.85	23.0 (22.1)	0.65	12.7 (12.5 ± 0.2)
Inverted	ternary	0.83	21.9 (21.0)	0.64	11.6 (11.3 ± 0.3)
Top MoO_3	ternary	0.83	19.9 (19.0)	0.63	10.4 (10.2 ± 0.2)
Flexible	ternary	0.79	21.7 (20.6)	0.57	9.8 (9.4 ± 0.4)

Note: a. Device configuration: ITO/PEDOT:PSS/active layer/PDINO/Ag. b. ITO/ZnO/active layer/MoO_3/Ag. c. AgNWs/PH1000/PEDOT:PSS/active layer/PDINO/Ag. d. PBDB-TF-T10:Y6 = 1:1.2. e. PBDB-TF-T10:Y6:$PC_{71}BM$ = 1:1:0.2. f. Calculated photocurrent densities are listed in the parentheses. g. The maximal PCE. h. The average PCE from 15 PSCs are given in the parentheses. i. ST-PSCs with a 15 nm Ag cathode.

were measured via a space charge limited current (SCLC) method (Fig. S7 and Table S4). Moderate and well-balanced hole mobilities (μ_h) and electron mobilities (μ_e) on the order of 1×10^{-4} $cm^2 \cdot V^{-1} \cdot s^{-1}$ were recorded from all PSCs, exhibiting small μ_e/μ_h within 1.5. Recombination losses of PSCs were evaluated by measuring the dependence of J_{sc} and V_{oc} upon the intensity (I) of incident light[76]. All PSCs showed high and comparable α of about 0.98 under short-circuit conditions (Fig. S8a and S8c), irrespective of material components and device structures. The results disclose that mobility and recombination were not critical limiting factors for the photovoltaic performance of PSCs. In addition, the slope (nkT/q) of fitting lines ranged from 1.09 kT/q to 1.32 kT/q (Fig. S8b and S8d). It implies that the bimolecular mechanism dominated the recombination losses of PSCs under open-circuit conditions[76]. Based on similar blend morphologies, mobilities and recombination losses, the enhanced J_{sc} and PCEs of the ternary PSCs mainly stem from the synergistic effects of broader absorption spectra and improved exciton dissociation compared to the binary PSCs[72,73,77]. On the other hand, the reduced exciton dissociation and increased recombination losses limited J_{sc} and PCEs of inverted PSCs in both binary and ternary systems.

Based on the best-performing ternary blend (PBDB-TF-T10:Y6:PC$_{71}$BM = 1:1:0.2), flexible and ITO-free PSCs were fabricated using a device configuration of polyethylene terephthalate (PET)/silver nanowires (AgNWs)/PH1000 (160 nm)/PEDOT:PSS (40 nm)/active layer/PDINO/Ag. AgNWs and PH 1000 have been regarded as promising alternatives to rigid electrodes (i.e., ITO) in ST-PSCs due to their high conductivity, good flexibility and transparency. AgNWs and PH 1000 were employed to fabricate a composite and flexible electrode in this study. The ratio (mass fraction) of PEDOT to PSS is 1:2.5 in PH 1000, leading to a higher electrical conductivity (0.2–1 $S \cdot cm^{-1}$) than PEDOT:PSS (4083, 10^{-4}–10^{-2} $S \cdot cm^{-1}$). Furthermore, the upper PEDOT:PSS layers can reduce the surface roughness of the AgNWs electrode and enhance the compatibility between the active layer and AgNWs electrode. The flexible PET/AgNWs/PH1000/PEDOT:PSS (4083) electrode was prepared via a modified method based on previous reports[73,78-80]. Scanning electron microscope (SEM) images show that AgNWs featured an average diameter of about 30 nm and an average length of about 25 μm, exhibiting a high aspect ratio of >800 (Fig. S9). The detailed fabrication of the flexible electrode is described in Section S4. Flexible PSCs afforded a V_{oc} of 0.82 V, a J_{sc} of 27.0 $mA \cdot cm^{-2}$, a FF of 0.71, reaching a high PCE of 15.7% (Fig. 2f). Flexible PSCs showed broad EQE similar to the rigid PSCs (Fig. S10), efficient exciton dissociation (P_{diss} = 96%) (Fig. S11) and bimolecular-dominated and small recombination losses (α = 0.98, slope: 1.08 kT/q) (Fig. S12). Compared to ITO-based PSCs, the performance of flexible PSCs was limited by the higher sheet resistance ($R_{sh} \approx 30$ Ω) and lower transparency (transmittance of 74.8% in the wavelength range of 400−900 nm) of the PET/AgNWs/PH1000/PEDOT:PSS electrode (Table S5 and Fig. S13). The rigid glass/ITO/PEDOT:PSS electrode afforded a higher R_{sh} of about 10 Ω and transmittance of 81.3% in the wavelength range of 400−900 nm.

A bending test was performed to assess the durability of flexible PSCs against deformation. J–V curves of flexible PSCs were measured under different bending cycles at a bending radius of 5.0 mm. It is noted that less than 2% of PCE decay was recorded after 100 cycles of continuous bending. About 85% of PCE can be retained after 500 cycles of bending (Fig. 2g). Its outstanding photovoltaic and mechanical performance is among the best-performing flexible PSCs[73]. Scanning electron microscopy (SEM) images were taken to evaluate the morphological variations of the flexible PET/AgNWs/PH 1000/PEDOT:PSS electrode and ternary active layer before and after bending. The bending strain generally forms small cracks on brittle ITO electrodes and the upper active layers, increasing the sheet resistance of PSCs[81]. On the other hand, no evident wrinkles or cracks were observed on the PET/AgNWs/PH 1000/PEDOT:PSS electrode and active layer after 100 and 500 cycles of bending (Fig. S14). The PET/AgNWs/PH 1000/PEDOT:PSS electrode can retain a R_{sh} of about 30 Ω after 100 and 500 cycles of bending. The results are in good agreement with the stable photovoltaic performance during mechanical bending, highlighting the excellent flexibility and durability of the PET/AgNWs/PH 1000/PEDOT:PSS layer used as the deformable and robust electrode.

▶ *Fabrication and characterization of ST-PSCs*

Conventional, inverted, electrode-modified and flexible ST-PSCs were fabricated based on the best-performing ternary system (PBDB-TF-T10:Y6:$PC_{71}BM$ = 1:1:0.2) and the optimized thickness (15 nm) of Ag electrode. Conventional and inverted ST-PSCs attained high PCEs of 12.7% and 11.6%, respectively (Fig. 3a). For optical modulation, a molybdenum trioxide (MoO_3) layer was deposited on the Ag anode of inverted ST-PSCs, forming a dielectric/metal/dielectric (D/M/D) structure as the top electrode. The top MoO_3 layer reduced the transmittance of the MoO_3/Ag/MoO_3 electrode in the wavelength range of 350−650 nm, whereas it improved the transmittance in the wavelength range of 650−1000 nm (Fig. S15). The reflectance of the Ag layer can be selectively reduced via the D/M/D structure[82,83]. The thickness of the top MoO_3 layer was increased from 10 nm to 50 nm. Notably, 30 nm of the MoO_3 layer attained a maximum average transmittance (22.4%) in the wavelength range of 350−900 nm among the D/M/D electrodes. As a result, the MoO_3 (8 nm)/Ag (15 nm)/MoO_3 (30 nm) layer was employed as the top electrode to improve the transmittance and color-rendering properties of ST-PSCs. The MoO_3 layer slightly lowered J_{sc} of inverted ST-PSCs from 21.9 mA·cm^{-2} to 19.9 mA·cm^{-2} and retained a moderate PCE of 10.4%. Based on the PET/AgNWs/PH1000/PEDOT:PSS electrode, flexible ST-PSCs in a conventional structure showed a good PCE of 9.8%. For all ST-PSCs, the photocurrent density calculated from the EQE spectrum agreed

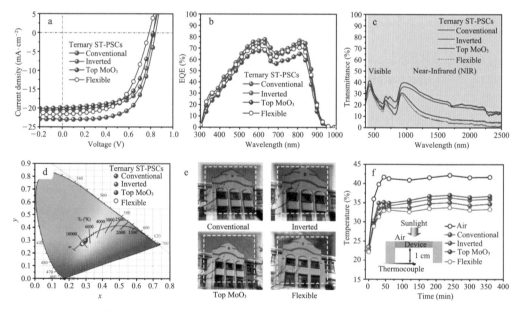

a. J–V curves; b. EQE spectra; c. Transmittance spectra; d. Color coordinates; e. Digital photographs of ST-PSCs with different device structures; f. Temperature changes of surrounding air below different ST-PSCs under solar irradiation.

Fig. 3　Fabrication and characterization of ST-PSCs

well with J_{sc} (Fig. 3b). The exciton dissociation and recombination characterization reveal that the exciton dissociation was the main limiting factor for the photovoltaic performance of ST-PSCs (Fig. S16 and Fig. S17).

Transparency and color perception were critical parameters of ST-PSCs. Transmittance spectra of four ST-PSCs (active area: 1 cm^2) in a broad wavelength range of 350−2500 nm were measured (Fig. 3c). Their transmittance showed three high transmission bands ranging from 350−625 nm, 625−810 nm and 810−2500 nm, which agreed well with the absorbance spectrum of the active layer. Weighted with the AM 1.5G solar spectrum, the effective transmittances of ST-PSC in the wavelength ranges of 350−1000 nm (T_1, absorption band of the active layer), 380−780 nm (T_2, wavelengths of visible light) and 780−2500 nm (T_3, wavelengths of NIR light) were calculated and summarized in Table S6. Portions of the photons above the absorption band of the active layer (350−1000 nm) can be converted into electricity *via* ST-PSCs. The conventional and inverted ST-PSCs showed comparable T_1 of 29.02% and 26.37%, respectively. Introduction of the top MoO$_3$ layer on inverted ST-PSC successfully improved T_1 to 30.24%. The lowest T_1 of 24.02% was recorded in flexible ST-PSCs, which mainly stem from the lower transmittance of the PET/AgNWs/PH1000/PEDOT:PSS electrode compared to the glass/ITO/PEDOT:PSS electrode.

VLT of four ST-PSCs was determined by the human eye photopic response and T_2, which refers to the brightness of an object as perceived by the human eyes. Conventional and inverted ST-PSCs showed comparable VLTs of 21.61% and 21.68%, respectively. Transmittance profiles of ST-PSCs in the visible light wavelengths play a vital role in the transparency and color perception of solar panels. The D/M/D electrode reduced the transmittance spectrum in the short-wavelength range of 350−650 nm, yet increased the transmittance in the long-wavelength range of 650−1000 nm. As a result, the MoO$_3$-modified ST-PSC attained a maximum VLT of 24.43%. The PET/AgNWs/PH1000/PEDOT:PSS electrode led to the lowest VLT of 18.75% in flexible ST-PSC. Moreover, color coordinates (*x*, *y*) of four ST-PCSs are depicted in the Commission Internationale de l'Eclairage (CIE) 1931 chromaticity diagram (Fig. 3d). The color coordinates of conventional, inverted and flexible were located at (0.28, 0.27), (0.29, 0.29) and (0.28, 0.28), respectively. Their strong transmittance in the wavelength range of 380−650 nm contributed to the light bluish color. The color coordinate of MoO$_3$-modified ST-PSC was located at (0.30, 0.29), which is the nearest to the color-neutral point of the AM 1.5G solar spectrum (0.33, 0.34). Under real sky, a slight blue tint was observed through conventional and inverted ST-PSCs (Fig. 3e). MoO$_3$-modified ST-PSC showed better color reproduction of buildings and afforded a high color rendering index (CRI) of 94. The flatter transmittance of MoO$_3$-modified ST-PSC was suggested to be the key factor for its negligible color aberration[72,83,84]. Despite the inevitable PCE loss, the result highlights the superior transparency and color rendering ability of MoO$_3$-modified ST-PSC compared to other ST-PSCs.

Most below-bandgap photons in the wavelength range of 1000−2500 nm were not captured via ST-PSCs. Some photons were reflected on the surface of the electrodes. Others were converted to heat and lost in the surrounding air via convection[85]. The heatproof properties of ST-PSCs were characterized based on an artificial setup. The solar intensity on the top surface of ST-PSCs (active area: 1 cm^2) was set to be 1 sun. The surface temperatures of ST-PSCs were monitored by an infrared (IR) camera (Fig. S18). The transmitted light gradually warmed the air below ST-PSCs. K-type thermocouples were placed 1 cm under ST-PSCs and measured the air temperatures (Fig. 3f). With no covering, the air temperature rapidly increased under 1 sun and reached a steady state of about 42 °C in 1 h. Without significant heat loss through convective airflow, ST-PSCs slowed the increase in air temperatures and reduced the air temperatures by about 6 °C under 1 sun. Among four ST-PSCs, MoO_3-modified and inverted ST-PSCs showed relatively higher transmittance, specifically in the wavelength range of 1000−2500 nm. The stronger transmitted light led to their lower surface temperatures of about 48 °C and higher air temperatures (about 36 °C) below the two devices. On the other hand, the relatively lower transmittance of flexible ST-PSC contributed to its higher surface temperature of about 53 °C and lower air temperature of about 33 °C. As a result, the surface temperatures of ST-PSCs and the air temperatures below ST-PSCs can be well correlated with the transmittance and device structures of ST-PSCs. The results highlight the good thermal insulating properties of ST-PSCs for window-integrated applications. Window insulation based on ST-PSC/TENG hybrid systems was characterized in the next sections.

▶ *Fabrication and characterization of ST-PSC/TENG hybrid devices*

Based on the high photovoltaic performance, good optical properties and excellent thermal insulation, MoO_3-modified and AgNW-based ST-PSCs were selected and integrated with TENGs, forming two ST-PSC/TENG hybrid systems with rigid and flexible properties. Device configurations of the ST-PSC/TENG hybrid systems are illustrated in Fig. 4a. TENG was employed as the upper electric generator in each hybrid device, converting the mechanical energy of water droplets to electricity. A Pt wire (diameter: about 100 μm) was used as the top electrode in TENG. A transparent layer of fluorinated ethylene propylene (FEP) served as the friction layer[86]. The output voltage of TENG based on different FEP thicknesses (50−160 μm) was measured (Fig. S19). The output voltage gradually raised when the FEP thickness increased from 50 to 130 μm. The voltage reached a steady value when the thickness was higher than 130 μm. As a result, a 130 μm FEP layer was used as the friction layer in this study. The FEP film featured a smooth and water-repellant surface as characterized by SEM images and water contact angle measurements (Fig. S20 and S21). Furthermore, a water wettability experiment was conducted by recording a water droplet when it was contacting with and then separating from the FEP surface. No clear water residue can be seen on the surface of the FEP film as depicted via a high-resolution camera (Fig. S22).

The equivalent circuit and working principle of TENG are illustrated in Fig. S23 and S24, respectively. Negative charges were generated on the FEP surface due to the triboelectric effect between water droplets and FEP. Meanwhile, positive charges were generated on the ITO or AgNWs electrode, forming an equivalent capacitor (named C_{FEP}). The interlayer between the water droplet and FEP surface formed an electric double layer (EDL) and acted as another capacitor (named C_{EDL}). The water droplet can be considered as a conductor, since ions were always found in natural raindrops. As a result, water droplets can be regarded as low resistors (named R_w). Another resistor was the external resistance (named R_L). When a water droplet arrived at the Pt wire, CFEP can charge CEDL. Positive charges on the ITO or AgNWs electrode were transferred to the water droplet through the external circuit and Pt electrode. When a water droplet left the Pt electrode, a reverse electron flow moving from the water droplet to the ITO or AgNWs electrode was recorded[87]. All ST-PSCs were fabricated and encapsulated with ultraviolet-cured resin/glass in inert atmosphere. For the fabrication of rigid ST-PSC/TENG hybrid devices, TENG was placed on top of MoO_3-modified ST-PSC and both devices shared the same glass/ITO electrode. For the fabrication of flexible ST-PSC/TENG hybrid devices, the bottom AgNWs-based ST-PSC shared the PET/AgNW electrode with TENG.

The experimental setup for the characterization of solar and mechanical energy conversion is illustrated in Fig. 4b. All hybrid devices were tested in ambient conditions. For each hybrid device, the electrical output of TENG was measured via a source measure unit (SMU). Water droplets (volume: 35 μL) were

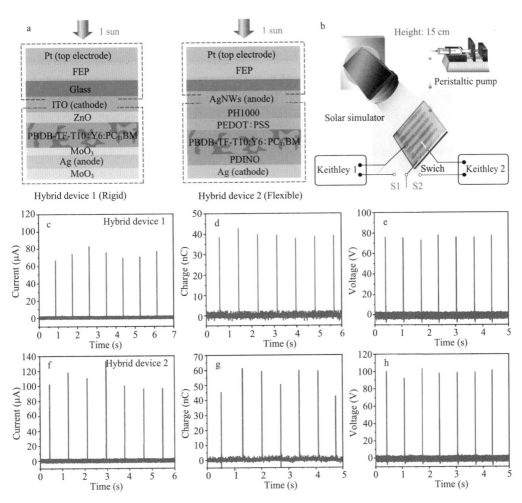

a. Device structures of rigid and flexible ST-PSC/TENG hybrid devices. b. Schematic illustration of dual-energy conversion characterization of ST-PSC/TENG hybrid devices. c. Output current; d, e. Transferred charges and Voltage of TENG in hybrid device 1; f–h. Current, transferred charges and voltage of TENG in hybrid device 2.

Fig. 4　Fabrication and characterization of ST-PSC/TENG hybrid devices

continuously dropped onto the surface of the hybrid device from a height of 15 cm. In hybrid device 1, high pulsed current peaks of about 80 μA were recorded due to the transient charging process, exhibiting an average duration time as short as about 50 μs (Fig. 4c and Fig. S25). The transferred charge and voltage of hybrid device 1 were about 40 nC and about 80 V, respectively (Fig. 4d and Fig. 4e). In hybrid device 1, the electrical power output (power density) of TENG was calculated to be 1.47 W·m^{-2} based on a standard external load of 1 MΩ (Fig. S26). Apart from device performance, reproducibility is another critical factor in the fabrication of TENG and hybrid devices. Hybrid device 1 (7 devices) was made in batch production and characterized under the same conditions. All TENG in these hybrid devices showed comparable voltage output of about 80 V (Fig. S27). The result highlights the fabrication accuracy and reproducibility of the hybrid devices.

When measured under the same conditions, the current of about 120 μA (Fig. 4f), transferred charges of about 60 nC (Fig. 4g) and a voltage of about 100 V (Fig. 4h) were synergistically enhanced in hybrid device 2, leading to a higher electrical power output of 2.62 W·m^{-2}. Similar results have been emerged in recent reports[88,89]. The reduced thickness of the intermediate dielectric layer from (0.7 mm, the glass layer) to (0.188 mm, the PET layer) can increase the induced charges on the PET/AgNWs electrode and enhance the charge-transfer efficiencies and overall power output of TENGs. Notably, one waterdrop (volume: 35 μL, height: 15 cm) can light more than 20 light-emitting diodes in hybrid device 2 (Supporting Video S1). The results demonstrate that the hybrid devices can efficiently convert the mechanical energy of water droplets into electricity.

Transmittance spectra of TENG, MoO$_3$-modified ST-PSC, hybrid device 1 and hybrid device 2 were measured and summarized in Fig. 5a. A glass slide (thickness: 1.1 mm) was also measured as a control group. Weighted with the AM 1.5G solar spectrum, the effective transmittances of all devices in the wavelength ranges of 350−1000 nm (T_1, absorption band of the active layer), 380−780 nm (T_2, wavelengths of visible light) and 780−2500 nm (T_3, wavelengths of NIR light) were summarized in Table S7. The glass slide showed high and steady transmittance of >90% in a broad wavelength range of 350−2500 nm. TENG afforded a high effective transmittance of 87.75% in the wavelength range of 350−1000 nm. It reveals that the majority of above-bandgap photons can pass through TENG and can be absorbed by MoO$_3$-modified ST-PSC in hybrid device 1. TENG provided high and steady transmittance of 88.76% in the visible wavelength range, exhibiting a comparable VLT of 90.28% to the glass slide (VLT = 91.76%). Furthermore, TENG showed a color coordinate (x, y) of (0.34, 0.34), approaching the white point in the CIE chromaticity diagram (Fig. S28). Based on the outstanding transparency and color-neutrality of TENG, hybrid device 1 attained a high VLT of 23.49%, a high CRI of 92 and a good color coordinate (x, y) of (0.30, 0.30), which are comparable to MoO$_3$-modified ST-PSC. Lower VLT of the PET/AgNWs/PH1000/PEDOT:PSS electrode suppressed VLT of hybrid device 2 (17.73%) compared to hybrid device 1. Hybrid device 1 can reproduce the background building colors without significant color aberration (Fig. 5b). Notably, a gradual decline in the transmittance of TENG (T_3 = 75.12%) was observed in the NIR wavelength range (780−2500 nm), indicating the great potential of TENG for window insulation.

Apart from water droplet-to-electricity conversion, sunlight passed through TENG and triggered the light-to-electricity conversion of ST-PSCs in hybrid devices. The photovoltage and photocurrent were characterized via another source measure unit (SMU) (Fig. 4a). J–V and EQE curves of the hybrid devices are plotted in Fig. 5c and Fig. S29. The photovoltaic parameters are summarized in Table S8. ST-PSC in hybrid device 1 attained a V_{oc} of 0.83 V, a J_{sc} of 19.4 mA·cm^{-2}, a FF of 0.63 and a high PCE of 10.1% under 1 sun, exhibiting nearly the same PCE with neat MoO$_3$-modified ST-PSC (10.4%). ST-PSC in hybrid device 2 showed a good PCE of 8.4%, which is comparable to AgNW-based ST-PSCs (9.8%). The high transmittance of TENG in the wavelength range of 350−1000 nm played a role in the limited PCE decay of ST-PSCs in the hybrid devices.

In order to evaluate the complementary electricity generation, a capacitor (0.47 μF) and a two-way switch were used to form an RC charging circuit and stored the free charges generated in hybrid device 1 (Fig. 5d). The inset figure depicts the two-way switch connection diagram of the charging circuit. When switch S1 was on, ST-PSC in hybrid device 1 can charge up the capacitor under 1 sun. A sharp increase in voltage was recorded, reaching a steady value of 0.83 V within seconds (Fig. S30a). The voltage was consistent with V_{oc} of MoO$_3$-modified ST-PSC. This charging process (Stage 1) indicates the efficient light-to-electricity conversion in ST-PSC on sunny days. When switch S2 was on, the capacitor was charged up via TENG in hybrid device 1. The voltage rise was steep since the charging rate was fast at the start of charging

(Fig. S30b). The enlarged view of the charging curve shows the gradual voltage accumulation along with the duration of water drops. After the transient period, the voltage tapered off exponentially as the capacitor took on additional charges at a reduced rate, approaching the supply voltage of over 13 V. For potential practical applications, several capacitors in series could be used as the power supply for electronic devices. This charging process (Stage 2) implies the efficient water droplet-to-electricity conversion in TENG when it rains. TENG and ST-PSC were independent parts in the hybrid system, preventing the high voltage of TENG from breaking ST-PSC down.

In order to examine the device performance in different weather conditions, varied sizes (10–60 μL) and frequencies (0.5–7.5 Hz) of water droplets were used to simulate different types of precipitation (i.e., drizzle, light and heavy rains) and corresponding electrical output of TENG in hybrid device 1 was measured. A lower height of 3 cm was set to avoid possible interferences (i.e., airflow) from the environment. The mechanical energy from tiny water droplets as small as 10 μL can be monitored and converted to electrical signals (Fig. 5e). A larger water droplet can lead to a more extensive spread area of water droplets on the surface of TENG, generating more frictional charges. As a result, the transferred charges and voltage of TENG gradually increased along with the larger water droplets. In addition, the output voltage of TENG in hybrid device 1 showed good response in a broad frequency range (0.5–7.5 Hz) of water droplets (Fig. S31). Specifically, TENG can retain a high voltage of about 80 V under a relatively high frequency of 3.5 Hz, corresponding to heavy rains. The results highlight the great potential of the ST-PSC/TENG hybrid device for green electricity generation under

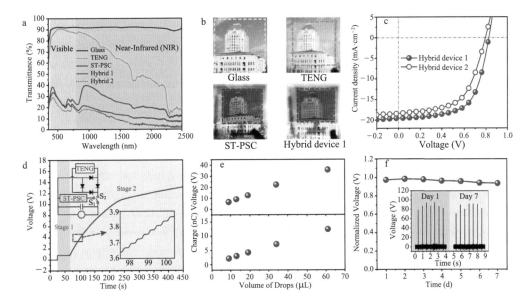

a, b. Transmittance spectra and digital photographs of glass, TENG, ST-PSC and hybrid devices. c. J–V curves of ST-PSCs in hybrid devices. d. A capacitor charging graph based on ST-PSC and TENG in hybrid device 1 (the left inset figure shows the two-way switch connection diagram of the charging circuit and the right insert image shows the voltage accumulation along with the number of water droplets). e. Voltage and transferred charges of TENG in hybrid device 1 measured under different volumes of water droplets. f. Stability test of TENG in hybrid device 1 (the insert image shows the voltage of TENG measured on different days).

Fig. 5 Transmittance spectra of TENG, MoO_3-modified ST-PSC, hybrid device 1 and hybrid device 2

different precipitation events.

The long-term stability of TENG and ST-PSC in hybrid device 1 was characterized. Under a constant drip of water drops (volume: 35 μL, height: 15 cm), no apparent reduction in the voltage of TENG was observed in 7 days (Fig. 5f). The illumination stability of ST-PSC was characterized by exposing hybrid device 1 under continuous illumination (1 sun). PCEs of ST-PSC suffered from a burn-in period and rapidly decayed below 90% in the first 6 h. After that, PCEs went into a relatively steady state. Encouragingly, ST-PSC can retain over 85% of the initial PCE after 12 h (Fig. S32). Furthermore, the shelf life testing was carried out by storing hybrid devices in a glovebox and periodically measuring the J–V curves of ST-PSCs. For ST-PSC in hybrid device 1, a relatively rapid reduction in PCEs was observed in the first 15 days. After that, PCEs reached a relatively steady state. ST-PSC can retain about 93% of initial PCE after 60 days (Fig. S33). The inverted device structure of ST-PSC removed several hygroscopic and acidic interlayers (i.e. PEDOT:PSS and PDINO), preventing ST-PSC from several moisture-induced defects (i.e., interfacial passivation and electrode corrosion). On the other hand, conventional ST-PSC in hybrid device 2 showed more significant PCE decay. Only about 70% of initial PCE was retained after 30 days (Fig. S34). The hydrophilic properties of PEDOT:PSS and PDINO inevitably led to moisture-related defects (interfacial passivation and electrode corrosion), lowering the long-term stability of conventional ST-PSC in hybrid device 2. The results indicate the encouraging performance stability of hybrid device 1. We agree that the inherent instability remains a major concern in the OPV community. Advanced material strategies and encapsulation barriers are explicitly required for the potential applications in moist environments (i.e., semitransparent roofing and walling).

Apart from dual-energy conversion and transparency characterization, thermal-insulating properties of TENG, MoO_3-modified ST-PSC and hybrid device 1 (active area: 1 cm^2) were evaluated based on an artificial setup (Fig. 6a). Under 1 sun, the surface temperatures of all devices rapidly increased and reached steady states in 30 min. Hybrid device 1 attained the maximum surface temperature of 50.9 °C, which was higher than the surface temperatures of ST-PSC (48.3 °C) and TENG (39.8 °C) (Fig. 6b). The glass slide showed the lowest surface temperature of 37.1 °C. Notably, the air temperatures below the devices showed a reverse trend relative to their surface temperatures (Fig. 6c). Hybrid device 1 showed the lowest air temperature of about 33.0 °C after 1 h, which was lower than the air temperature measured below ST-PSC (about 35.2 °C) and TENG (about 42.0 °C). The highest air temperature of >43.0 °C was recorded under the glass slide. Notably, hybrid device 1 significantly reduced the air temperature by about 10 °C compared to the typical glass slide. Similar to the thermal insulating properties of different ST-PSCs, the surface and air temperatures agreed well with the transmittance of different devices. Weighted with the AM 1.5G solar spectrum, MoO_3-modified ST-PSC showed efficient transmittance (T_2 = 29.52% and T_3 = 29.02%) in the visible and NIR wavelength ranges of 350−780 and 780−2500 nm, respectively (Table S7). TENG showed slightly lower effective transmittance (T_2 = 88.76%) to the glass slide (T_2 = 91.37%) in the visible wavelength range. Notably, a significantly lower effective transmittance (T_3 = 75.12%) was recorded in the NIR wavelength range of TENG. As a result, integration of TENG on top of ST-PSC can synergistically reduce T_2 to 26.23% and T_3 to 17.00% in hybrid device 1. In the whole energy of the solar spectrum, the visible and NIR light constitutes 43% and 53% of solar energy. Hybrid device 1 further suppressed the transmittance of both visible and NIR light compared to single ST-PSC and TENG, resulting in the lowest air temperature below the device. Meanwhile, more incident photons can be converted to heat and warmed hybrid device 1, leading to its highest surface temperature compared to other devices. The results highlight the use of ST-PSC/TENG hybrid systems as passive window thermal barriers, efficiently reducing heat

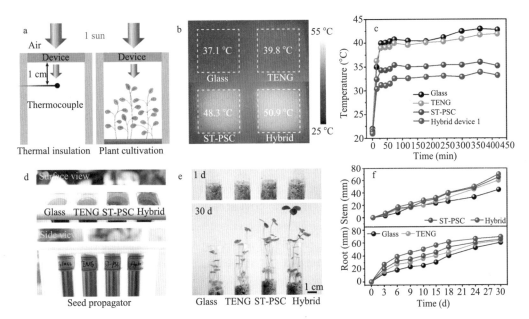

a. Schematic illustration of thermal insulation and plant cultivation experiments. b. Surface IR images of glass, TENG, ST-PSC and hybrid device 1 under solar irradiation. c. Temperature changes of surrounding air below glass, TENG, ST-PSC and hybrid device 1 over time. d. Surface and side view of the 3D-printed seed propagator. e, f. Digital photographs of plant growth and summary of stem and root length below glass, TENG, ST-PSC and hybrid device 1 over days.

Fig. 6　Thermal insulation and plant cultivation experiments

conduction from outdoor to interior environments.

In order to assess potential agricultural applications of the hybrid systems, a crop growth factor was defined based on the transmittance spectra of devices and the plant action spectrum[36,90]. Hybrid device 1 afforded a high crop growth factor of 25.32%, comparable to the MoO$_3$-modified ST-PSCs (27.39%). The result indicates the transmitted light through both devices can sustain plant growth. A 3D-printed seed propagator was designed for plant growth characterization (Fig. 6d). TENG, MoO$_3$-modified ST-PSC and hybrid device 1 (active area: 1 cm^2) were placed 1 cm higher than the surface of the growing medium, which simulated the roofs of greenhouses. A glass slide (thickness: 1.1 mm) was measured as a control group. Seeds of three-leaf clover (a staple crop for silage in the temperature northern hemisphere) were sowed into test tubes (nursery pots) with the same growing medium. The glass slide was also measured as a control group. Seeds were germinated and cultivated at constant light intensity (1 sun) and photoperiod (10 h). Our previous reports have shown that the absorbance spectra of common photosynthetic pigments (i.e., chlorophyll, carotenoid and phycobilin) partially overlapped with the transmittance spectra of PBDB-TF:Y6 derived ST-PSCs, specifically in the wavelength ranges of 380−530 nm and 620−700 nm[72,91]. Owing to the high and broad transmittance of TENG, the photons transmitted through hybrid device 1 can participate in the photosynthetic process of plants, sustaining and even regulating plant growth.

Time-lapse photographs of clover growth under different devices were taken over days and are depicted in Fig. 6e and Fig. S35. The length of stems and roots were measured over time and summarized in Fig. 6f. All clover started to germinate after 1 day and similar germination rates of >90% were found after 3 days. In the primary growth stage, a gradual increase in the lengths of stems and roots was observed in all samples. Interestingly, the clover below hybrid device 1 showed small branches and grew to about 20 mm long on day 7, exhibiting the fastest growth rate of 4.6 mm·d^{-1} on average. The clover continued to grow to adult plants, showing an average length of about 70 mm on day 30. Notably, the clover under the glass slide showed the lowest growth rate of 3.5 mm·d^{-1} on average, which only reached a half-height of the clover below hybrid device 1 after 30 days. The plant growth results indicate that all clovers were in a photo-inhibiting state. In this case, the reduced transmission of hybrid device 1 can provide positive shading effects and also decrease the ambient temperature (<33 °C) in the seed propagator. On the other hand, the ambient temperature under glass increased from about 22 to about 42 °C, which lies beyond the normal temperature fluctuations that common plants can tolerate. As a result, the more proper light and air temperature provided by hybrid device 1 led to the fastest growth rate. The plant cultivation characterization highlights the potential to use the ST-PSC/TENG hybrid system as translucent roof panels and walls for self-powering and energy-saving greenhouses. It would be more interesting if the transmittance (i.e., wavelength and intensity) of hybrid systems could be modulated to target the growth of specific plants in future research.

Conclusions

In summary, we have developed the first example of ST-PSC/TENG hybrid systems and have demonstrated their use in efficient solar and waterdrop energy conversion, complementary electricity generation and promising window-integrated applications. Rational integration of TENGs and ST-PSCs (TENGs on top and ST-PSCs on bottom) into a monolithic structure delivered several unique advantages, which are rationalized as follows: (1) Efficient solar power generation: Based on high-performing opaque PSCs (PCEs of >17%) and ST-PSCs (PCEs of >12%), TENGs provided high and even light transmittance of 87.75% in a broad wavelength range of 350−1000 nm, ensuring efficient light absorption of ST-PSCs. The device architecture ensured well-balanced power generation and transparency. As a result, the rigid and flexible ST-PSC/TENG hybrid systems can attain high PCEs of 10.1% and 8.4% under 1 sun, respectively, indicating efficient light-to-electricity conversion (a maximum electrical power output: 101 W·m^{-2}) on sunny days. (2) Efficient waterdrop energy conversion: Under water droplet stimuli, the hybrid devices can generate a current of about 120 μA, transferred charges of about 60 nC and a voltage of about 100 V, leading to a maximum electrical power output of 2.62 W·m^{-2} and indicating complementary electricity generation in the rain. (3) Good transparency, color rendering and window insulation: The hybrid devices showed good color neutrality of visible light (VLT of 23.49% and CRI of 92). Introduction of TENG significantly lowered the effective transmittance of ST-PSCs in the NIR wavelength range. The shading effects efficiently hampered the heat transfer and decreased the air temperature by about 10 °C compared to the typical glass slide, ensuring both visual and temperature comfort. (4) Sustainable gardening: the specific transmittance windows and controlled light intensities provided by the hybrid devices can sustain the germination and growth of clover and show a high plant growth factor of 25.3%, highlighting their great potential in energy-efficient greenhouse applications. Our work paves an intriguing prospect of developing ST-PSC/TENG hybrid systems for solar and raindrop energy conversion, not merely scaling up the green electricity production under different weather conditions, but also evaluating their integrability, transparency, amenity and sustainability for versatile window-integrated applications.

Experimental Section

▶ *Fabrication of ST-PSCs, TENG and hybrid devices*

The fabrication of different ST-PSCs and electrode modulations are described in Section S4, Supporting Information. For the fabrication of TENG, a piece of glass/ITO or PET/AgNWs substrate was cleaned in pure water and ethanol. A copper wire (diameter: 100 μm) was connected to the glass/ITO or PET/AgNWs electrode via conductive copper foil tapes. A FEP film (about 35 μm) was used as the friction layer. The FEP film was mounted onto the glass/ITO or PET/AgNWs electrode via a thin layer of optically clear adhesive. Finally, a Pt wire (diameter: 100 μm) was placed on the FEP layer as the counter electrode. For the fabrication of the ST-PSC/TENG hybrid devices, the FEP film was mounted onto ST-PSCs via a layer of optically clear adhesive. The glass/ITO or PET/AgNWs electrode acted as the shared electrode for ST-PSCs and TENGs. Finally, a Pt wire was placed on the FEP layer as the counter electrode.

▶ *Solar power conversion characterization*

J–V curves were recorded in backward scan direction by using a Keithley 2400 source meter under a solar simulator (Oriel Sol3A, 69920, Newport). The light intensity was calibrated by using a Si-based power meter (PT-SI-SRC, Pharos). For the EQE measurements, the photocurrent was measured by using a Keithley 485 picoammeter under monochromatic light (MS257) illumination across the PSCs. The current was recorded as the voltage over a 50 Ω resistance and converted to a EQE profile by comparing the data with a calibrated Si reference cell.

▶ *Water droplet energy conversion characterization*

Ordinary tap water with the total dissolved solids (TDS) of 1120 mg·L^{-1} was used. A microinjector and a syringe pump generated continuous water droplets with controlled volume (10–60 μL) and volumetric flow rates. All hybrid devices were projected at 45° to the ground. The heights of water drops can be varied from 3 cm to 20 cm. The current of TENGs in hybrid devices was measured via a low-noise current amplifier (Stanford Research System Model SR570) and an oscilloscope (Rohde and Schwarzrte, RTA4004). Charge transfer and voltage of TENGs in hybrid devices were measured via a Keithley 6514 source meter and an oscilloscope RTA4004 equipped with a high-impedance (10 MΩ) probe, respectively.

Supporting information

Supplementary data associated with this article can be found online: Materials, optical and electrochemical measurements, optimization of opaque PSCs and ST-PSCs, AFM and SEM images, exciton dissociation, mobility and charge recombination characterization, device stability and plant growth measurements.

Acknowledgements

We acknowledge the financial support from the Taishan Scholar Program of Shandong Province, China (Grant No. tsqn201812026), the Fundamental Research Funds for the Central Universities, China (Grant No. 201941011, 202161055 and 202242001), the Natural Science Foundation of Shandong Province, China (Grant No. ZR2019QB014, ZR2020ZD33 and ZR2021QE043) and the National Natural Science Foundation of China (Grant No. 52101390).

Author Contributions

Device fabrication, evaluation and investigation were carried out by Tong Liu and Yang Zheng. Optical, thermal insulation and plant growth characterization were conducted by Tong Liu. Material synthesis was performed by Yunxiang Xu. Xiaofeng Xu conceived, coordinated and supervised the project. Design of the experiments, analysis of the results and manuscript writing were done by Xiaofeng Xu, Xianjie Li, Chuanfei Wang and Petri Murto. Mats Fahlman, Junwu Chen, Liangmin Yu and X. Liu involved in the discussion and data analysis. All the authors discussed the results and contributed to the review and revision of the manuscript.

Competing Interests

The authors declare that they have no known competing financial interests or personal relationships that could have appeared to influence the work reported in this paper.

References

1. Meddeb H, Götz-Köhler M, Neugebohrn N, et al. Tunable photovoltaics: Adapting solar cell technologies to versatile applications[J]. Adv. Energy Mater., 2022, 12: 2200713.
2. Zhao Y, Zhu Y, Cheng H W, et al. A review on semitransparent solar cells for agricultural application[J]. Mater. Today Energy, 2021, 22: 100852.
3. Xue Q, Xia R, Brabec C J, et al. Recent advances in semi-transparent polymer and perovskite solar cells for power generating window applications[J]. Energy Environ. Sci., 2018, 11: 1688-1709.
4. Jiang Y, Dong X, Sun L, et al. An alcohol-dispersed conducting polymer complex for fully printable organic solar cells with improved stability[J]. Nat. Energy, 2022, 7: 352-359.
5. Liu Y, Liu B, Ma C Q, et al. Recent progress in organic solar cells (part i material science) [J]. Sci. China Chem., 2022, 65: 224-268.
6. Liu Y, Liu B, Ma C Q, et al. Recent progress in organic solar cells (part ii device engineering) [J]. Sci. China Chem., 2022, 65: 1457-1497.
7. Kan B, Kan Y, Zuo L, et al. Recent progress on all-small molecule organic solar cells using small-molecule nonfullerene acceptors[J]. InfoMat, 2021, 3: 175-200.
8. Riede M, Spoltore D, Leo K. Organic solar cells: The path to commercial success[J]. Adv. Energy Mater., 2021, 11: 2002653.
9. Feng K, Guo H, Sun H, et al. N-type organic and polymeric semiconductors based on bithiophene imide derivatives[J]. Acc. Chem. Res., 2021, 54: 3804-3817.
10. He Y, Li N, Heumüller T, et al. Industrial viability of single-component organic solar cells[J]. Joule, 2022, 6: 1160-1171.
11. Cui Y, Xu Y, Yao H, et al. Single-junction organic photovoltaic cell with 19% efficiency[J]. Adv. Mater., 2021, 33: 2102420.
12. Zhu L, Zhang M, Xu J, et al. Single-junction organic solar cells with over 19% efficiency enabled by a refined double-fibril network morphology[J]. Nat. Mater., 2022, 21: 656-663.
13. Gao W, Qi F, Peng Z, et al. Achieving 19% power conversion efficiency in planar-mixed heterojunction organic solar cells using a pseudo-symmetric electron acceptor[J]. Adv. Mater., 2022, 34: 2202089.
14. Chong K, Xu X, Meng H, et al. Realizing 19.05% efficiency polymer solar cells by progressively improving charge extraction and suppressing charge recombination[J]. Adv. Mater., 2022, 34: 2109516.
15. Sun R, Wu Y, Yang X, et al. Single-junction organic solar cells with 19.17% efficiency enabled by introducing one asymmetric guest acceptor[J]. Adv. Mater., 2022, 34: 2110147.
16. Zhan L, Li S, Li Y, et al. Desired open-circuit voltage increase enables efficiencies approaching 19% in symmetric-asymmetric molecule ternary organic photovoltaics[J]. Joule, 2022, 6: 662-675.
17. Wei Y, Chen Z, Lu G, et al. Binary organic solar cells breaking 19% via manipulating vertical component distribution[J]. Adv. Mater., 2022, 34: 2204718.
18. Zuo L, Jo S B, Li Y, et al. Dilution effect for highly efficient multiple-component organic solar cells[J]. Nat. Nano., 2022, 17: 53-60.
19. Wang D, Qin R, Zhou G, et al. High-performance semitransparent organic solar cells with excellent infrared reflection and see-through functions[J]. Adv. Mater., 2020, 32: 2001621.
20. Li X, Xia R, Yan K, et al. Semitransparent organic solar cells with vivid colors[J]. ACS Energy Lett., 2020, 5: 3115-3123.
21. Xu C, Jin K, Xiao Z, et al. Wide bandgap polymer with narrow photon harvesting in visible light range enables efficient semitransparent organic photovoltaics[J]. Adv. Funct. Mater., 2021, 31: 2107934.
22. Li Y, He C, Zuo L, et al. High-performance semi-transparent organic photovoltaic devices via improving absorbing selectivity[J]. Adv. Energy Mater., 2021, 11: 2003408.
23. Zhang Y, He X, Babu D, et al. Efficient semi-transparent organic solar cells with high color rendering index enabled by self-assembled and knitted agnps/mwcnts transparent top electrode via solution process[J]. Adv. Opt. Mater., 2021, 9: 2002108.
24. Jiang T, Zhang G, Xia R, et al. Semitransparent organic solar cells based on all-low-bandgap donor and acceptor materials and their performance potential[J]. Mater. Today Energy, 2021, 21: 100807.
25. Huang X, Zhang L, Cheng Y, et al. Novel narrow bandgap terpolymer donors enables record performance for semitransparent organic solar cells based on all-narrow bandgap semiconductors[J]. Adv. Funct. Mater., 2022, 32: 2108634.
26. Zhao Y, Cheng P, Yang H, et al. Towards high-performance semitransparent organic photovoltaics: Dual-functional p-type soft interlayer[J]. ACS Nano, 2022, 16: 1231-1238.
27. Yuan X, Sun R, Wu Y, et al. Simultaneous enhanced device efficiency and color neutrality in semitransparent organic photovoltaics employing a synergy of ternary

28. Jing J, Dong S, Zhang K, et al. Semitransparent organic solar cells with efficiency surpassing 15%[J]. Adv. Energy Mater., 2022, 12: 2200453.
29. Liu W, Sun S, Xu S, et al. Theory-guided material design enabling high-performance multifunctional semitransparent organic photovoltaics without optical modulations[J]. Adv. Mater., 2022, 34: 2200337.
30. Liu W, Sun S, Zhou L, et al. Design of near-infrared nonfullerene acceptor with ultralow nonradiative voltage loss for high-performance semitransparent ternary organic solar cells[J]. Angew. Chem. Int. Ed., 2022, 61: e202116111.
31. Guan S, Li Y, Yan K, et al. Balancing the selective absorption and photon-to-electron conversion for semitransparent organic photovoltaics with 5.0% light utilization efficiency[J]. Adv. Mater., 2022, 34: 2205844.
32. Liu S, Li H, Wu X, et al. Pseudo-planar heterojunction organic photovoltaics with optimized light utilization for printable solar windows[J]. Adv. Mater., 2022, 34: 2201604.
33. Duan L, Hoex B, Uddin A. Progress in semitransparent organic solar cells[J]. Solar RRL, 2021, 5: 2100041.
34. Sun C, Xia R, Shi H, et al. Heat-insulating multifunctional semitransparent polymer solar cells[J]. Joule, 2018, 2: 1816-1826.
35. Xia R, Brabec C J, Yip H L, et al. High-throughput optical screening for efficient semitransparent organic solar cells[J]. Joule, 2019, 3: 2241-2254.
36. Wang D, Liu H, Li Y, et al. High-performance and eco-friendly semitransparent organic solar cells for greenhouse applications[J]. Joule, 2021, 5: 945-957.
37. Wang D, Li Y, Zhou G, et al. High-performance see-through power windows[J]. Energy Environ. Sci., 2022, 15: 2629-2637.
38. Emmott C J M, Röhr J A, Campoy-Quiles M, et al. Organic photovoltaic greenhouses: A unique application for semi-transparent pv?[J]. Energy Environ. Sci., 2015, 8: 1317-1328.
39. Li Z, Ma T, Yang H, et al. Transparent and colored solar photovoltaics for building integration[J]. Solar RRL, 2021, 5: 2000614.
40. Shockley W, Queisser H J. Detailed balance limit of efficiency of p-n junction solar cells[J]. J. Appl. Phys., 1961, 32: 510-519.
41. Eisner F, Tam B, Belova V, et al. Color-tunable hybrid heterojunctions as semi-transparent photovoltaic windows for photoelectrochemical water splitting[J]. Cell Rep. Phys. Sci., 2021, 2: 100676.
42. Wang Z L. Triboelectric nanogenerator (teng): Sparking an energy and sensor revolution[J]. Adv. Energy Mater., 2020, 10: 2000137.
43. Zheng L, Lin Z H, Cheng G, et al. Silicon-based hybrid cell for harvesting solar energy and raindrop electrostatic energy[J]. Nano Energy, 2014, 9: 291-300.
44. Liu Y, Sun N, Liu J, et al. Integrating a silicon solar cell with a triboelectric nanogenerator via a mutual electrode for harvesting energy from sunlight and raindrops[J]. ACS Nano, 2018, 12: 2893-2899.
45. Cao R, Wang J, Xing Y, et al. A self-powered lantern based on a triboelectric–photovoltaic hybrid nanogenerator[J]. Adv. Mater. Technol., 2018, 3: 1700371.
46. Cho Y, Lee S, Hong J, et al. Sustainable hybrid energy harvester based on air stable quantum dot solar cells and triboelectric nanogenerator[J]. J. Mater. Chem. A, 2018, 6: 12440-12446.
47. Guo Q, Yang X, Wang Y, et al. Dielectric hole collector toward boosting charge transfer of cspbbr3 hybrid nanogenerator by coupling triboelectric and photovoltaic effects[J]. Adv. Funct. Mater., 2021, 31: 2101348.
48. Jung S, Oh J, Yang U J, et al. 3d cu ball-based hybrid triboelectric nanogenerator with non-fullerene organic photovoltaic cells for self-powering indoor electronics[J]. Nano Energy, 2020, 77: 105271.
49. Lee J W, Jung S, Jo J, et al. Sustainable highly charged c60-functionalized polyimide in a non-contact mode triboelectric nanogenerator[J]. Energy Environ. Sci., 2021, 14: 1004-1015.
50. Yoo D, Park S C, Lee S, et al. Biomimetic anti-reflective triboelectric nanogenerator for concurrent harvesting of solar and raindrop energies[J]. Nano Energy, 2019, 57: 424-431.
51. Liu X, Cheng K, Cui P, et al. Hybrid energy harvester with bi-functional nano-wrinkled anti-reflective pdms film for enhancing energies conversion from sunlight and raindrops[J]. Nano Energy, 2019, 66: 104188.
52. Wang L, Wang Y, Wang H, et al. Carbon dot-based composite films for simultaneously harvesting raindrop energy and boosting solar energy conversion efficiency in hybrid cells[J]. ACS Nano, 2020, 14: 10359-10369.
53. Zhao L, Duan J, Liu L, et al. Boosting power conversion efficiency by hybrid triboelectric nanogenerator/silicon tandem solar cell toward rain energy harvesting[J]. Nano Energy, 2021, 82: 105773.
54. Yang D, Ni Y, Su H, et al. Hybrid energy system based on solar cell and self-healing/self-cleaning triboelectric nanogenerator[J]. Nano Energy, 2021, 79: 105394.
55. Xie L, Yin L, Liu Y, et al. Interface engineering for

56. Zheng Y, Liu T, Wu J, et al. Energy conversion analysis of multilayered triboelectric nanogenerators for synergistic rain and solar energy harvesting[J]. Adv. Mater., 2022, 34: 2202238.
57. Guo Q, Guo Q, Geng Y, et al. Recent advances in pm6:Y6-based organic solar cells[J]. Mater. Chem. Front., 2021, 5: 3257-3280.
58. Zhang M, Guo X, Ma W, et al. A large-bandgap conjugated polymer for versatile photovoltaic applications with high performance[J]. Adv. Mater., 2015, 27: 4655-4660.
59. Yuan J, Zhang Y, Zhou L, et al. Single-junction organic solar cell with over 15% efficiency using fused-ring acceptor with electron-deficient core[J]. Joule, 2019, 3: 1140-1151.
60. Yuan J, Huang T, Cheng P, et al. Enabling low voltage losses and high photocurrent in fullerene-free organic photovoltaics[J]. Nat. Commun., 2019, 10: 570.
61. Wei Q, Liu W, Leclerc M, et al. A-da'd-a non-fullerene acceptors for high-performance organic solar cells[J]. Sci. China Chem., 2020, 63: 1352-1366.
62. Yin Y, Zhan L, Liu M, et al. Boosting photovoltaic performance of ternary organic solar cells by integrating a multi-functional guest acceptor[J]. Nano Energy, 2021, 90: 106538.
63. Liang J, Pan M, Chai G, et al. Random polymerization strategy leads to a family of donor polymers enabling well-controlled morphology and multiple cases of high-performance organic solar cells[J]. Adv. Mater., 2020, 32: 2003500.
64. Song J, Ye L, Li C, et al. An optimized fibril network morphology enables high-efficiency and ambient-stable polymer solar cells[J]. Adv. Sci., 2020, 7: 2001986.
65. Wu J, Li G, Fang J, et al. Random terpolymer based on thiophene-thiazolothiazole unit enabling efficient non-fullerene organic solar cells[J]. Nat. Commun., 2020, 11: 4612.
66. Guo X, Fan Q, Wu J, et al. Optimized active layer morphologies via ternary copolymerization of polymer donors for 17.6% efficiency organic solar cells with enhanced fill factor[J]. Angew. Chem. Int. Ed., 2021, 60: 2322-2329.
67. Jiang H, Han C, Li Y, et al. Rational mutual interactions in ternary systems enable high-performance organic solar cells[J]. Adv. Funct. Mater., 2021, 31: 2007088.
68. Ma R, Zhou K, Sun Y, et al. Achieving high efficiency and well-kept ductility in ternary all-polymer organic photovoltaic blends thanks to two well miscible donors[J]. Matter, 2022, 5: 725-734.
69. Xu Y, Ji Q, Yin L, et al. Synergistic engineering of substituents and backbones on donor polymers: Toward terpolymer design of high-performance polymer solar cells[J]. ACS Appl. Mater. Interfaces, 2021, 13: 23993-24004.
70. Zhang W, Huang J, Xu J, et al. Phthalimide polymer donor guests enable over 17% efficient organic solar cells via parallel-like ternary and quaternary strategies[J]. Adv. Energy Mater., 2020, 10: 2001436.
71. Yu R, Yao H, Cui Y, et al. Improved charge transport and reduced nonradiative energy loss enable over 16% efficiency in ternary polymer solar cells[J]. Adv. Mater., 2019, 31: 1902302.
72. Zhang N, Jiang T, Guo C, et al. High-performance semitransparent polymer solar cells floating on water: Rational analysis of power generation, water evaporation and algal growth[J]. Nano Energy, 2020, 77: 105111.
73. Yan T, Song W, Huang J, et al. 16.67% rigid and 14.06% flexible organic solar cells enabled by ternary heterojunction strategy[J]. Adv. Mater., 2019, 31: 1902210.
74. Xie Y, Yang F, Li Y, et al. Morphology control enables efficient ternary organic solar cells[J]. Adv. Mater., 2018, 30: 1803045.
75. Mihailetchi V D, Koster L J A, Hummelen J C, et al. Photocurrent generation in polymer-fullerene bulk heterojunctions[J]. Phys. Rev. Lett., 2004, 93: 216601.
76. Cowan S R, Roy A, Heeger A J. Recombination in polymer-fullerene bulk heterojunction solar cells[J]. Phys. Rev. B, 2010, 82: 245207.
77. Zhang G, Ning H, Chen H, et al. Naphthalenothiophene imide-based polymer exhibiting over 17% efficiency[J]. Joule, 2021, 5: 931-944.
78. Song W, Fan X, Xu B, et al. All-solution-processed metal-oxide-free flexible organic solar cells with over 10% efficiency[J]. Adv. Mater., 2018, 30: 1800075.
79. Zhang W, Song W, Huang J, et al. Graphene:Silver nanowire composite transparent electrode based flexible organic solar cells with 13.4% efficiency[J]. J. Mater. Chem. A, 2019, 7: 22021-22028.
80. Sun Y, Chang M, Meng L, et al. Flexible organic photovoltaics based on water-processed silver nanowire electrodes[J]. Nat. Electron., 2019, 2: 513-520.
81. Zhang N, Xu Y, Zhou X, et al. Synergistic effects of copolymerization and fluorination on acceptor polymers for efficient and stable all-polymer solar cells[J]. J. Mater. Chem. C, 2019, 7: 14130-14140.
82. Upama M B, Wright M, Elumalai N K, et al. High-efficiency semitransparent organic solar cells with non-fullerene acceptor for window application[J]. ACS

Photonics, 2017, 4: 2327-2334.
83. Xie Y, Xia R, Li T, et al. Highly transparent organic solar cells with all-near-infrared photoactive materials[J]. Small Methods, 2019, 3: 1900424.
84. Zhang J, Xu G, Tao F, et al. Highly efficient semitransparent organic solar cells with color rendering index approaching 100[J]. Adv. Mater., 2019, 31: 1807159.
85. Ji Q, Li N, Wang S, et al. Synergistic solar-powered water-electricity generation via rational integration of semitransparent photovoltaics and interfacial steam generators[J]. J. Mater. Chem. A, 2021, 9: 21197-21208.
86. Xu W, Zheng H, Liu Y, et al. A droplet-based electricity generator with high instantaneous power density[J]. Nature, 2020, 578: 392-396.
87. Zhang Q, Li Y, Cai H, et al. A single-droplet electricity generator achieves an ultrahigh output over 100 v without pre-charging[J]. Adv. Mater., 2021, 33: 2105761.
88. Li X, Lau T H, Guan D, et al. A universal method for quantitative analysis of triboelectric nanogenerators[J]. J. Mater. Chem. A, 2019, 7: 19485-19494.
89. Li X, Tao J, Wang X, et al. Networks of high performance triboelectric nanogenerators based on liquid–solid interface contact electrification for harvesting low-frequency blue energy[J]. Adv. Energy Mater., 2018, 8: 1800705.
90. Shi H, Xia R, Zhang G, et al. Spectral engineering of semitransparent polymer solar cells for greenhouse applications[J]. Adv. Energy Mater., 2019, 9: 1803438.
91. Yin L, Zhou Y, Jiang T, et al. Semitransparent polymer solar cells floating on water: Selected transmission windows and active control of algal growth[J]. J. Mater. Chem. C, 2021, 9: 13132-13143.

Experimental investigation of a novel OWC wave energy converter

Tongshun Yu[1,2,*], Qiyue Guo[1], Hongda Shi[1,2], Tingyu Li[1], Xiaoyu Meng[1], Shoukun He[1], Peixuan Li[1]

1 College of Engineering, Ocean University of China, Qingdao, 266100, China
2 Shandong Provincial Key Laboratory of Ocean Engineering, Qingdao, 266100, China
* Corresponding author: Tongshun Yu (tshyu707@ouc.edu.cn).

Abstract

Wave energy, one of the safest and currently growing forms of renewable energy, requires the development of the appropriate wave energy converters (WECs). This study presents a recently patented WEC; and proves the feasibility of this concept using a series of experiments. To achieve this, two rectangular chambers of an oscillating water column (OWC) system were arranged in series, and the wave with a propagation direction that forms a relatively small angle of 30° with the front walls was referred to as obliquely incident wave. The effects of the wave amplitude, front wall draughts, and opening ratios were investigated in relation to the overall hydrodynamic performance and the respective performances of the two sub-chambers. It was found that the overall efficiency of the system was greatly reduced as the front wall draught in the high-frequency zone increased. Furthermore, the opening ratios of the individual chambers significantly affected the overall hydrodynamic performance for all frequency bandwidths under obliquely incident waves, and the optimal efficiency achieved at the opening ratio of $\varepsilon = 1\%$. The rear chamber is inferior to the front chamber for most of the wave frequencies.

Keywords: OWC wave energy converter; obliquely incident wave; wave energy; hydrodynamic performance; model testing

Introduction

As the worldwide energy demand continues to grow and global environmental issues become increasingly apparent, renewable energy sources have been actively studied by an increasing number of researchers. Among the various renewable energy resources, ocean energy, and in particular wave energy, has the potential to provide a substantial amount of renewable energy worldwide. Owing to its high power density, wave energy has long been considered as one of the most promising renewable energy[1]. More than 1000 wave energy converters (WECs) have been developed and patented worldwide[2]. Among them, the oscillating water column (OWC) WEC is considered to be one of the most successful devices owing to its structural and mechanical simplicity[3-6].

A typical OWC device mainly consists of two components: a hollow pneumatic chamber with a large opening below the water surface and a turbine coupled with an electric generator. An incident wave

---

① 本文于2022年8月发表在 *Ocean Engineering* 第257卷，https://doi.org/10.1016/j.oceaneng.2022.111567。

excites the water column inside of the chamber such that it oscillates and forces the trapped air above the water column to flow into and out of the turbine. The electric generator is driven by the reciprocating air flow to realize the conversion of pneumatic energy to electrical energy. Owing to their simple structure in which the only moving mechanical component is a turbine, OWC devices can be flexibly adapted to variable ocean environments.

In recent years, a large number of OWC devices have been deployed to investigate their hydrodynamic efficiencies. These full-sized OWC prototypes include Tofteshallen in Norway (500 kW)[7]; the Pico plant in Portugal (400 kW)[8]; a shoreline OWC demonstration plant in Guangdong Province, China (2001) (100 kW)[9]; and a bottom-standing OWC in Yongsoo (500 kW)[5]. Although varying levels of success have been achieved, further investigation of OWC devices is required to improve their hydrodynamic performances and wave energy absorption efficiencies.

Extensive studies have been carried out to investigate and optimize OWC devices numerically[10-12] and experimentally[13-15,27]. The majority of these studies have focused on single-chamber OWC devices, which are recommended for operation at near-resonance conditions to achieve high-efficiency performances[16]. Dual-chamber OWC devices have been proposed to improve the hydrodynamic efficiency and broaden the range of operating conditions of these devices.

Dual-chamber OWCs operate similarly to single-chamber OWCs except that there are two chambers and two power take-off (PTO) components in the former devices. The hydrodynamic performances of dual and single-chamber OWCs have been numerically studied and compared using potential flow theory[17]. It was observed that the performance curve of the dual-chamber device exhibited more than one peak. Rezanejad et al.[18], also investigated a dual-chamber OWC placed over a stepped seabed. They compared their results with a typical single-chamber device with and without a stepped seabed and found that the existence of at least one step on the seabed in front of the chamber had a significant effect on the hydrodynamic efficiency of the dual-chamber device. Ning et al.[19] used potential flow theory to estimate the hydrodynamic efficiency of a dual-chamber OWC with a shared single orifice and studied the effects of geometric parameters on the air pressure and free water surface elevation in each chamber. The hydrodynamic performances of single and dual-chamber offshore OWCs using the STAR-CCM + CFD package have been studied[20]. It was found that the dual-chamber improves the overall power output compared to the a single-chamber. Ning et al.[21] also analyzed a dual-chamber device with two independent orifices using potential flow theory. They used the air pressure and free surface elevation of the chambers to validate the numerical model results with their experimental data. It was found that the dual-chamber OWC had a broader effective frequency bandwidth and an improved efficiency compared to the single-chamber device. By keeping the length of the total chamber constant and examining three different lengths for each chamber, it was found that the overall efficiency of such a system is insensitive towards the variation of the sub-chamber, while the efficiency of two sub-chamber systems increases with the breadth of the system when the total chamber breadth is kept constant.

To improve the hydrodynamic efficiency of OWCs, various OWC systems configurations have been studied. He et al.[22] carried out a set of experiments using a dual-chamber OWC system in which two chambers were attached to either side of a floating breakwater. Likewise, a novel dual-chamber OWC system that consists of two concentric cylindrical chambers was previously proposed[23]. The cylindrical chambers were bounded by a central cylinder and two outer rigid shells, and they were connected at the top of the system.

Scaled-model preliminary testing using the WaveCat[24] patent system, which is an offshore floating WEC in the presence of oblique waves, was conducted. These experiments provided a proof of concept using WaveCat as a viable technology for converting wave energy into electricity using the novel approach of oblique overtopping, and it was found that the derived power greatly depended on the angle between hulls.

Inspired by a previous study of a dual-chamber OWC and concept of WaveCat, a novel three-dimensional multi-chamber OWC wave energy converter (MCOW) is proposed in this study. This study considers the MCOW, which is a recently patented WEC (Fig. 1), is an offshore floating WEC system. The overall length of the prototype of MCOW is 90 m. It consists of two hulls that converging at the stern by a hinge, thereby forming a wedge in the plan view that allows the angle between the hulls to be varied depending on the sea state. An equal number of OWCs are installed in each of the two hulls. Additionally, the MCOW is intended for offshore deployment, in water depths between 50 m and 100 m where the wave energy resources are greater than those in nearshore or onshore locations.

The MCOW is moored to a catenary-buoy using a single point mooring system[25]. It rotates around a fixed object on the sea along with the changes in the wave directions by using the mooring system such that the opening angle of the two hulls always face the incident wave (Fig. 2). Additionally, the opening angle of the hinged hulls can be varied in the range of 0° to 120° according to the size of the wave. Under extreme sea conditions, the opening angle of the hulls may be reduced to 0° to ensure the survival of the system. As discussed above, the direction of wave propagation forms a relatively small angle with the side of the inner hull.

The objectives of this study are to test the performance of MCOW, and prove the feasibility of this concept using a series of experiments. To achieve this, the effects of the wave amplitude, front wall draughts, and opening ratios were investigated in relation to the overall hydrodynamic performance

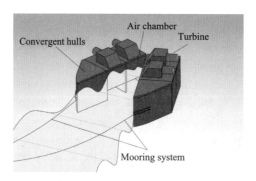

Fig. 1　Schematic of the multi-chamber device concept

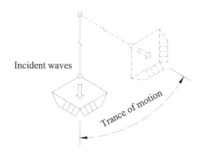

Fig. 2　Plan view of the mooring system

and the respective performances of the two sub-chambers over a wide range of wave periods. The wave with a propagation direction that forms a relatively small angle of 30° with the front walls of the chambers was referred to as obliquely incident wave. The process of oblique wave propagation towards the MCOW differs from the mechanism of traditional OWCs in which the wave propagation directly impacts the front wall. Therefore, specific physical model tests should be conducted.

The remainder of this paper is organized as follows. Section "Model Design" describes the model design of the MCOW, and Section "Experiments" presents the descriptions of the experimental setup and data analysis method. In Section "Results and Discussion", the effects of the geometrical parameters (i.e., the front wall draught and the orifice diameter) and incoming wave conditions on the performances of the entire device and the two sub-chambers are systematically investigated. Finally, the conclusions of this study are provided in Section "Conclusions".

Model Design

According to the principle of symmetry, the experimental model was simplified to half of the MCOW using a 1:30 scale as shown in Fig. 3(a). The model was made of acrylic plates with thicknesses of $C = 0.01$ m, and a photo of the OWC model installed in the wave-current flume is shown in Fig. 3b. In Fig. 3a, the geometric parameters of the model include the total length of the model $L = 1.036$ m, the width of the model $L_W = 0.32$ m, and the height of the model $L_H = 0.52$ m. The model consists of two air chambers, and the inner geometric parameters of each chamber include: the length of the chamber $L_{CL} = 0.3$ m and the width of the chamber $L_{CW} = 0.3$ m. The draught of the front wall (i.e., $d = 0.1, 0.12$, and 0.14 m) can be changed by adjusting between three front wall lengths (i.e., $L_{FH} = 0.3, 0.32$, and 0.34 m).

To simulate the power take-off (PTO) system, a circular orifice is drilled into the center of the top cover of each chamber. Three orifices with different opening ratios ε, defined by the ratio of the orifice area ($S_1 = \pi D^2/4$, where D is the diameter of orifice) to the chamber area ($S = L_{CL} \times L_{CW}$), were manufactured[12]. Three opening ratios of $\varepsilon = 0.35\%$ ($D = 0.02$ m), $\varepsilon = 0.68\%$ ($D = 0.028$ m), and $\varepsilon = 1\%$ ($D = 0.034$ m) were utilized in this study. It is noteworthy that the incompressibility of the air inside of the chamber will result in

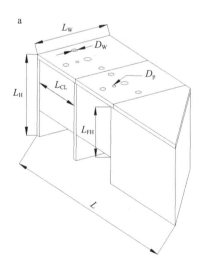 

a. Simplified model; b. The physical model in the wave tank.

Fig. 3 Schematic of the OWC model

the overestimation of the maximum conversion efficiency of the full-scale device, especially at the occurrence of resonance. As a small-scale model test and with the purpose of performing a feasibility analysis of the MCOW device, the effect of air compressibility on the performance of the OWC system is not considered.

For each chamber, three circular holes with a diameter of D_w = 27 mm were drilled on each top cover for installation of the wave gauges. One was located 60 mm from the front wall, while the other two were the same distance from the side walls. A circular hole with a diameter of D_p = 8 mm was placed near the center orifice for the installation of an air pipe to measure the air pressure inside of the chamber.

Experiments

▶ *Experimental set-up*

Preliminary tests were conducted with a fixed model in the wave flume at the Engineering Hydrodynamics Laboratory, Ocean University of China. A piston-type wave-maker was installed at one end of the flume, and an artificial beach was constructed at the other end to absorb the waves. The parameters of A_i (incident wave amplitude), h (static water depth), D (diameter of the orifice), d (immergence of the front wall), and h_c (height of the air chamber; i.e., the distance between the still water surface and the ceiling) were used in this study.

The experimental setup of the wave flume is illustrated in Fig. 4. The water depth was fixed at h = 0.6 m. The plan view of the experimental setup is shown in Fig. 5. The OWC model was placed at a distance of 15 m from the piston wave maker to eliminate as many reflecting waves as possible. Three wave gauges (Gs) were installed in the Ø27 holes for each chamber. One gauge was adjacent to the front wall and the other two gauges were deployed near the side walls to evaluate the free-surface wave elevations inside of the chambers. Two pressure sensors (S_1 and S_2) were used to measure the air pressures inside of the two chambers. Both the surface and pressure signals were sampled at 50 Hz. All installation points were sealed to prevent air leakage.

Serial sets of experiments were carried out to investigate the effects of the incident wave amplitude, front wall draught, and opening ratios on the hydrodynamic performance of the OWC. The chamber height h_c = 0.2 m remained constant in the experiments (i.e. by keeping constant the hydrostatic depth of h = 0.6 m). The characteristic waves of the South China Sea were chosen as the reference for the test conditions. Wang et al.[26] studied and simulated the wind and wave characteristics in the South China Sea, and found that the wave energy in the South China Sea is mainly concentrated in waves with periods of 6–11 s and effective wave heights of 1–6 m. Therefore, combining with former studies[21]

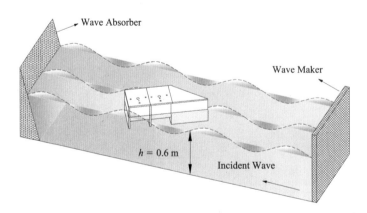

Fig. 4　Schematic of experimental model

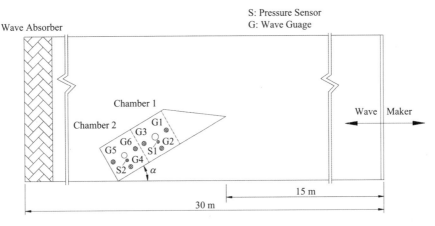

Fig. 5 Plan view of experimental model

and scale, 11 wave periods T with an interval of 0.1 s in the range of 1.0 s to 2.0 s were considered for each set of experiments and three wave amplitudes of 0.02 m, 0.03 m and 0.04 m considered in the experiments. The draughts of the two front walls were the same in this study. The parameters A_i = 0.04 m, d = 0.1 m, and D = 0.028 m were chosen as reference values. Only one corresponding parameter was varied in each set of experiments while the others were kept constant. The geometric parameters chosen for the experiment are shown in Table 1. A total of 77 tests were carried out to study the hydrodynamic performance of the MCOW device.

Table 1 Detailed parameters of the MCOW models in each set of experiments to obtain 11 wave periods T with an interval of 0.1 s in the range of 1.0 s to 2.0 s

Experiment	A_i (m)	d (m)	D (m)	T (s)
Set 1	0.02	0.10	0.028	1.0-2.0
Set 2	0.03	0.10	0.028	1.0-2.0
Set 3	0.04	0.10	0.028	1.0-2.0
Set 4	0.04	0.12	0.028	1.0-2.0
Set 5	0.04	0.14	0.028	1.0-2.0
Set 6	0.04	0.10	0.020	1.0-2.0
Set 7	0.04	0.10	0.034	1.0-2.0

▶ *Data analysis*

The experimental procedure for each set of experiments can be simply described as follows. First, the wave-maker is initialized to generate incident regular waves. After the wave train arrives at the device, data recording is triggered by the water level change inside of the chamber due to the first wave entering the chamber. The spatially-averaged vertical velocities of the water surface discharges are then calculated using the free water surface elevations in the chambers. Each case is tested at least three times to determine the repeatability and accuracy of the processed results. And the averaged values from these three repetitions were used as the representative values for this text. The standard deviations of all data processed were less than 0.05, which indicated that all measured data were repeatable enough. All the data were recorded for over 10 wave cycles after the regular wave arrived in the chamber and the initial effects disappeared.

A full-scale (FS) types of Gs were used in the experiments, and the accuracy of the Gs is ±0.5% of the FS. The pressure sensor was used to measure the differential pressure inside and outside the chamber, produced by the Japanese company Keyence, model KEYENCE AP-10S, range selection 3 kPa, and the accuracy of sensor is ±0.5% of FS_{max}. All the voltage signals from all measurement instruments were acquired and processed by a data acquisition (DAQ) system, which was developed by the Ocean University of China research team that is based on the CompactDAQ® system. A program compiled in LABVIEW ™ was used for the data processing and recording.

Inside of the OWC chamber, the periodic motion of the water column forces the airflow to enter and exit through the orifice, which in turn leads to air pressure fluctuations inside of the chambers. The work associated with the air pressure due to the motion of the water surface inside of the OWC chamber produces pneumatic power that is extracted by the OWC device. The relative pneumatic power, $P_{P(1,2)}$ (i.e., the period-averaged power captured by each chamber), can be calculated as follows:

$$P_{P(1,2)} = \frac{1}{T}\int_{t_0}^{t_0+T} p_{(1,2)}(t) \cdot Q_{(1,2)}(t) dt \quad (1)$$

where $p(t)$ is the instantaneous value of the air pressure inside of the chamber, and $Q(t)$ is the air flow rate into the orifice that is calculated by:

$$Q_{(1,2)}(t) = U_{(1,2)}(t) \cdot S \quad (2)$$

where $U_{(1,2)}$ represents the spatially averaged vertical velocities of the water surfaces inside of the chambers, and it is approximated using the velocities of the water surfaces measured by six wave gauges (i.e., the first order derivatives of the free surface elevation time-series measured by the Gs). S is the cross-sectional area of the water surface in the chambers, and subscripts "1" and "2" denote that these variables were measured and calculated for chambers 1 and 2, respectively.

The corresponding input energy flux per unit wave crest length, P_W, is calculated as follows:

$$P_W = \frac{1}{2}\rho_w g A_i^2 C_g \quad (3)$$

where ρ_w is the gravitational acceleration, g is the water density, A_i is the incident wave amplitude, and C_g is the group velocity of the incident waves defined as,

$$C_g = \frac{\omega}{2k}\left(1 + \frac{2kh}{\sinh 2kh}\right) \quad (4)$$

where k is the wave number and ω is the wave angular frequency.

Therefore, the relative hydrodynamic efficiency ξ of each chamber may be defined as

$$\xi_{(1,2)} = \frac{P_{P(1,2)}}{P_W \cdot L_{CW} \cdot \theta} \quad (5)$$

where θ is the angle between the model and the tank (i.e., $\theta = 30°$ in this study); The overall hydrodynamic efficiency of the model is:

$$\xi = \xi_1 + \xi_2 \quad (6)$$

Results and Discussion

In this study, the MCOW was exposed to the wave conditions in which the waves formed an angle of 30° with respect to the front wall. The main objective of this study is to verify the feasibility of the MCOW device and to investigate its preliminary hydrodynamic performance.

▶ *Time histories of the measured data*

Two typical cases were utilized to illustrate the time histories of the measured data obtained during the test. Fig. 6 shows the time series of the relative wave amplitudes a/A_i for $T = 1.8$ and 1.0 s, where a represents the free water surface amplitude inside of the chambers. The variation in air pressure inside of each chamber is illustrated in Fig. 7. All measured free water surface elevations in the chambers were approximately in the same phase shown in Fig. 6. The positive and negative values of the water elevations were all less than those of the incident wave amplitude. The free water surface elevations at different gauge points were almost identical for large wave periods, as shown in Fig. 6a. For $T = 1.0$ s, the negative values at G2 (Chamber 1) and G4 (Chamber 2) near the front wall were smaller than the average values, which are shown in Figs. 6c and 6d, respectively. The positive and negative values at G6 (in Chamber 2), which is located near the wall that separates the two chambers, is greater than the average values. The uneven distribution characteristics of the free water surface elevations in the chambers were observed in Chamber 2 at small wave periods in Fig. 6d. Furthermore, it is shown that the motion of the wave column trapped inside of the chambers is sensitive to the incident wave length, and a smaller wave length will cause an oscillation of the free surface inside of the chambers, especially in Chamber 2.

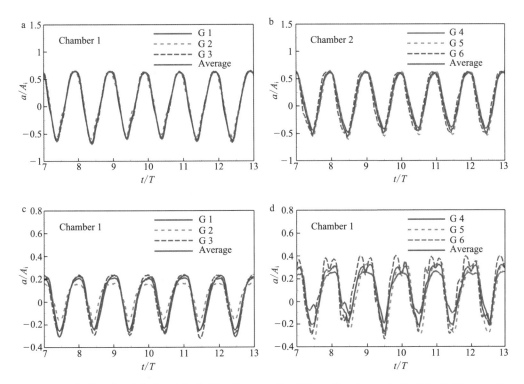

a. $T = 1.8$ s in Chamber 1. b. $T = 1.8$ s in Chamber 2. c. $T = 1.0$ s in Chamber 1. d. $T = 1.0$ s in Chamber 2. $A_i = 0.04$ m; $d = 0.1$ m; $\varepsilon = 0.68\%$.

Fig. 6 Time series of the measured instantaneous free water surface elevations inside of Chambers 1 and 2 for $T = 1.8$ s and 1.0 s

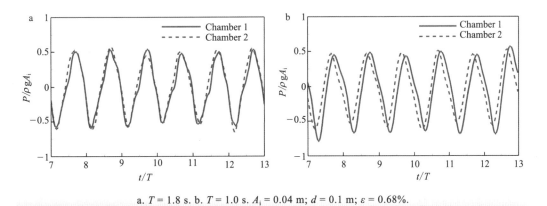

a. $T = 1.8$ s. b. $T = 1.0$ s. $A_i = 0.04$ m; $d = 0.1$ m; $\varepsilon = 0.68\%$.

Fig. 7 Time series of the measured instantaneous air pressures for $T = 1.8$ s and 1.0 s

▶ *Effect of the incident wave amplitude*

The experiments were conducted using varying incident wave amplitudes while keeping the other parameters constant at: $h = 0.6$ m, $d = 0.1$ m and $D = 0.028$ m. Figs. 8a and 8b show the variation of the hydrodynamic efficiency with kh under the conditions of three incident wave amplitudes (i.e., $A_i = 0.02$, 0.03, and 0.04 m) for Chambers 1 and 2, respectively. For a certain wave condition kh, it is observed that the hydrodynamic efficiency decreases as the wave amplitudes near the peak of the curves increases. Furthermore, the wave amplitude has little influence between the curves of $A_i = 0.03$ and 0.04 m, and this effect is more obvious in Chamber 1. It should be noted that two peaks are present in the efficiency curves. For convenience, we referred the larger peak as the main-peak, and the smaller peak as the sub-peak.

Fig. 9 shows the variation of the overall hydrodynamic efficiency. It is observed that the trend of the overall hydrodynamic efficiency curve is same in each sub-chamber. While the main-peak for all three wave amplitudes was observed at $kh = 1.78$, the hydrodynamic efficiencies achieved maximum values for $A_i = 0.02$, 0.03, and 0.04 m of 0.56, 0.49, and 0.46, respectively. Therefore, the maximum efficiency is achieved at $A_i = 0.02$ m.

The effects of the relative wave amplitude on the free water surface elevations and air pressure in Chambers 1 and 2 are given in Figs. 10 and 11, respectively. As shown, the free surface elevations in each sub-chamber decrease with increasing kh. In addition, it can be observed that the relative wave amplitude decreases with increasing wave amplitude A_i, and the variation is more obvious in the low frequency region. The variation of the relative air pressure is not sensitive to the variation of wave amplitude A_i almost the entire frequency range expect near $kh = 0.98$, at which point the minimum overall efficiency is achieved. The relative air pressure was observed to be higher in Chamber 1 than in Chamber 2.

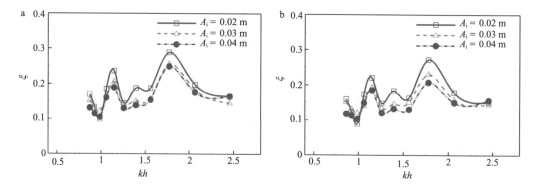

a. Hydrodymatic efficiency of Chamber 1. b. Hydrodymatic efficiency of Chamber 2. $d = 0.1$ m; $\varepsilon = 0.68\%$.

Fig. 8　Variations of the efficiencies of Chambers 1 and 2 with kh for three wave amplitudes

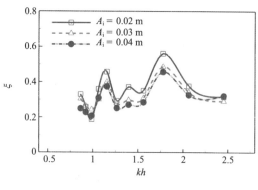

$d = 0.1$ m; $\varepsilon = 0.68\%$.

Fig. 9　Variations of the overall hydrodynamic efficiency with kh for three wave amplitudes

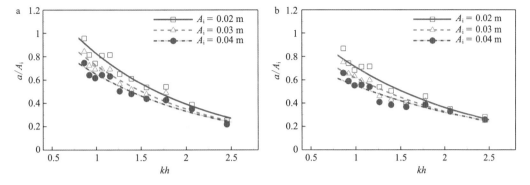

a. Relative wave amplitude of Chamber 1. b. Relative wave amplitude of Chamber 2. $d = 0.1$ m; $\varepsilon = 0.68\%$.

Fig. 10　Variations of the average relative wave amplitude for Chambers 1 and 2 with kh

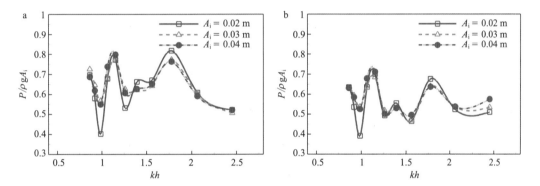

a. Air pressure amplitude of Chamber 1. b. Air pressure amplitude of Chamber 2. $d = 0.1$ m; $\varepsilon = 0.68\%$.

Fig. 11　Variations of the air pressure amplitudes for Chambers 1 and 2 with kh

▶ *Effect of the front wall draught*

The effect of the front wall draught is investigated in this section, in which three different front wall draughts (i.e., $d = 0.1$, 0.12, and 0.14 m) are considered. The wave amplitude and opening ratio are maintained at $A_i = 0.04$ m and $\varepsilon = 0.68\%$, respectively. It should be noted that the draughts of the two front walls are kept the same in this study.

Figs. 12a and 12b show the hydrodynamic efficiencies for Chambers 1 and 2, respectively. In the high-frequency region (at approximately $kh > 1.56$), the variations of the front wall draught have a more obvious influence in Chamber 1. Additionally, it can be seen that the efficiency of Chamber 1 is greater than that of Chamber 2 in the majority of the high-frequency range, especially considering the frequencies near the main-peak. It is worth noting a lower limit for the front wall draught is utilized to avoid the leakage of air from the chamber.

Fig. 13 shows the variation of the overall efficiency of the model for the three front wall draughts (i.e., $d = 0.1$, 0.12, and 0.14 m). It can be observed that at most frequencies, the efficiency (especially the peak efficiency) decreases as the front wall draught d increases. The main-peak occurred at $kh = 1.78$ and corresponds to the hydrodynamic efficiencies of 0.46, 0.4, and 0.36 for $d = 0.1$, 0.12, and 0.14 m, respectively. It can be observed that the two sub-chambers exhibit a greater performance as the front wall draught is reduced in high-frequency region (at approximately $kh > 1.56$), while the changing front wall draught has little influence on

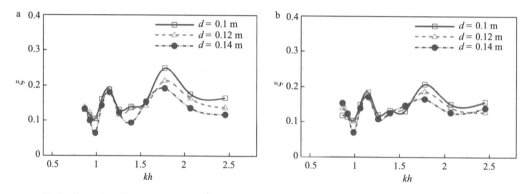

a. Hydrodynamic efficiency of Chamber 1. b. Hydrodynamic efficiency of Chamber 2. $A_i = 0.04$ m; $\varepsilon = 0.68\%$.

Fig. 12　Variations of the efficiencies for Chambers 1 and 2 with kh for the three front wall draughts

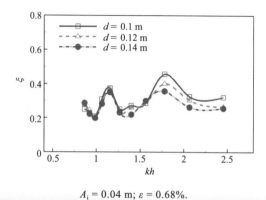

$A_i = 0.04$ m; $\varepsilon = 0.68\%$.

Fig. 13　Variations of the overall efficiency with kh for the three front wall draughts

the sub-chamber performance in the low-frequency region (at approximately $kh < 1.56$). This is because in the low-frequency long wave region, the front wall draught is adequately small relative to the wave length such that changes in the front wall draught have a lower effect on the propagation of long waves. In contrast, in the high-frequency region, the existence of the front wall draught cannot be ignored relative to the wave length, and therefore the variation of the short wave length significantly affects the front wall draught.

　　The effect of the front wall draught on the averaged free surface elevation and the air pressure inside of the sub-chamber were also investigated, as shown in Figs. 14 and 15, respectively. The averaged free surface elevation is shown to be insensitive towards the front wall draught. Considering both the averaged free surface elevation and air pressure, chamber 1 performs better than Chamber 2. Considering the relative air pressure, Chamber 2 is also not sensitive towards the variation of the front wall draught, while this factor has a significant effect on Chamber 1. In Chamber 1, limited changes occur in the relative air pressure as the front wall draught is varied from 0.1 to 0.12 m. However, when the front wall draught changes from 0.12 to 0.14 m, the relative air pressure decreases drastically. Therefore, the general trend of the overall efficiency curve is dominated by the change of the air pressure in Chamber 1.

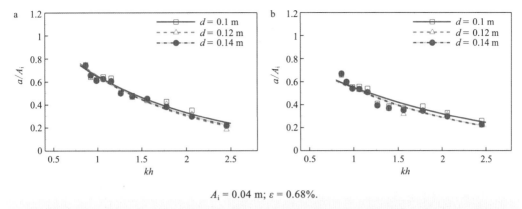

$A_i = 0.04$ m; $\varepsilon = 0.68\%$.

Fig. 14　Variations of the average relative wave amplitudes for Chambers 1 and 2 with kh

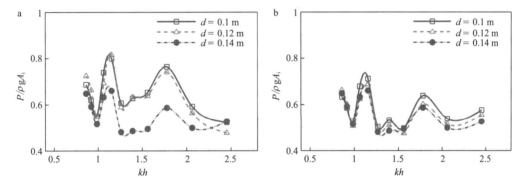

a. Air pressure amplitude of Chamber 1. b. Air pressure amplitude of Chamber 2. $A_i = 0.04$ m; $\varepsilon = 0.68\%$.

Fig. 15　Variations of the air pressure amplitudes for Chambers 1 and 2 with kh

▶ *Effects of the opening ratio*

In this section, three opening ratios (i.e., $\varepsilon =$ 0.35%, 0.68%, and 1%, which correspond to D = 0.02, 0.028, and 0.034 m, respectively) were tested via experiment. In this set of experiments, the incident wave amplitude was $A_i = 0.04$ m and the front wall draught was $d = 0.01$ m. It should be noted that the independent orifices used in this study were changed in the same manner.

Figs.16a and 16b show the hydrodynamic efficiencies of Chambers 1 and 2, respectively, for the opening ratios of $\varepsilon = 0.35\%$, 0.68%, and 1%. Chamber 2 is shown to be inferior compared to the Chamber 1 over the entire frequency region considered in this study.

Fig. 17 shows the variation of the overall efficiency of the model for the three orifice diameters. It can be seen that a higher orifice diameter results in a higher efficiency. The value of the main-peak is greatly influenced by the opening ratio with $\xi = 0.26$ ($\varepsilon = 0.35\%$), 0.46 ($\varepsilon = 0.68\%$), and 0.52 ($\varepsilon = 1\%$). Therefore, the optimal opening ratio orifice diameter considered in this study that corresponds to the aforementioned ideal orifice diameter is $D = 0.034$ m. It is worth noting that when the opening ratio increases from $\varepsilon = 0.35\%$ to 0.68%, the overall hydrodynamic efficiency changes significantly, while when the opening ratio increases from $\varepsilon = 0.68\%$ to 1%, the variation is less significant.

To further investigate the aforementioned phenomena, Figs. 18 and 19 present comparisons of the averaged relative wave amplitudes and air pressures in the chambers at different opening

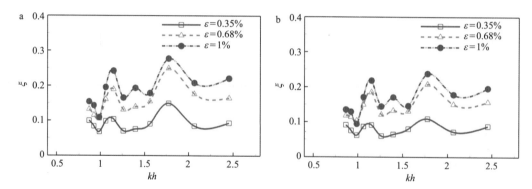

a. Hydrodynamic efficiency of Chamber 1. b. Hydrodynamic efficiency of Chamber 2. $A_i = 0.04$ m; $d = 0.1$ m.

Fig. 16　Variations of the efficiency for Chambers 1 and 2 with kh

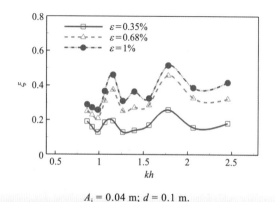

$A_i = 0.04$ m; $d = 0.1$ m.

Fig. 17　Variations of the hydrodynamic efficiency with kh for three orifice diameters

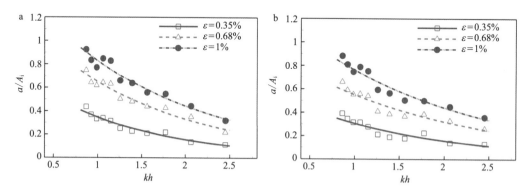

a. Relative wave amplitude of Chamber 1. b. Relative wave amplitude of Chamber 2. $A_i = 0.04$ m; $d = 0.1$ m.

Fig. 18　Variations of the average relative wave amplitudes for Chambers 1 and 2 with kh

ratios, respectively. Chamber 1 performs better than Chamber 2 in terms of both the relative wave amplitude and air pressure. The water column motion is influenced by the oscillation of the air pressure inside of the chambers. The air pressure inside of the chambers decreases as the opening ratio

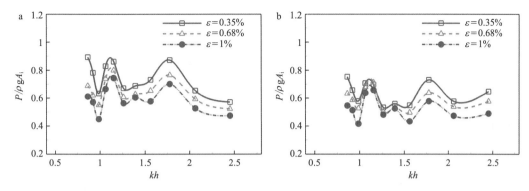

a. Air pressure amplitude of Chamber 1. b. Air pressure amplitude of Chamber 2. $A_i = 0.04$ m; $d = 0.1$ m.

Fig. 19 Variations of the air pressure amplitudes for Chambers 1 and 2 with kh

is increased, while the averaged relative free water surface elevation exhibits an opposite trend. These results may aid in the investigation of the turbine damping of OWC devices such that optimal energy extraction may be achieved.

Conclusions

In this study, the hydrodynamic performance of a fixed multi-chamber OWC model was experimentally investigated using a series of wave-flume tests. We will optimize the MCOW device in a further study based on the results of the experiment. We attempted to optimize the performance of the model via preliminary tests. The effects of the incident wave amplitude, front wall draught, and opening ratio of the orifice on the hydrodynamic performance of the MCOW model were examined. The following conclusions may be drawn from this study.

1. The MCOW is a feasible technology for extracting energy from waves, as evidenced by the satisfactory hydrodynamic performances achieved in this study. Further investigations are required to optimize the performance of MCOWs.

2. The hydrodynamic efficiency of this system increased with the decreasing of wave amplitude and reached a maximum at $A_i = 0.02$ m. In addition, the effects of the incident wave amplitude on the hydrodynamic efficiency were more apparent in air Chamber 2.

3. The hydrodynamic efficiency was significantly reduced with increasing front wall draught d in the high-frequency zone (at approximately $kh > 1.56$), while it was not sensitive to changes of draught d in the low-frequency zone (at approximately $kh < 1.56$).

4. The opening ratios of the individual chambers have a significant effect on the overall hydrodynamic performance of the system across the entire investigated frequency range under obliquely incident waves, and the optimum hydrodynamic efficiency was achieved at an opening ratio of $\varepsilon = 1\%$ for the same orifice diameter in the two sub-chambers.

5. The inter-comparisons of the two sub-chambers revealed that the rear chamber (the stern-side Chamber 2) was inferior to the front chamber (the bow-side Chamber 1) for most of the wave frequencies considered in this study, especially at frequencies near the main resonance frequency of the system. It is therefore recommended that the design of rear Chamber 1 should be carefully considered during the design of these systems.

Considering the actual marine environment, we make the following recommendations for further research:

1. The effect of the variation angle between the incident wave and the front wall on the hydrodynamic efficiency should be investigated.

2. Further study of the force variation in the front wall under obliquely incident waves compared with

traditional OWCs should be investigated.

Acknowledgments

This work was financially supported by the National Natural Science Foundation of China (52071304, 52071303); the Taishan Scholars Program of Shandong Province (No. ts20190914).

Author Contributions

Tongshun Yu: Conceptualization, methodology, validation, writing (review and editing). Qiyue Guo: Methodology, formal analysis, investigation, writing (original draft). Hongda Shi: Supervision. Tingyu Li: Investigation. Xiaoyu Meng: Investigation. Shoukun He: Data curation, validation. Peixuan Li: Investigation.

Competing Interests

The authors declare that they have no known competing financial interests or personal relationships that could have appeared to influence the work reported in this paper.

References

1. Pelc R, Fujita R M. Renewable energy from the ocean[J]. Mar. Policy, 2002, 26: 471-479.
2. Falcão A F O. Wave energy utilization: A review of the technologies[J]. Renew. Sustain. Energy Rev., 2010, 14: 899-918.
3. Delauré Y M C, Lewis A. 3D hydrodynamic modelling of fixed oscillating water column wave power plant by a boundary element methods[J]. Ocean Eng., 2003, 30: 309-330.
4. Esteban M D, López-Gutiérrez J S, Negro V, et al. A new classification of wave energy converters used for selection of devices[J]. J. Coast. Res., 2018, 85: 1286-1290.
5. Falcão A F O, Henriques J C C. Oscillating-water-column wave energy converters and air turbines: A review[J]. Renew. Energy, 2016, 85: 1391-1424.
6. Wu B J, Li M, Wu R K, et al. Experimental study on primary efficiency of a new pentagonal backward bent duct buoy and assessment of prototypes[J]. Renew. Energy, 2017, 113: 774-783.
7. Teixeira P R F, Davyt D P, Didier E, et al. Numerical simulation of an oscillating water column device using a code based on Navier–Stokes equations[J]. Energy, 2013, 61: 513-530.
8. Falcão A F O, Sarmento A J N A, Gato L M C, et al. The Pico OWC wave power plant: Its lifetime from conception to closure 1986–2018[J]. Appl. Ocean Res., 2020, 98: 102104.
9. Zhang D, Li W, Lin Y. Wave energy in China: Current status and perspectives[J]. Renew. Energ, 2009, 34: 2089-2092.
10. Deng Z, Wang C, Wang P, et al. Hydrodynamic performance of an offshore-stationary OWC device with a horizontal bottom plate: Experimental and numerical study[J]. Energy, 2019, 187: 115941.
11. Liu Z, Xu C, Qu N, et al. Overall performance evaluation of a model-scale OWC wave energy converter[J]. Renew. Energy, 2020, 149: 1325-1338.
12. Ning D Z, Wang R Q, Zou Q P, et al. An experimental investigation of hydrodynamics of a fixed OWC Wave Energy Converter[J]. Appl. Energy, 2016, 168: 636-648.
13. Elhanafi A, Fleming A, Macfarlane G, et al. Underwater geometrical impact on the hydrodynamic performance of an offshore oscillating water column–wave energy converter[J]. Renew. Energy, 2017, 105: 209-231.
14. Kharati-Koopaee M, Fathi-Kelestani A. Assessment of oscillating water column performance: Influence of wave steepness at various chamber lengths and bottom slopes[J]. Renew. Energ, 2020, 147: 1595-1608.
15. Ning D Z, Wang R Q, Gou Y, et al. Numerical and experimental investigation of wave dynamics on a land-fixed OWC device[J]. Energy, 2016, 115: 326-337.
16. De A F. Wave-power absorption by a periodic linear array of oscillating water columns[J]. Ocean Eng., 2002, 29: 1163-1186.
17. Rezanejad K, Bhattacharjee J, Soares C G. Analytical and numerical study of nearshore multiple oscillating water columns[J]. J. Offshore Mech. Arct. Eng., 2016, 138: 1-7.
18. Rezanejad K, Bhattacharjee J, Guedes Soares C. Analytical and numerical study of dual-chamber oscillating water columns on stepped bottom[J]. Renew. Energy, 2015, 75: 272-282.
19. Ning D, Wang R, Zhang C. Numerical simulation of a dual-chamber oscillating water column wave energy converter[J]. Sustain., 2017, 9: 1-12.
20. Elhanafi A, Macfarlane G, Ning D. Hydrodynamic performance of single–chamber and dual–chamber offshore–stationary Oscillating Water Column devices using CFD[J]. Appl. Energy, 2018, 228: 82-96.
21. Ning D Z, Wang R Q, Chen L F, et al. Experimental investigation of a land-based dual-chamber OWC wave energy converter[J]. Renew. Sustain. Energy Rev.,

2019, 105: 48-60.
22. He F, Leng J, Zhao X. An experimental investigation into the wave power extraction of a floating box-type breakwater with dual pneumatic chambers[J]. Appl. Ocean Res., 2017, 67: 21-30.
23. Ning D, Zhou Y, Zhang C. Hydrodynamic modeling of a novel dual-chamber OWC wave energy converter[J]. Appl. Ocean Res., 2018, 78: 180-191.
24. Fernandez H, Iglesias G, Carballo R, et al. The new wave energy converter WaveCat: Concept and laboratory tests[J]. Mar. Struct., 2012, 29: 58-70.
25. JohanningL, Smith G H, Wolfram J. Measurements of static and dynamic mooring line damping and their importance for floating WEC devices[J]. Ocean Eng., 2007, 34: 1918-1934.
26. Wang Z, Duan C, Dong S. Long-term wind and wave energy resource assessment in the South China sea based on 30-year hindcast data[J]. Ocean Eng., 2018, 163: 58-75.
27. Ning D Z, Shi J, Zou Q P, et al. Investigation of hydrodynamic performance of an OWC (oscillating water column) wave energy device using a fully nonlinear HOBEM (higher-order boundary element method)[J]. Energy, 2015, 83: 177-188.

Numerical investigation of spudcan penetration under partially drained conditions[①]

Kehan Liu[1], Dong Wang[1,2*], Jingbin Zheng[1]

1 Shandong Provincial Key Laboratory of Marine Environment and Geological Engineering, Ocean University of China, Qingdao, China
2 Pilot National Laboratory for Marine Science and Technology, Qingdao, China
* Corresponding Author: Dong Wang (dongwang@ouc.edu.cn)

Abstract

Estimation of spudcan penetration resistance during jack-up rig installation is critical to determine its final resting depth under full pre-load. The penetration resistance profile is conventionally calculated by assuming undrained conditions for clay and fully drained conditions for sand in current practice, but partial drainage in intermediate soils, such as in silts and sandy silts, leads to the resistance in between. This paper reports the results from a series of numerical investigations conducted to explore spudcan penetration response, which spans from fully undrained to fully drained conditions. The investigation is carried out using a large deformation finite-element approach within the framework of effective stress analysis. The numerical approach is validated by simulating existing centrifuge tests. Parametric studies are then performed using a wide range of normalized penetration velocity. The transition point where partially drainage occurs is identified, and is compared with those obtained from centrifuge tests. The variation of normalized penetration resistance with normalized velocity is compared with the published experimental data on UWA kaolin and silt. The backbone and consolidation index curves for Malaysian kaolin are subsequently provided to quantify the influence of different drainage conditions. Finally, the partial drainage effects for spudcan and penetrometers is compared and discussed.

Keywords: spudcan; bearing capacity; partial drainage; silts; large deformation analysis; finite element methods

Introduction

Mobile jack-up rigs are employed extensively by the offshore oil and gas industry in water depths up to about 120 m. Most jack-ups consist of a buoyant triangular platform and three independent legs with each conical spudcan footing at the bottom. Installation of the jack-up rig involves the penetration of spudcan into the seabed under the weight of the structure and then the pre-loading provided by water ballast. The penetration resistance profile of the spudcan footing needs to be estimated accurately to ensure the stability of mobile jack-up rig during installation. In almost all the existing approaches, the penetration resistance is calculated by assuming undrained conditions in clay and fully drained conditions in sand[1,2]. However, the soil

---

① 本文于2022年1月发表在 *Ocean Engineering* 第244卷，https://doi.org/10.1016/j.oceaneng.2021.110425。

response induced by spudcan penetration may fall somewhere in between the two extremes conditions, especially for silty soils[3]. The consolidation of the soil under partially drained conditions may result in a zone of increased strength of the soil and then an enhancement in the penetration resistance of spudcan. The drainage condition around an advancing spudcan depends not only on the soil permeability, but also on the penetration rate, v, and the spudcan diameter, D. The normalized velocity, $V = vD/c_v$, has thus been used widely as an indicator, where c_v is the coefficient of consolidation[4,5]. Typically, $vD/c_v > 30$ represents the nearly undrained conditions and $vD/c_v < 0.01$ the drained conditions for shallow circular foundations, while the range between the two threshold values refers to various degrees of partial drainage[4].

The effects of partial drainage on penetration resistance for cone, T-bar and ball penetrometers have been explored in centrifuge tests conducted at different penetration rates[5-7]. These tests were performed in normally consolidated kaolin for rates varying over two orders of magnitude. Numerical analysis of the effect of partial drainage on the cone tip resistance was conducted by Yi et al.[8]. The penetration resistance first declines as the rate decreases because of the reduced viscous effect, followed by an increase in resistance with a further reduction of rate[5,6,9]. This is attributed to the dissipated excess pore pressure around the penetrometer owing to the reduced rate. Similar to cone, T-bar and ball penetrometers, the penetration resistance of spudcan is increased with the decreasing penetration rate given that the soil around the spudcan is under partially drained conditions. This was evidenced by the centrifuge tests in studies by Barbosa-Cruz[10] and Cassidy[3]; however, the number of the existing centrifuge tests is limited. A typical trend obtained from spudcan resistances at different penetration depths (i.e. different magnitudes of c_v) in kaolin clay, reported by Barbosa-Cruz[10], was that the resistance at about $V = 1.75$ was about 2.5 times the undrained resistance. In addition, Cassidy[3] noted that the penetration resistance of spudcan at about $V = 0.8$ in calcareous silt was about 6 times greater than that under undrained conditions. Unfortunately, the penetration rate in the reported tests were not slow enough to capture the full range of partially drained conditions.

Most of the existing numerical analyses of spudcan penetration assumed either undrained or fully drained conditions[11-16]. The consolidation response during the pauses in spudcan penetration was investigated using numerical modelling[17,18]. The short consolidation contributed to an increase in soil shear strength and then an enhancement in the spudcan resistance. Nevertheless, the effect of partial drainage (or consolidation) on penetration resistance for an advancing spudcan has rarely been studied.

In this paper, the spudcan penetration resistance under partially drained conditions is investigated using a large deformation finite-element (LDFE) approach within the framework of effective stress analysis. The key aim of this study is to propose the backbone and consolidation index curves for assessing the penetration response of spudcan under different drainage conditions in Malaysian kaolin clay. The numerical model is validated by comparing with the existing centrifuge tests. The normalized penetration velocity of spudcan is varied to achieve partially drained conditions in normally consolidated Malaysian kaolin. Based on the numerical results, the change in normalized penetration resistance with normalized velocity is presented, compared with those obtained experimentally. A theoretical backbone curve is adopted to assess the penetration resistance under partial drainage according to the normalized velocity. The transition point from undrained to partially drained conditions is proposed based on the numerical results. Finally, the consolidation index is used to determine the parameters of backbone curve, followed by a comparison with those for penetrometers.

Methodology

The spudcan footing has to penetrate by a large

distance in soil under the designed pre-loading, which results in a mesh distortion of adjacent soil and hence computational convergence problems in traditional finite-element analyses[19]. Here, the spudcan penetration in cohesive soil was investigated using a LDFE approach, termed remeshing and interpolation technique with small strain (RITSS)[20,21]. The penetration of spudcan was divided into a sequence of incremental steps in the large deformation analysis. The spudcan displacement in each step must be sufficiently small to avoid excessive distortion of the soil element, and the updated Lagrangian calculation was conducted in each step. At the beginning of next step, the deformed geometry was remeshed, followed by the mapping of field variables (e.g. effective stresses, pore pressures and void ratio) from the old mesh to the new one. Python files were programmed to establish the finite element model and to extract necessary data from the result files.

In order to capture the penetration-induced accumulation and re-distribution of pore-pressures in cohesive soil, the LDFE approach was programmed for coupled pore fluid-effective stress analysis. The mesh generation and Lagrangian calculation in each step were completed through commercial software Abaqus, and the Lagrangian calculation was for effective stress – pore pressure coupled analysis. The modified Cam-clay (MCC) model was incorporated into the RITSS, simulating spudcan penetration and dissipation of excess pore pressure in cohesive soil. The details of the LDFE approach used can be found in Wang et al.[21,22].

▶ *Geometry strategy*

Spudcan penetration is an axisymmetric problem and hence a two-dimensional axisymmetric model was created. The spudcan-soil interaction was assumed as smooth since the interface roughness has minimal effect on the penetration resistance[23], even in consolidation problems considering the pore pressure, has been justified that frictional contact yielded only slightly higher results[18]. The stiffness of the spudcan is much higher than soil, and the spudcan was thus idealized as an impermeable rigid body. Soil was discretized with quadrilateral elements with quadratic displacements, linear pore pressure and reduced integration, i.e. CAX8RP in Abaqus.

For spudcan penetration from the soil surface, the soil surface geometry around the spudcan will change extremely, which results in distorted soil elements, and the computational convergence is challenging. Hence, the spudcan was pre-embedded at a depth of $0.6D$ before the penetration to ensure convergence, as suggested by Ragni et al.[17] and Wang and Bienen[18]. The horizontal and vertical dimensions of the soil domain were $20D$, which is sufficient to minimize boundary effects. The horizontal and vertical dimensions of the model were shown to be sufficiently large as the failure mechanisms were well within the mesh domain, and the changes in soil and pore pressures were not observed[18]. The zones of penetration-induced excess pore pressure were confined close to the spudcan so that flow across the sides or base of the soil region was not expected. The minimum element size of soil around the spudcan was chosen as $0.023D$. For each incremental step of the LDFE analyses, the vertical displacement of spudcan was selected as $0.01D$. This was shown not to affect the results, which proves that the selected incremental step of $0.01D$ is sufficiently small to ensure the accuracy of the Lagrangian calculation in each step. The spudcan diameter simulated in the prototype was 12.5 m, unless otherwise stated.

As shown in Fig. 1a, horizontal and vertical constraints were applied on the vertical and base boundaries of soil, respectively. The drainage of excess pore pressure was allowed only at the soil surface. The model spudcan has the same shape as that of the centrifuge test by Purwana et al.[24], as shown in Fig. 1b. The shoulder angle at the bottom face of the spudcan was 9°. However, round corners were used for the spudcan shoulder to ease convergence in numerical analyses. The filter element of pore pressure was fitted into the conical

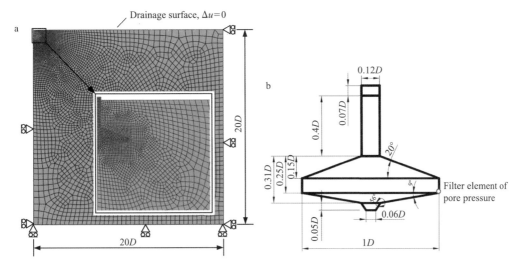

a. Soil domain and mesh strategy; b. Spudcan dimensions.

Fig. 1　Finite element model

underside of the spudcan shoulder to monitor the computed pore pressure variation during penetration.

▶ *Material properties*

The cohesive soils were simulated by the MCC model. The soils considered in the LDFE analyses include: (1) UWA kaolin, in order to make comparison with the centrifuge tests at various penetration rates performed by Barbosa-Cruz[10], and (2) the Malaysian kaolin, with permeability around ten times of UWA kaolin[18]. The higher permeability of the Malaysian kaolin was to benefit formation of partial drainage during spudcan penetration. The MCC parameters of both soils are listed in Table 1. The coefficient of earth pressure at rest was $K_0 = 1 - \sin \varphi'$, where φ' was the effective stress friction angle. The effective unit weights of both kaolin clays were considered to be constant, i.e. $\gamma' = 6$ kN/m³. The elastic shear modulus, G, can be expressed in terms of ν and κ as

$$G = \frac{3(1-2\nu)}{2(1+\nu)} \frac{p'(1+e)}{\kappa} \quad (1)$$

where p' is the mean effective stress, i.e. $p' = (1 + 2K_0)\sigma'_v/3$, and σ'_v is the vertical effective stress. The soil permeability k governs the rate of generation and dissipation of excess pore pressures. The permeability was assumed as isotropic, which was calculated as

$$k = \gamma_w m_v c_v \quad (2)$$

where γ_w is the unit weight of water and m_v is the one-dimensional compressibility of soil. Following Mahmoodzadeh et al.[25,26], the permeability was expressed as a function of the void ratio. The initial void ratio, e_0 and the undrained strength, s_u of normally consolidated soil can be deduced from the MCC model[27].

Table 1　Parameters of kaolin soils

Property	Malaysian kaolin[28]	UWA kaolin[36]
Angle of internal friction, φ': degrees	23	23
Void ratio at $p' = 1$ kPa on virgin consolidated line, e_N	2.35	2.25
Slope of normal consolidation line, λ	0.244	0.205
Slope of swelling line, κ	0.053	0.044
Poisson ratio, ν	0.3	0.3

Validation Against Centrifuge Tests

For the rapid penetration of spudcan footings

under nearly undrained conditions ($vD/c_v > 30$) reported in Wang and Bienen[18], the resistance predicted by the LDFE approach has already been validated thoroughly by simulating the centrifuge tests conducted by Purwana[28]. As the MCC model cannot allow for soil strength degradation, the penetration resistance was overestimated by about 10%–15% compared with the centrifuge tests[29].

To validate the numerical model under partially drained conditions, analyses were carried out simulating the centrifuge tests reported by Barbosa-Cruz[10] for spudcan penetration in UWA kaolin under an acceleration of 200g. In order to make a straightforward comparison with the tests, the spudcan model rather than prototype was considered in LDFE analyses, while the gravitational acceleration (g-forces) on the soil domain has also been simulated. In this way, the complex scaling rules for dimensions and material parameters between the model and prototype scales can be avoided. The diameter of the spudcan model was 60 mm, and the penetration rate was varied between 0.003 mm/s and 0.05 mm/s. The coefficient of consolidation of UWA kaolin can be estimated as[30]

$$c_v = 0.032\sqrt{1+14\sigma'_v/p_a} \qquad (3)$$

where p_a is atmospheric pressure. The spudcan penetration depth, z is measured from the mudline to the lowest point of the largest cross-section of spudcan. The normalized penetration velocity, vD/c_v, thus varied between 1.4 and 23.8 at the normalized penetration depth, $z/D = 1.5$, which fell within the range of partially drained conditions. In addition, a pre-embedment depth of $0.5D$ was adopted in this validation.

The undrained shear strength profile measured by Barbosa-Cruz[10] through the T-bar tests was $s_u = 1z$ kPa, which is different from $s_u = 1.46z$ kPa deduced from the MCC model. Therefore, the numerical and experimental penetration profiles are plotted in Fig. 2 in terms of the normalized bearing capacity F/s_uA, where F is the vertical resistance on spudcan, A is the bearing area and su is taken as the local undrained shear strength at spudcan penetration depth. The divergence at the beginning of penetration is attributed to the pre-embedment of spudcan in the LDFE analyses. More LDFE analyses with a pre-embedment depth of $0.7D$ for $v = 0.05$ and 0.03 mm/s were performed, as plotted in Fig. 2. The comparison with centrifuge tests shows that the bearing capacity factor in the LDFE analyses becomes independent of the embedment at $z/D > 0.8$ for $v = 0.05$, 0.03 and 0.01 mm/s and at about $z/D >$

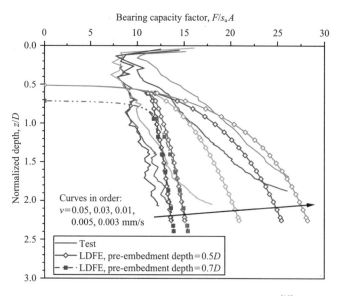

Fig. 2　Comparison of bearing capacity profiles from centrifuge tests[10] and numerical analyses

1.2 for v = 0.005 and 0.003 mm/s, respectively. The numerical normalized bearing capacity at a given depth increases with decreasing penetration rate, which is consistent with the tendency reported by Barbosa-Cruz[10].

The normalized penetration velocities at the normalized depth of z/D = 1.5, i.e. $V_{z/D=1.5}$, were 14.3 and 23.8 for v = 0.03 and 0.05 mm/s, respectively. This suggests that the soil response was close to undrained conditions. The numerical resistance increased sharply once the spudcan started penetration from the pre-embedment. The resistance at deep depths (e.g. z/D > 1) was independent of the pre-embedment. It is observed that the numerical resistance is around 10%–15% higher than the measured in deep soil. This may be due to the strain-softening of soil which was neglected in the LDFE analyses. With a further reduction of penetration rate to v = 0.01 mm/s (i.e. $V_{z/D=1.5}$ = 4.8), the difference between the normalized bearing capacities from centrifuge test and LDFE analysis was mostly more than 30% at z/D > 1.5. Whereas, this big divergence between numerical and experimental penetration profiles merely exists in the case for v = 0.01 mm/s.

The values of $V_{z/D=1.5}$ are decreased to 1.4 and 2.4 when the rates are 0.003 mm/s and 0.005 mm/s, respectively. The quick nonlinear increase of the computed resistance at the early penetration stage was also due to the pre-embedment of spudcan adopted in the LDFE analysis. For v = 0.005 mm/s, the dramatic increase of the measured resistance at z/D = 1.7 was caused by the boundary effect of strong box (sample thickness was approximately 2.8D). Consequently, the numerical and measured penetration profiles intersect at the subsequent penetration depth. A larger penetration distance of spudcan was required to diminish the influence of pre-embedment on penetration resistance than the faster rate. However, the normalized resistance increases almost linearly at subsequent depth. Generally, the computed capacity factors achieve reasonable agreement with the measured results in deep penetration, except for the rate of 0.01 mm/s.

▶ *Resistance profiles for various penetration rates*

Additional LDFE analyses were performed to quantify the influence of partial drainage on the penetration resistance of spudcan footing in Malaysian kaolin. The effect of partially drained condition can be achieved by either increasing the permeability of soil or reducing the penetration rate of spudcan. This study adopted the latter approach, as routinely performed by Barbosa-Cruz[10] and Cassidy[3]. Trial analyses have been performed by varying c_v and varying v to achieve the different drainage conditions, which indicates that the effect of reducing v and increasing c_v on the penetration response is equivalent provided that the dimensionless velocity vD/c_v is used for description.

The transition from fully drained to fully undrained ranges at least requires three orders of magnitude increase in penetration rate, and normalized velocities of 0.01 and 30 were suggested to confirm the limits of fully drained and fully undrained conditions[4]. Therefore, the penetration rates were varied between a sufficiently wide range, i.e. v = (9 × 10^{-8} – 1.8) mm/s, to cover the cases from fully drained to fully undrained conditions. The coefficient of consolidation for Malaysian kaolin can be fitted as[18]

$$c_v = 1.26(\sigma'_v/p_a)^{0.45} \quad (4)$$

As c_v increases with increasing σ'_v, the normalized penetration velocity decreases with the penetration depth for a constant penetration rate. The normalized velocities, V, were calculated as 0.0008 and 16,767 at the penetration depth of 1.5D for v = 9 × 10^{-8} mm/s and 1.8 mm/s, respectively. The order of magnitude for V is generally constant during penetration. Therefore, the results of numerical analyses are believed to cover fully drained to fully undrained conditions according to the range suggested by Finnie and Randolph[4]. The calculated spudcan resistance profiles at the considered penetration rates are presented in Fig. 3. q is the net penetration resistance, which represents the total vertical resistance provided by the soil deducting the submerged weight of the spudcan in

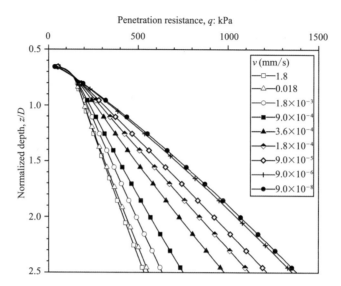

Fig. 3　Penetration resistance profiles for a series of penetration rates in Malaysian kaolin

soil. The resistance increases almost linearly with depth except for the initial penetration stage due to the pre-embedment, which is shown not to affect the results after a penetration depth of approximately $0.2D$.

From Fig. 3, the lower penetration rate results in more dissipation of excess pore pressures and higher penetration resistance, which is consistent with the observations of Barbosa-Cruz[10] and Cassidy[3]. The penetration resistance with $v = 9 \times 10^{-8}$ mm/s is about 2.4 times larger than that with $v = 0.018$ mm/s at $z/D = 1.5$, and it is noted that higher values were recorded at subsequent penetration depths. If no account is taken of partially drained condition for relatively permeable silts, the penetration resistance will tend to be underestimated.

The resistances corresponding to $v = 0.018$ and 1.8 mm/s ($V_{z/D=1.5}$ = 168 and 16,767) are close to each other, suggesting the undrained response was approached at $v = 0.018$ mm/s. The resistance at $v = 1.8$ mm/s is referred as the undrained penetration resistance, q_{un}, which is used as the base case to quantify the increased penetration resistance under partial drainage. The penetration resistance at $v = 9.0 \times 10^{-6}$ mm/s ($V_{z/D=1.5}$ = 0.08) is almost identical to that at $v = 9.0 \times 10^{-8}$ mm/s ($V_{z/D=1.5}$ = 0.0008), indicating fully drained conditions prevailed at these slow rates. Further, this indicates that the transition from fully undrained to fully drained condition occurs over a thousandfold decrease in penetration rate.

Backbone Curves

The variation of normalized penetration resistance, q/q_{un}, with the normalized velocity at several typical penetration depths is shown in Fig. 4. The normalized penetration resistance is expected to avoid, at least partially, the influence of strain-softening. As the pre-embedment in the numerical analysis with slower penetration rate resulted in a slightly underestimation of penetration resistance at $z/D < 1.2$, the penetration resistances at $z/D > 1.5$ are only presented.

As shown in Fig. 4, the normalized penetration resistances at $v = 0.018$ mm/s ($V_{z/D=1.5}$ = 168) are consistent with the results for $v = 1.8$ mm/s ($V_{z/D=1.5}$ = 16,767), providing further confidence that their penetration conditions are undrained. The increased degree of drainage around spudcan, as normalized velocity decreases, attributes a higher q/q_{un}. In addition, it can be seen in Fig. 4 that the normalized resistance shows consistent trends at different

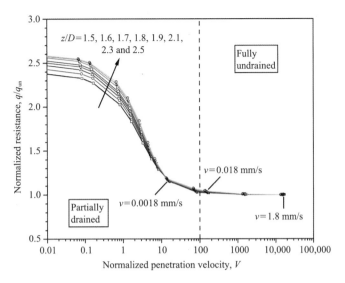

Fig. 4 Variations of normalized penetration resistance with normalized velocity at different depths in Malaysian kaolin

depths, but the variation in normalized penetration resistance among different penetration depths exists at penetration rate below 0.0018 mm/s.

▶ *Transition of drainage conditions*

The different ranges of the normalized velocity can be corresponding to fully undrained and partially drained conditions. According to Fig. 4, when the rate reduces to 0.018 mm/s, the increased penetration resistance due to partial drainage occurs. Therefore, the partially drained condition is achieved with V less than 100. Based on centrifuge tests, Barbosa-Cruz[10] suggested that for spudcan and plate footings in UWA kaolin, the transition from fully undrained to partially drained conditions occurs at $V = 80$. Additionally, the transition is at about $V = 10$ for spudcan penetration in calcareous silt, as suggested by the centrifuge tests conducted by Cassidy[3], in which $c_v = 1.55$ mm²/s of calcareous silt at $\sigma'_v = 120$ kPa is larger than 1.37 mm²/s of Malaysian kaolin and 0.14 mm²/s of UWA kaolin. Overall, the transition points of V for different drainage conditions in the literatures[3-5,10], agree reasonably with the results obtained from the numerical analyses in this study.

The transition value of V for partially drained behavior is useful to predict the spudcan penetration resistance profile in relatively permeable silts. Based on the numerical results, the partial drainage may occur when $c_v > 2.78$ mm²/s for large spudcans with $D = 10$ m and a practical penetration rate between 0.1 m/h and 4 m/h. Estimation of the spudcan penetration resistance assuming undrained conditions may be reasonable for field soils with $c_v < 2.78$ mm²/s.

Based on numerical analyses, the fully drained condition is likely to be achieved with V less than 0.01. In practice, the normalized velocity at which the penetration response changes from partially to fully drained condition is difficult to be clearly identified for clays and silts. This is because it is challenging to establish fully drained conditions in these soils. Hence, the penetration rate in the reported tests[3,10] were not slow enough to capture the full range of drained conditions.

When the viscous effect dominates in the undrained region, the normalized penetration resistance of spudcan shows a maximum increase by approximately 20% with the rise of rate, as reported by Barbosa-Cruz[10]. More centrifuge tests reported by Chung et al.[6] suggested a hypothetical relationship to quantify the viscous effect on the penetration resistance of T-bar penetrometers. Lehane et al.[9] extended the research of undrained viscous rate effect for ball penetrometers. However, this effect was not observed in the spudcan penetration tests performed by Cassidy[3] in calcareous silt and the numerical study using Malaysian kaolin in this work.

▶ *Comparison with centrifuge tests*

The experimental data of normalized penetration resistance reported in Barbosa-Cruz[10] and Cassidy[3] are compared with the LDFE results in Fig. 5. The region of partial drainage at $1 < V < 100$ is focused. The normalized penetration depths for calcareous silt corresponding to the normalized resistance are not indicated in the legend of the figure, since the majority of the penetration resistance profiles reported by Cassidy[3] are not clearly given.

Regardless of the normalized penetration depth, the LDFE results are broadly in a good agreement with the centrifuge test data in Barbosa-Cruz[10] for $V > 3$, while are approximately 10%–25% lower for $1 < V < 3$. It is noted that friction angle and other soil parameters may lead to the variation of this ratio of normalized resistance[6,8,10]. Generally, the higher friction angle of the soil around spudcan contributed to a higher q/q_u at the same drained condition, as suggested by Barbosa-Cruz[10]. Furthermore, Yi et al.[8] found that higher modulus ratio (ratio of shear modulus to initial mean effective stress), G/p', will lead to a higher q/q_u in terms of the study of cone penetration at partially drained conditions. The modulus ratio of UWA kaolin, G/p' = 25.1 is higher than 18.7 of Malaysian kaolin at z/D = 1.5, which may result in a higher q/q_u of UWA kaolin. Therefore, the divergence for $V > 3$ between UWA kaolin and Malaysian kaolin can be tolerated.

Experimental results in calcareous silt give a broader undrained condition when compared to the Malaysian kaolin, as shown in Fig. 5. For partially drained condition, the normalized penetration resistances of LDFE results achieve a reasonable agreement with the tests in calcareous silt for $2 < V < 10$, but a higher q/q_u in calcareous silt can be observed for $V \approx 1.4$.

▶ *Fitting of backbone curve*

House et al.[31] and Randolph and Hope[5] suggested that the normalized penetration resistance, q/q_{un}, can be fitted using the backbone curve expressed as a hyperbolic function:

$$\frac{q}{q_{un}} = 1 + \frac{b}{1 + cV^m} \quad (5)$$

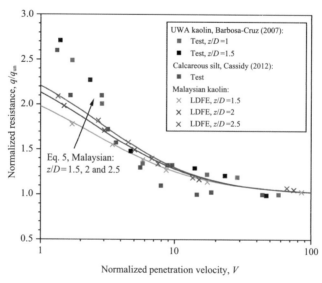

Fig. 5　Comparison of normalized resistance from numerical analyses and centrifuge tests[3,10]

where b, c and m are correlation parameters. At high V, the penetration resistance approaches to q_{un}, while the penetration resistance under fully drained conditions, q_{dr}, is achieved as V approaches zero. Therefore, the ratio of fully drained to fully undrained penetration resistances, q_{dr}/q_{un}, is expressed as, $q_{dr}/q_{un} = 1 + b$.

The experimental data of Barbosa-Cruz[10] and numerical results obtained in this study are both fitted by the Eq. 5. The corresponding b, c and m values are listed in Table 2 for comparison. As illustrated in Fig. 5, the fitted backbone curves can effectively quantify the influence of partial drainage on penetration resistance. The enhanced penetration resistance caused by the increase of soil shear strength as a result of the consolidation can be directly predicted through Eq. 5. An increasing value of q_{dr}/q_{un} (i.e. $1 + b$) with penetration depth can be observed in Fig. 4, 5, and q_{dr}/q_{un} tends to be united for a deeper depth. The maximum value of q_{dr}/q_{un} derived from the LDFE results for Malaysian kaolin at deep penetration (i.e. $z/D > 2.5$) is recommended as 2.6. The method reported by Hossain and Randolph[32] is recommended to calculate q_{un} in single layer clay, so that the prediction of q for partially drained conditions can be achieved using the established backbone curves.

Table 2 Values of correlation parameters for the backbone curves

Source	z/D	Parameters		
		b	c	m
Barbosa-Cruz (2007) UWA kaolin	1	3.35	0.75	1
	1.5	3.5	0.75	1
LDFE results Malaysian kaolin	1.5	1.37	0.4	1.04
	2	1.54	0.36	1.09
	2.5	1.57	0.3	1.15
	>2.5	1.6	0.3	1.15

▶ *Consolidation index curve*

A consolidation index (CI) is defined to normalize q/q_{un} by q_{dr}/q_{un}, expressed as[33]

$$CI = \frac{q/q_{un} - 1}{q_{dr}/q_{un} - 1} = \frac{1}{1 + cV^m} \quad (6)$$

where CI is derived from the backbone curve to quantify the degree of drainage, the values of CI vary between 0 for undrained condition and 1 for fully drained condition. The interval of normalized velocity is somewhere in $0 < CI < 1$, represents the partially drained condition. With increasing m, the interval for partially drained condition becomes narrower, whereas, the increase of c results in a shift of the interval to the left[33].

The numerical CI curve of spudcan for Malaysian kaolin is compared with that derived from centrifuge tests of Barbosa-Cruz[10] in UWA kaolin, as shown in Fig. 6. The coefficient of consolidation of Malaysian kaolin in LDFE analyses is about 10 times that of UWA kaolin. Correspondingly, the numerical CI curve of Malaysian kaolin is located on the right side of that for UWA kaolin derive from Barbosa-Cruz[10], which results in a lower value of parameter c for the numerical results. Furthermore, the values of m fitted for the LDFE analyses achieve a good agreement with the centrifuge tests in UWA kaolin at $z/D = 1.5$, implying that the magnitudes of interval at which partially drained conditions dominate for both soils appear to be uniform.

Although small differences in the values of c and m obtained from LDFE analyses at different penetration depths were recorded, their fitted CI curves are very close to each other (Fig. 6). This indicates that parameters c and m are independent of the penetration depth. Therefore, the average values of c and m corresponding to different penetration depths were adopted, and thus c and m for the backbone curve of spudcan in Malaysian kaolin are recommended as 0.35 and 1.1, respectively. With the identified parameters, the predicted CI curve is also included in Fig. 6 for comparison with the LDFE results for different penetration depths.

▶ *Comparison with CI curves of penetrometers*

The reported CI curves for cone, plate and ball penetrometers are also illustrated in Fig. 6. The transition from undrained to partially drained

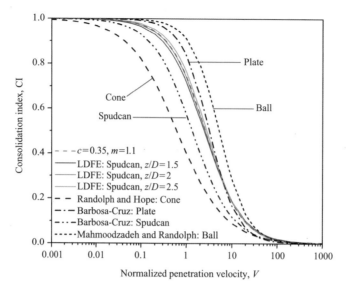

Fig. 6　CI curves for spudcan and penetrometers

conditions is almost independent of the type of penetrometers. Whereas, as indicated in Fig. 6, the transition value of V for fully drained behavior for cone penetrometer, obtained by Finnie and Randolph[4], is obviously larger than spudcan and other penetrometers. Yi et al.[8] gave the drained and undrained limits of cone for V of 0.01 and 100 based on numerical analysis, and a rather narrower range with V of 0.05 and 10 was suggested by Kim et al.[34] through cone tests. Therefore, the values of V of 0.01 and 100 are sufficiently small and large to define the limits of fully drained and fully undrained regions (i.e. $0 < CI < 1$) for spudcan, cone, plate and ball penetrometers.

The CI curve of cone plots to the left of the spudcan curves in Fig. 6, suggesting a lower rate of drainage around the cone than around the spudcan. By contrast, the comparison of CI curves between ball and spudcan is the other way around. This may be because the soil can flow around the ball but is displaced by the cone, resulting in a more extensive field of excess pore pressure around the cone than the ball during penetration[6]. Therefore, a lower penetration rate is required for the cone to achieve the same effect of partial drainage when compared to the ball. This implies that the mobility of soil around the advancing spudcan and plate is somewhere in between, since their CI curves are bracketed by those of cone and ball.

According to Fig. 6, the parameter c of CI curve is influenced by the mobility of soil around the penetrometers, which may be determined by the shape of penetrometer. The easier flow of soil around the penetrometer leads to a lower value of parameter c.

▶ *Δu generated during penetration under partially drained conditions*

Figs. 7a–c show the excess pore pressure contours in the spudcan vicinity for $V = 1676.7$, 8.4 and 0.8, respectively, at the penetration depth of $1.5D$ (18.75 m). As shown in Fig. 7a, the highest excess pore pressure around the spudcan is associated with the undrained condition, and decreases with V (Fig. 7b and 7c). The excess pore pressure, Δu generated during spudcan penetration under different drained conditions was measured at the spudcan shoulder base (Fig. 1b). A slower penetration rate of spudcan leads to lower pore pressures. For example, the value of Δu is reduced from 251 kPa at $V = 1,676.7$ to 7.9 kPa at $V = 0.17$ while $z/D = 1.5$. The influence of partial drainage on pore pressure ratio for cone penetrometer can be interpreted using the

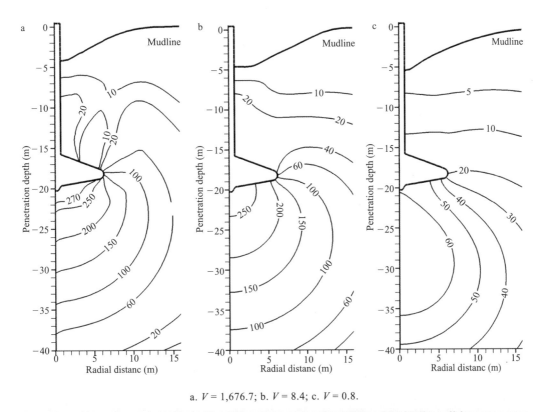

a. $V = 1,676.7$; b. $V = 8.4$; c. $V = 0.8$.

Fig. 7 Computed excess pore pressure contours for different drained conditions

normalized backbone curve proposed by DeJong and Randolph[35]. Similarly, the effect of dissipation during spudcan penetration on Δu may be effectively captured by using the normalized pore pressure ratio expressed as a function of the normalized velocity, suggested as

$$\frac{\Delta u}{\Delta u_{un}} = 1 - \frac{1}{1+cV^m} \quad (7)$$

where Δu_{un} is the referenced value of Δu obtained from undrained condition. The relationship between $\Delta u/\Delta u_{un}$ and V is presented in Fig. 8. When the Δu values for different penetration depths (i.e. z/D = 1.5, 2 and 2.5) are predicted and plotted in the dimensionless space of "$\Delta u/\Delta u_{un} - V$", all data points approximately form a unique backbone curve with $c = 0.15$ and $m = 1.06$. The threshold value of $V = 100$ indicates the transition of the drainage state, and the threshold value agrees with those in the backbone curves of normalized penetration resistance presented in Fig. 5. The consolidation index, CI can also be derived from the ratio of excess pore pressure, i.e. $CI = 1 - \Delta u/\Delta u_{un}$.

Conclusions

This paper has established the backbone curve of normalized resistance for spudcan footing, in an effort to quantify the effect of partially drained condition on penetration resistance. The study was achieved by performing coupled LDFE analyses incorporating the MCC model, which has been validated against centrifuge tests at different drainage conditions. The effect of partially drained condition is achieved by reducing the penetration rate of spudcan. The increased degree of drainage around an advancing spudcan, as penetration rate decreases, is contributed to a higher penetration resistance. The transition point from undrained to partially drained condition is determined as $V = 100$, which agrees well with those suggested by centrifuge tests. The variation of normalized

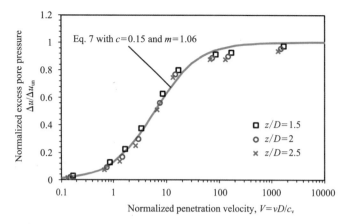

Fig. 8 Backbone curve of the normalized excess pore pressure generated during penetration for different penetration depths

penetration resistance with normalized velocity, focusing for the common situation of spudcan penetration under partial drainage, also agrees well with the experimental results. Based on the numerical results, the parameters for the backbone curve of spudcan penetration in Malaysian kaolin have been investigated. Finally, the consolidation index derived from the backbone curve is used to quantify the degree of drainage, followed by a comparison with those for penetrometers. The established backbone curve and CI curve may be used to predict the spudcan penetration resistance under partially drained conditions in offshore practice.

Acknowledgments

This research is supported by the National Natural Science Foundation of China (through the grants of No. U1806230, 42025702, 41772294, 51809247).

Author Contributions

Kehan Liu: Conceptualization, validation, writing (original draft), formal analysis, methodology. Dong Wang: Software, supervision, writing (review and editing), funding acquisition. Jingbin Zheng: Investigation, writing (review and editing), visualization, funding acquisition.

Competing Interests

The authors declare that they have no known competing financial interests or personal relationships that could have appeared to influence the work reported in this paper.

References

1. ISO (International Organization for Standardization). ISO/FDIS 19905-1: Petroleum and natural gas industries—site-specific assessment of mobile offshore unit—Part 1: jack-ups[S]. Geneva, Switzerland: ISO, 2016.
2. SNAME (Society of Naval Architects and Marine Engineers). Guidelines for site specific assessment of mobile jack-up units, SNAME Technical and Research Bulletin 5-5A, Rev. 3[Z]. Jersey City, NJ, USA: Society of Naval Architects and Marine Engineers, 2008.
3. Cassidy M J. Experimental observations of the penetration of spudcan footings in silt[J]. Géotechnique, 2012, 62(8): 727-732.
4. Finnie I M S, Randolph M F. Punch-through and liquefaction induced failure of shallow foundations on calcareous sediments[M]. Proc., Int. Conf. on Behaviour of Offshore Structures, 1994: 217-230.
5. Randolph M F, Hope S. Effect of cone velocity on cone resistance and excess pore pressures[J]. Proceedings of the IS Osaka - Engineering Practice and Performance of Soft Deposits, 2004: 147-152.
6. Chung S F, Randolph M F, Schneider J A. Effect of penetration rate on penetrometer resistance in clay[J].

Journal of Geotechnical and Geoenvironmental Engineering, 2006, 132(9): 1188-1196.
7. Mahmoodzadeh H, Randolph M F. Penetrometer testing: Effect of partial consolidation on subsequent dissipation response[J]. Journal of Geotechnical and Geoenvironmental Engineering, 2014, 140(6): 04014022.
8. Yi J, Goh S, Randolph M F. A numerical study of cone penetration in fine-grained soils allowing for consolidation effects[J]. Géotechnique, 2012, 62(8): 707-719.
9. Lehane B M, O'Loghlin C D, Gaudin C, et al. Rate effects on penetrometer resistance in kaolin[J]. Géotechnique, 2009, 59(1): 41-52.
10. Barbosa-Cruz. Partical consolidation and breakthrough of shallow foundations in soft soil[D]. Perth: University of Western Australia, 2007.
11. Li Y, Yi J, Lee F, et al. Effect of lattice leg and sleeve on the transient vertical bearing capacity of deeply penetrated spudcans in clay[J]. Journal of Geotechnical and Geoenvironmental Engineering, 2018, 144(5): 04018019.
12. Qiu G, Grabe J. Numerical investigation of bearing capacity due to spudcan penetration in sand overlying clay[J]. Canadian Geotechnical Journal, 2012, 49(12): 1393-1407.
13. Tho K K, Leung C F, Chow Y K, et al. Eulerian finite-element technique for analysis of jack-up spudcan penetration[J]. International Journal of Geomechanics, 2012, 12(1): 64-73.
14. Yi J, Zhao B, Li Y, et al. Post-installation pore-pressure changes around spudcan and long-term spudcan behaviour in soft clay[J]. Computers and Geotechnics, 2014, 56: 133-147.
15. Zheng J, Hossain M S, Wang D. Numerical modeling of spudcan deep penetration in three-layer clays[J]. International Journal of Geomechanics, 2015, 15(6): 04014089.
16. Zheng J, Hossain M S, Wang D. Estimating spudcan penetration resistance in stiff-soft-stiff clay[J]. Journal of Geotechnical and Geoenvironmental Engineering, 2018, 144(3): 04018001.
17. Ragni R, Wang D, Masin D, et al. Numerical modelling of the effects of consolidation on jack-up spudcan penetration[J]. Computers and Geotechnics, 2016, 78: 25-37.
18. Wang D, Bienen B. Numerical investigation of penetration of a large diameter footing into normally consolidated kaolin clay with a consolidation phase[J]. Géotechnique, 2016, 66(11): 947-952.
19. Hossain M S, Hu Y, Randolph M F, et al. Limiting cavity depth for spudcan foundations penetrating clay[J]. Géotechnique, 2005, 55(9): 679-690.
20. Hu Y, Randolph M F. A practical numerical approach for large deformation problems in soil[J]. International Journal for Numerical and Analytical Methods in Geomechanics,1998, 22(5): 327-350.
21. Wang D, Bienen B, Nazem M, et al. Large deformation finite element analyses in geotechnical engineering[J]. Computers and Geotechnics, 2015, 65: 104-114.
22. Wang D, Hu Y, Randolph M F. Three-dimensional large deformation finite element analysis of plate anchor in uniform clay[J]. Journal of Geotechnical and Geoenvironmental Engineering, 2010, 136(2): 355-365.
23. Hossain M S, Randolph M F. Deep-penetrating spudcan foundations on layered clays: Numerical analysis[J]. Géotechnique, 2010, 60(3): 171-184.
24. Purwana O A, Leung C F, Chow Y K, et al. Influence of base suction on extraction of jack-up spudcans[J]. Géotechnique, 2005, 55(10): 741-753.
25. Mahmoodzadeh H, Randolph M F, Wang D. Numerical simulation of piezocone dissipation test in clays[J]. Géotechnique, 2014, 64(8): 657-666.
26. Mahmoodzadeh H, Wang D, Randolph M F. Interpretation of piezoball dissipation testing in clay[J]. Géotechnique, 2015, 65(10): 831-842.
27. Wroth C P. The interpretation of in situ soil tests[J]. Géotechnique, 1984, 34(4): 449-489.
28. Purwana O A. Centrifuge model study on spudcan extraction in soft clay[D]. Singapore: National University of Singapore. 2006.
29. Zhang Y, Wang D, Cassidy M J, et al. Effect of installation on the bearing capacity of a spudcan under combined loading in soft clay[J]. Journal of Geotechnical and Geoenvironmental Engineering, 2014, 140(7): 04014029.
30. Richardson M. Rowe cell test on kaolin clay[R]. COFS internal report. Crawley, Australia: the University of Western Australia, 2007.
31. House A R, Oliveira J R M S, Randolph M F. Evaluating the coefficient of consolidation using penetration tests[J]. International Journal of Physical Modelling in Geotechnics, 2001, 1(3): 17-26.
32. Hossain M S, Randolph M F. Effect of strain rate and strain softening on the penetration resistance of spudcan foundations on clay[J]. International Journal of Geomechanics, 2009, 9(3): 122-132.
33. Lee J, Randolph M F. Penetrometer-Based assessment of spudcan penetration resistance[J]. Journal of Geotechnical and Geoenvironmental Engineering, 2011, 137(6): 587-596.
34. Kim L, Prezzi M, Salgado R, et al. Effect of penetration

rate on cone penetration resistance in saturated clayey soils[J]. Journal of Geotechnical and Geoenvironmental Engineering, 2008, 134(8): 1142-1153.
35. DeJong J T, Randolph M. Influence of partial consolidation during cone penetration on estimated soil behavior type and pore pressure dissipation measurements[J]. Journal of Geotechnical and Geoenvironmental Engineering, 2012, 138(7): 777-788.
36. Stewart D P. Lateral loading of pile bridge abutments due to embankment construction[R]. Perth: University of Western Australia, 1992.

Comparative study on metaheuristic algorithms for optimizing wave energy converters[①]

Feifei Cao[1,2,3], Meng Han[1], Hongda Shi[1,2,3,4*], Ming Li[1,2,3], Zhen Liu[1,2,3]

1 Department of Ocean Engineering, College of Engineering, Ocean University of China, 238, Songling Road, Qingdao 266100, China
2 Shandong Provincial Key Laboratory of Ocean Engineering, Ocean University of China, 238, Songling Road, Qingdao 266100, China
3 Qingdao Municipal Key Laboratory of Ocean Renewable Energy, Ocean University of China, 238, Songling Road, Qingdao 266100, China
4 Pilot National Laboratory for Marine Science and Technology (Qingdao), 1, Wenhai Road, Qingdao 266237, China
* Corresponding author: Hongda Shi (hongda.shi@ouc.edu.cn)

Abstract

The optimisation of WECs is a focus current research. With the increase in optimisation variables, traditional optimisation methods are insufficient to find the best design in a wide and complex solution space, and more advanced algorithms are required. In addition, the iterative optimisation of the WEC requires a reliable and accurate WEC simulation method to ensure the effectiveness of optimisation. Therefore, a MATLAB-APDL-AQWA united simulation system (MAA-system) is developed to calculate the WEC generation power accurately. Then, a memory mechanism is designed to avoid calculating the fitness of duplicate individuals in the search process by creating a memory base to store individual coding information and fitness information. Seven metaheuristic algorithms—the genetic algorithm, differential evolution, immune algorithm (IA), covariance matrix adaptation evolution strategy (CMA-ES), particle swarm optimisation, ant colony optimisation, and simulated annealing—with memory mechanisms are utilised to optimise the draft, power take-off parameter, and layout of WEC globally in random wave conditions. The performances of these algorithms in different search spaces are analysed. The results show CMA-ES is the best choice for solving the unimodal optimisation problem of WEC, and IA is more suitable for multimodal optimisation such as the WEC array layout.

Keywords: metaheuristic algorithm; configuration optimisation; wave energy converter; numerical simulation; power take-off; array layout

---

① 本文于2022年3月发表在 *Ocean Engineering* 第247卷，https://doi.org/10.1016/j.oceaneng.2021.110425。

Nomenclature

c_1, c_2	Acceleration coefficient
g	Gravitational acceleration, m·s^{-2}
$m^{(g)}$	The gth generation population average fitness
p_c	Crossing probability
p_m	Mutation probability
p_r	Proportion of population refresh
p_s	Proportion of immune selection
q	Interaction factor
v_{max}	Maximum velocity
z	Heave displacement, m
$x_k^{(g+1)}$	The kth individual in the $(g+1)$st generation population
$\dot{z}$	Heave velocity, m·s^{-1}
z_i	Heave displacement matrices of floaters i, m
z_j	Heave displacement matrices of floaters j, m
C	Radiation damping, N·s·m^{-1}
$C^{(g)}$	Evolutionary covariance matrix of the gth generation population
C_m	Covariance matrix
C_{PTO}	Linear damping coefficient, N·s·m^{-1}
Cr	Crossover operator
C_i	Radiation damping matrices by the motion of float i itself, N·s·m^{-1}
C_{iPTO}	Linear damping coefficient matrices of the ith float, N·s·m^{-1}
C_{ij}	Radiation damping matrices of float i increased by the motion of float j, N·s·m^{-1}
D	Float draft, m
F	Mutation operator
F_0	Initial mutation operator
F_{ex}	Excitation force, N
H	Float height, m
H_s	Annual significant wave height, m
I	Current evolution algebra
K_T	Attenuation coefficient
K	Static force coefficient, N·m·s^{-2}
K_i	Static force coefficient matrices of the ith float, N·m·s^{-2}
L_{dist}	Euclidean distance, m
L_{safety}	Safe distance, m
M	Mass of the float, kg
M_a	Added mass of the float, kg
MaxIt	Maximum evolution algebra
M_{ai}	Added mass matrices by the motion of float i itself, kg
M_{aij}	Added mass matrices of float i increased by the motion of float j, kg
M_i	Mass matrices of the ith float, kg
N	Floats number
N_{cl}	Multiples of antibody clone amplification
N_{ps}	Potential solutions number
NP	Number of individuals
NS	Total number of time steps
P	Instantaneous power, W
P_0	State transition probability constant
$\bar{P}$	Average power, W
R	Float radius, m
R_{ho}	Evaporation coefficient of pheromone
T	Simulation duration, s
$\bar{T}$	Average period, s
T_0	Initial control temperature, °C
T_I	Temperature of I iteration, °C
T_s	Time step, s

Greek

λ_L	Length scale
λ_P	Power scale
λ_T	Time scale
σ	Dynamic step parameters
$\sigma^{(g)}$	Evolutionary step size of the gth generation population

ω	Inertia weight
ω_{max}	Maximum inertia weight
ω_{min}	Minimum inertia weight

Abbreviation

(1+1)-EA	A simple evolutionary algorithm
ACO	Ant colony optimisation
BEM	Boundary element method
CFD	Computational fluid dynamics
CMA-ES	Covariance matrix adaptation evolution strategy
COBYLA	Constrained optimisation by linear approximations
DE	Differential evolution
GA	Genetic algorithm
GOA	Grasshopper optimisation algorithm
GSO	Glowworm swarm optimisation
GWO	Grey wolf optimizer
HG-PSO	Hybrid Genetic PSO
HHO	Harris hawks optimisation
HPTO	Hydraulic power take-off
IA	Immune algorithm
ISWEC	Inertial sea wave energy converter
LOCE	Levelized cost of energy
MAA-system	MATLAB-APDL-AQWA united simulation system
MFO	Moth–Flame optimisation algorithm
MVO	Multiverse optimisation algorithm
NLPQL	Non-linear programming by quadratic Lagrangian
OWSC	Oscillating wave surge converter
PC	Personal computer
PI	Parabolic intersection
PS	Parameter sweep
PSO	Particle swarm optimisation
PTO	Power take-off
RAM	Random-access memory
SA	Simulated annealing
SPH	Smoothed particle hydrodynamics
UGEN	U-shaped interior oscillating water column
WEC	Wave energy converter
WOA	Whale optimisation algorithm

Introduction

Wave energy is one of the most abundant renewable energy resources. Because of its wide distribution, abundant reserves, and easy conversion, it is an important direction for clean energy development and utilisation. The global wave energy resources available for utilisation are approximately 29,500 TW·h/a[1], showing great development potential. Power generation is the main form of effective use of wave energy. By 2020, the cumulative installations of global wave energy had reached 23.3 MW[2]. It is conservatively estimated that installations will reach 178 MW by 2030[3]. The growing demand from the energy market and the huge development potential of waves require the development of wave energy converters (WECs) with higher energy acquisition efficiency. Three optimisation approaches are commonly used to improve the efficiency of the WEC: geometry, control strategy, and array positioning[4]. Recently, the improvement of computer performance and computing power has enabled metaheuristic algorithms, such as the genetic algorithm (GA) and particle swarm optimization (PSO) to achieve excellent results in solving the geometric structure, Power take-off (PTO) parameters, and array layout optimisation problems of WECs. A summary of recent studies has been classified in Table A.1.

The structure of the WEC is closely related to the construction cost; in particular, the geometric shape may significantly affect the hydrodynamic performance of the equipment. The purpose of geometric optimisation of the WEC is to maximise the performance and reduces the cost. McCabe et al.[5] established different WEC shapes with single symmetry (one symmetry plane) and bisymmetric (two symmetry planes) based on the bicubic B-spline surface parameter description, and they optimised them using the GA. Subsequently, they considered the structure size into the objective function and set three objective functions[6]. Sirigu et al.[7] pursued different objective functions to address the comprehensive techno-economic optimisation of a pendulum WEC via the GA, considering a wide multivariate design space. These researchers all observed that the optimisation objective has a significant influence on the optimised design. Garcia-Teruel et al.[8] improved the geometry definition method of the McCabe spline surface and applied the GA and PSO with different parameter combinations to optimise the WEC geometry. Compared with McCabe's results, the objective function value is increased by 11% when using the most suitable algorithm. Silva et al.[9] compared and analysed the floater geometry optimisations of a WEC with a U-shaped interior oscillating water column (UGEN) using constrained optimisation by linear approximations (COBYLA) and the GA. The results showed that the COBYLA algorithm converged faster, but the GA could find a better solution. The GA was also used to optimise the absorber shape of a submerged planar pressure differential WEC. The results showed that the optimal shape may have a significantly higher energy capturing capability than a circular absorber shape in the same area[10]. The GA was used to optimise the geometric shape in the above research. To find a more suitable algorithm, GA, PSO, and hybrid genetic PSO (HG-PSO) were considered to determine the optimal values of the PTO stiffness parameters and the angular velocity of the gyroscope's flywheel of ISWEC. The results proved that all three algorithms can provide reliable solutions, but only the HG-PSO maintains a robust optimisation of the parameters, even with a more rigid stop condition[11].

The PTO system is a component that has a large potential for optimisation, and many researchers have focused on enhancing the PTO system with diverse algorithms. M'zoughi et al.[12] adjusted the airflow in the turbine duct using a proportional–integral–derivative controller tuned with the water cycle algorithm to avoid the stalling behaviour of the Wells turbine. The study showed that the gains obtained by optimisation are better than those obtained by the traditional Ziegler–Nichols–tuning method. To maximise the average absorbed power, Jusoh et al.[13], Calvário et al.[14] and Gao and Xiao[15] optimised the geometrical and control parameters of the hydraulic PTO (HPTO) system through the GA. The performances of the non-evolutionary non-linear programming by quadratic Lagrangian (NLPQL) and GA in optimising HPTO were compared. The results showed that the NLPQL algorithm converged faster, but the optimisation results were less effective than those of the GA[16]. Amini et al.[17] and Neshat et al.[18] evaluated the optimisation performances of different algorithms under different maximum evaluation numbers. Amini et al.[17] believed that multiverse optimisation algorithm (MVO) and CMA-ES are the most effective methods for optimizing PTO parameters, compared with the grey wolf optimizer (GWO), grasshopper optimisation algorithm (GOA), and Harris hawks optimisation (HHO). Neshat et al.[18] pointed out that the improved moth–flame optimisation (IMFO) algorithm using their proposed diversification strategy can outperform other bio-inspired optimisation algorithms, including GWO, whale optimisation algorithm (WOA), standard MFO, PSO, and CMA-ES.

The hydrodynamic interaction between the floats in the WEC array could have a positive or negative impact on the total power of the array. Simulation and experimental investigations demonstrated that the wave energy capture efficiency can be improved by optimising the spacing and arrangement of WECs[19]. Their optimisation model of oscillating-array-buoys was practically applied to the next generation prototype. In addition, adjusting the WEC type in the array may produce a constructive mixing effect, thereby improving the performance of the array[20]. Budal[21] defined the interaction factor q to characterise the hydrodynamic interactions between floating bodies, where q is the ratio of the total energy gain of the array composed of N WECs to that of a single device, and $q > 1$ means that the hydrodynamic interaction between the two devices has a positive effect—otherwise, it has a negative effect. Child and Venugopal[22] first applied a metaheuristic algorithm to solve the optimisation problem of a WEC array and proposed a parabolic intersection (PI) method for optimising the layouts of truncated WEC cylinders. This method placed the float in the wave field in turn, and each float was located at the intersection of the parabolic wake. However, the performance of this method is not superior to that of the GA. A real-coded elitism GA optimisation method with the advantage of implementing non-discretized space was utilised to optimise the WEC array layouts with five heave-constrained cylinders in a random unidirectional sea state. It was concluded that the optimisation results of the real-coded GA are better than those of the binary GA[23]. Then, both the power and cost of the WEC array in the objective function were considered, and a binary GA was used to analyse the optimal array layout under different spacings[24]. The good performance of the GA in the field of array configuration optimisation has increased interest in the GA. It was utilised to optimise the array layout of 40 and 12 oscillating wave surge converters (OWSCs) respectively, by Sarkar et al.[25] and Tay and Venugopal[26]. Sarkar et al.[25] suggested that the clustering of devices should be avoided while designing layouts of WEC arrays. The GA adopted in the latter study did not introduce the elitism strategy, and the population size and iteration times were too small, which resulted in poor search abilities. Therefore, its optimisation results tend to converge with those of the conventional method, but they are not completely consistent. Giassi and Göteman[27,28] also applied an elitism GA to optimise the internal parameters of single-point-absorbing axisymmetric cylindrical WECs (buoy radius, draft, and generator damping). A parameter sweeps (PS) optimisation of a single device made it possible to validate the GA method and evaluate its accuracy. However, the power result of any other layout found by the GA did not exceed the arbitrarily simulated layouts provided by the author, indicating that the search ability of the GA cannot satisfy the need for the search space. Therefore, other metaheuristic algorithms are also used to optimise the WEC array configuration, and their array optimisation results are compared to evaluate their reliability and efficiency. A simple evolutionary algorithm called the (1+1)-EA and CMA-ES algorithm was utilised to optimise the layout of 25 and 50 fully submerged three-tether WECs. The results of 20 independent runs showed that the (1+1)-EA with a simple mutation operator outperformed the CMA-ES[4]. By combining a four-parameter layout description, the optimisation performance and computational cost of a CMA-ES, a GA, and a GSO algorithm were compared and analysed[29]. However, the results show that CMA-ES had no obvious advantage compared with the GA and GSO. Fang et al.[30] improved the DE algorithm by introducing an adaptive mutation operator and applied the algorithm to optimise three-, five-, and eight-float arrays. The results show that the improved DE algorithm used for array optimisation is superior to the traditional DE one. Faraggiana et al.[31] showed that the GA and PSO have the same optimisation effect on a metric reflecting the levelized cost of energy (LOCE) of a multi-float configuration of a WaveSub device. No one is completely outstanding the other.

Optimising the configuration of WECs with an optimisation algorithm requires calculating the energy capture of WECs with different parameters in several iterations. Almost all researchers have applied numerical models to predict the WEC power output. The boundary element method (BEM), computational fluid dynamics (CFD) and smoothed-particle hydrodynamics (SPH) are widely applied in the wave energy context[32]. While limited by the linear nature of the potential flow theory, the speed at which numerical simulation can be performed makes the BEM the most popular choice for WEC optimisations. Researchers in this field extensively utilise the commercial BEM solver WAMIT and open-source BEM solver NEMOH to estimate the WEC hydrodynamic parameters[5,6,8,10,23,24,26,29,33]. For some wave energy floats with simple structures and regular shapes, analytical methods[9,22,33-37] or semi-analytical methods[27] can also be used for hydrodynamic analysis. Then, the power output of the device is evaluated by frequency domain simulation. However, the preprocessing function of these methods is poor, which makes it more difficult to change the float structure parameters. Moreover, the accuracy of the WEC power calculation based on frequency-domain analysis is lower than that of time-domain analysis. Faraggiana et al.[31] developed a hybrid framework based on time-domain analysis to assess WEC capacitation. This hybrid framework utilizes Salome-Meca (for the generation of the mesh), NEMOH (for the hydrodynamic coefficient simulation), and WEC-Sim (for the dynamic system simulation). However, its complexity is high, and its generality is poor.

Although numerous beneficial and enlightening explorations have been conducted in the field of WEC optimisation, there are still some problems that must be solved urgently. The application of suitable optimisation algorithms can significantly improve the WEC optimisation efficiency. The iterative optimisation of WEC also requires a reliable and accurate WEC simulation method to ensure the effectiveness of optimisation. In summary, the improvement of the WEC optimisation lies in finding the appropriate metaheuristic algorithm and adopting the WEC simulation method with stronger generality and higher accuracy. This work proposes a MATLAB-APDL-AQWA united simulation system to calculate the WEC generation power. The system allows the modification of different optimisation parameters, such as the number of devices, layout position, geometric dimensions, and mooring conditions. Subsequently, the memory mechanism of avoiding repeated individuals is applied to seven common variable simultaneous optimisation algorithms, and the optimisation performances of different algorithms in different search spaces are evaluated and analysed.

This article is organized as follows. The MATLAB-APDL-AQWA united simulation system is introduced in detail in Section "MATLAB-APDL-AQWA United Simulation System". The different metaheuristic algorithms used in this study are outlined in Section "Optimisation Method". The specific optimisation settings are described in Section "Optimisation Settings". In Section "Results", the optimisation performances of different algorithms in different search spaces are analysed. Section "Conclusions" summarises the study and presents the conclusions. Finally, the future research directions are discussed in Section "Future Work".

MATLAB-APDL-AQWA United Simulation System

The Ansys Parametric Design Language (APDL) makes parametric modelling possible, as well as setting the preprocessor, solver, and postprocessor parameters. It provides the functionality of a generic programming language, including parameters, array parameters, expressions, functions, branches, loops, and macros. The users can realise the process of modeling, meshing, and material definition by composing APDL commands, and these commands are not limited by the Ansys software version. Compared with NEMOH and WAMIT, the commercial BEM solver, Ansys-AQWA has powerful preprocessing and postprocessing functions, and a good connection with the modelling software Ansys-APDL. The complete hydrodynamics analysis function enables AQWA to satisfy the demands of the WEC simulation. In this study, using MATLAB codes and Windows batch processing commands, APDL and AQWA are connected to build the MATLAB-APDL-AQWA united simulation system (MAA-system) to calculate the energy produced from WECs.

MATLAB is the central platform of the MAA-system, which processes input and output data, calls software through batch processing commands, and realises information transmission between different software packages. The software communication relationships are shown in Fig. 1. The system consists of four subsystems running in series: parameter input, model generation, AQWA solution, and data processing (Fig. 2).

In the parameter input module, some parameter variables that control the numerical model and simulation calculation are defined by MATLAB, such as float structural parameters (height, draft, etc.), mooring conditions, wave conditions, linear PTO load, number and position of floats, step size and step number, and sea state. Subsequent modules use the variables defined above.

In the model generation module, in light of the shape and dimensions of the floats, the modeling and meshing process of the WEC are transformed into APDL modelling command streams. These command streams are written to the TXT text. MATLAB calls APDL to read in the text, automatically completes the modelling and meshing, and then exports the AQWA text containing node and panel information. The logic command streams of modelling are different for floats with different geometries. This must be written in advance in MATLAB.

The AQWA solution module, composed of frequency domain calculations and time domain calculations, is the most critical part of the simulation system. In the frequency-domain part, MATLAB calculates the weight, inertia, and centre of gravity of the floats based on structural parameters. These parameters and AQWA text content are written into the frequency-domain calculation input text using

MATLAB transmits simulation control information to APDL and AQWA through codes and Windows batch commands, and reads their computation results.

Fig. 1 Communication relationships among MATLAB, AQWA and APDL

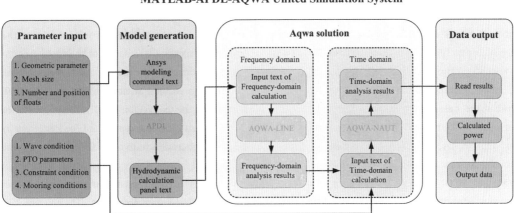

The MAA-system we proposed realizes the simulation process of WEC modelling, frequency-domain calculation, time-domain calculation and data output.

Fig. 2 MAA-system framework

MATLAB. Then, AQWA-LINE (a frequency-domain solver) is called by Windows batch commands to solve the frequency-domain problem. The results of the frequency-domain calculations are stored in the *.RES file. In the time-domain calculation, MATLAB compiles the time-domain calculation input text in accordance with time-domain control parameters, such as wave condition, step size, step number, and linear PTO load. The AQWA-NAUT is started by Windows batch commands for time-domain analysis and automatically reads the results of the frequency-domain analysis.

The calculation results of the time history motion response of floats are stored in the LIS text. Using MATLAB, the data processing module reads this text and calculates the WEC instantaneous power time history and average power according to the float velocity and PTO damping force.

The system can automatically calculate the generation power of WECs after all types of parameters preset, laying the foundation for the optimisation of the WEC configuration by a metaheuristic algorithm.

Optimisation Method

The mathematical and metaheuristic optimisation methods are two powerful tools for solving optimisation problems. The metaheuristic optimisation algorithm, which only needs to calculate the value of the objective function and has no continuous and differentiable requirements for the objective function, is more suitable for solving the configuration optimisation problem of WECs. Common metaheuristic algorithms can be divided into three categories: evolutionary algorithms inspired by Darwinian evolutionary theory (such as GA, DE, IA, and CMA-ES), swarm algorithms simulating social behavior of group animals (such as PSO and ACO) and physics algorithms simulating physical laws of nature (such as SA and MVO). GA, DE, IA, CMA-ES, PSO, ACO, and SA are the most widely used and representative algorithms in these categories. They have been successfully applied to solve many complicated optimisation problems. To find a suitable metaheuristic algorithm for solving the WEC optimisation problem, the average power of the WEC is taken as the objective function, and the above-mentioned seven metaheuristic algorithms are combined with the MAA-system to optimise the configuration of WECs. Then, their optimisation performance is compared and analysed. In this section, the basic ideas and settings of the algorithms used are introduced briefly.

▶ *Genetic algorithm (GA)*

The GA is a self-adaptive global optimisation search algorithm that simulates the genetic and evolutionary processes of organisms in the natural environment. It was first proposed by Holland[38]. In the GA, after the initial population is formed by binary coding, the genetic operation realises the evolutionary process of survival of the fittest through selection, crossover, and mutation. The GA used in this study utilizes roulette selection, single-point crossing with crossing probability p_c, and basic bit mutation with mutation probability p_m. The elitism strategy of replacing the worst individual of each generation with the optimum of history is applied to ensure the continuity of excellent genes. Fig. 3 shows the flowchart of the GA used in this study.

▶ *Differential evolution (DE)*

DE is an extraordinarily effective optimisation algorithm proposed by Storn and Price[39]. This algorithm uses the difference component of two individual vectors randomly selected from the population as the disturbance of the third random target vector to obtain the mutation vector. Then, the mutation vector and target vector are hybridised to generate the test vector. Finally, the target vector competes with the test vector, and the better one is the next-generation individual (Fig. 4). A discrete DE is used in this study. The adaptive mutation operator is introduced to adapt the mutation operation to the search progress[40]. The concept is as follows:

$$\begin{cases} F = F_0 \times 2^\lambda \\ \lambda = e^{1 - \frac{\text{maxIt}}{\text{maxIt} - I}} \end{cases} \quad (1)$$

The initial mutation operator F_0 is usually taken

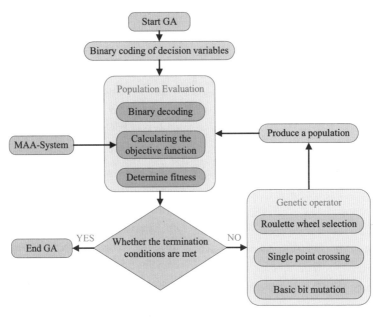

The elitism GA is utilized with crossing probability $p_c = 0.6$ and mutation probability $p_m = 0.01$.

Fig. 3   The flow chart of the GA

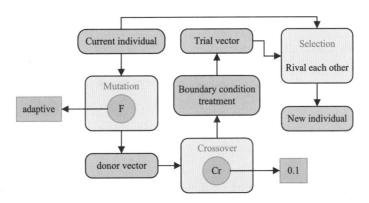

The mutation operator F is adaptive and the crossover operator Cr is 0.1.

Fig. 4   The process of the differential operation

as 0.4, MaxIt is the maximum evolution algebra, and I is the current evolution algebra. A better choice of crossover operator Cr is 0.1. Individuals that do not satisfy the boundary conditions in the process of mutation and hybridisation are replaced by random individuals in the feasible region.

▶ *Immune algorithm (IA)*

The IA is based on the learning mechanism of the biological immune system that recognises external pathogens and produces antibodies against pathogens[41]. Similar to the GA, the evolutionary optimisation process of the IA is realised by operators. Nevertheless, its ability to maintain population diversity is better than that of the GA. Immune operation simulates the biological immune system through immune selection, cloning, mutation, clonal inhibition, and population refresh. Moreover, the algorithm adds the calculation of the individual affinity, antibody concentration and stimulation degree, and it takes the antibody concentration as an index to evaluate individual quality. In this study, the proportion of immune selection p_s and population refresh p_r was set to 50%. The affinity is calculated by the Euclidean distance between antibody vectors, and the multiples of antibody clone amplification N_{cl} is 2–3.

▶ *Particle Swarm optimisation (PSO)*

PSO is an algorithm that simulates the mutual cooperation mechanism in the foraging behaviour of birds to find the optimal solution[42,43]. The velocities and positions of the particles are updated based on the fitness values of all particles in the swarm, and the swarm propagates toward optimal solutions. The discrete binary PSO[44] is suitable for solving WEC optimisation problems because it can map the discrete problem space to the continuous particle motion space. The control parameters of PSO are the acceleration coefficients c_1 and c_2, maximum velocity v_{max}, inertia weight ω, boundary treatment strategy, etc. This PSO takes $c_1 = c_2 = 1.5$, using the boundary absorption strategy. Here, ω determines the performance of the algorithm, and the weight of the linear change can better adapt to the search progress[45]. The change formula is

$$\omega = \omega_{max} - I\frac{\omega_{max} - \omega_{min}}{MaxIt} \quad (2)$$

where ω_{max} and ω_{min} represent ω upper and lower limits of change, respectively. Here, $\omega_{max} = 0.8$ and $\omega_{min} = 0.4$ are recommended. The optimisation process of PSO is shown in Fig. 5.

▶ *Ant Colony Optimisation (ACO)*

ACO is a probabilistic algorithm for metaheuristic search of the shortest path. It uses pheromones as the basis for ants to select follow-up behaviour[46]. As shown in Fig. 6, during foraging, the ants leave the bio-pheromone on the crawling path. Ants in ant colonies can sense the concentration of pheromones and always move in the direction of a high concentration of pheromones. Finally, the ant colony finds the optimal path. ACO sets the evaporation coefficient of pheromone (R_{ho}) to represent the evaporation degree of a realistic pheromone. The state transition probability constant (P_0) is used to determine the search range of the algorithm. When the state transition probability is less than P_0, a local search is performed, otherwise, a global search is conducted. Here, $R_{ho} = 0.9$ and

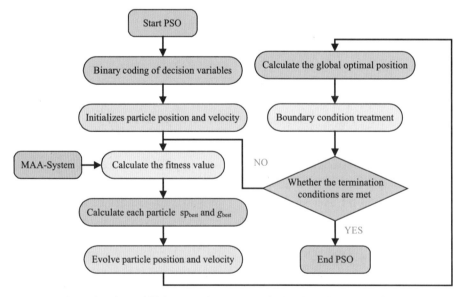

Both acceleration coefficient c_1 and c_2 are 1.5. The inertia weight varies linearly.

Fig. 5 The optimisation process of PSO

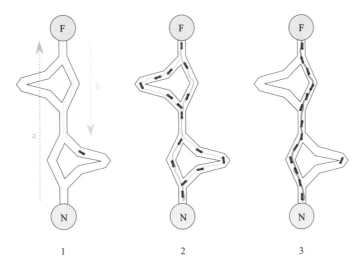

The evaporation coefficient of pheromone R_{ho} is 0.9 and the state transition probability constant P_0 is 0.2.

Fig. 6 Ant colony optimisation[47]

$P_0 = 0.2$ are recommended. An adaptive search step is defined to improve the local search, so that the local search range is reduced with an increase in iterations.

▶ *Simulated annealing (SA)*

Under the given initial values of control parameters, SA[48] starts from the feasible solution randomly and continues the iterative process of "generate new solution, judge, accept/abandon". A series of Markov chains are generated in the iterative process. The length of a Markov chain represents the number of searching times in an iteration. The Metropolis criterion is used to judge whether to accept the new solution. The search point gradually converges to the local optimum through calculating the time evolution process of the system. The initial control temperature $T_0 = 100\ °C$ and the temperature drop function $T_I = K_T^I \times T_0$, where K_T is the attenuation coefficient.

▶ *Covariance matrix adaptation evolution strategy (CMA-ES)*

The CMA learning mechanism is an adaptive learning method based on statistics. It generates new individuals by randomly sampling the probability distribution constructed during the evolution process. The probability distribution generated in the optimisation process describes the characteristics of the optimisation objective function. CMA-ES guides the direction of mutation evolution by adjusting the dynamic step parameter σ and dynamic positive definite covariance matrix C_m. It adaptively increases or decreases the search space for the next generation based on the shape of the covariance matrix. The basic equation is

$$x_k^{(g+1)} = m^{(g)} + \sigma^{(g)} N(0, C_m^{(g)}) \qquad (3)$$

here, $x_k^{(g+1)} \in R^n$ is the kth individual in the $(g + 1)$st generation population. $m^{(g)}$ is the average fitness of the gth generation population, $\sigma^{(g)}$ is the evolutionary step size of the gth generation population, and $C_m^{(g)}$ is the evolutionary covariance matrix of the gth generation population[49].

The running code of CMA-ES used in this study stems from the improvement introduced by Mostapha[50].

Optimisation Settings

In this section, the numerical model of the WEC used is introduced, and the discrete settings of optimisation variables are described in detail, especially the WEC layout method. A memory mechanism to avoid repeated calculations of the same individual fitness is proposed. This mechanism can significantly reduce the optimisation time of the evolutionary algorithm. Then, the stopping conditions of the algorithm are introduced, and the calculation times of fitness are used as the evaluation index of the algorithm optimisation time. Finally, the computational resources and memory mechanisms of all tests are presented.

▶ *Wave Energy Converter model*

The axisymmetric cylindrical float point absorbing WEC has a simple structure and only moves in the heave direction. Consequently, many studies on the optimisation of WEC structure parameters and array layouts are based on the device. Different metaheuristic algorithms have been applied to optimise the radius, draft, PTO damping and array arrangement of the cylindrical floats[22-24,27,28,34]. While the methodology of the metaheuristic algorithm combined with the proposed MAA-system is general and applies to any WEC design, the focus on the optimisation of the cylindrical heave float WEC. As the incoming waves reach the WEC, the cylindrical floats and the translator move upward (wave crest) and downward (wave through), as shown in Fig. 7. The PTO system converts this motion into an electrical output.

In this study, a time-domain simulation is applied. This type of simulation provides a higher accuracy than frequency-domain models, which makes them suitable for WEC optimisation studies. In the time domain, a linear PTO model is used. The external load added by the energy output is expressed by the linear damping coefficient, C_{PTO}. In addition, viscous forces and the mooring system are not modelled, and the kinematics are linearised. The equations of motion associated with a single cylindrical heave float when excited by a unit-amplitude plane wave with frequency ω follow from Newton's second law:

$$[-\omega^2(M + M_a) - i\omega(C + C_{PTO}) + K]z = F_{ex} \quad (4)$$

In Eq. 4, M and M_a are the mass and added mass of the float, respectively, C is the radiation damping, C_{PTO} is the linear damping coefficient, K is the static force coefficient, z is the heave displacement, and F_{ex} is the excitation force.

The interaction between floats should be considered when an array composed of N floats heaves. This is mainly reflected by the wave radiation force (i.e., added mass and radiation damping) generated by the float movement. The motion equation of the ith float in the array is expressed as

$$[-\omega^2(\boldsymbol{M}_i + \boldsymbol{M}_{ai}) - i\omega(\boldsymbol{C}_i + \boldsymbol{C}_{iPTO}) + \boldsymbol{K}_i]z_i + \sum_{j=1, j\neq i}^{n}(-\omega^2\boldsymbol{M}_{aij} + i\omega\boldsymbol{C}_{ij})z_j = \boldsymbol{F}_{ex} \quad (5)$$

where $\boldsymbol{M}_i$, $\boldsymbol{C}_{iPTO}$ and $\boldsymbol{K}_i$ are the mass, linear damping coefficient, and static force coefficient matrices of the ith float, respectively, $\boldsymbol{M}_{ai}$ and $\boldsymbol{C}_i$ are the added mass and radiation damping matrices, respectively, by the motion of float i. $\boldsymbol{M}_{aij}$ and $\boldsymbol{C}_{ij}$ are the added mass and radiation damping of float i increased by the motion of float j, and z_i and z_j are

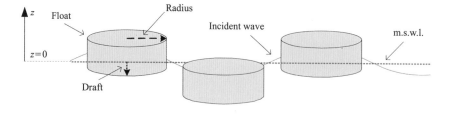

The dotted line indicates the mean surface water level (m.s.w.l). Float radius is constant and the incident wave is unidirectional irregular wave.

Fig. 7　Cylinder float WEC concept

the heave displacement matrices of floaters i and j, respectively.

In this research, the added mass, radiation damping, and other hydrodynamic parameters were calculated by MAA-system calling AQWA-LINE. AQWA-NUAT calculates the time-domain model based on the results of the frequency-domain analysis. The instantaneous wet surface integral method is used to calculate the static water restoring force and wave incident force, and the real-time motion response of the floats under PTO load is obtained.

In the time-domain analysis, the simulation of the WEC under irregular wave conditions may more accurately reflect the motion characteristics of a WEC in actual sea states. Zhaitang Island of China Qingdao is taken as the target sea area. According to the wave observation and statistical data, the annual significant wave height H_s is 0.61 m, and the average period $\bar{T}$ is 3.3 s. The ocean model uses irregular unidirectional waves together with the Joint North Sea Wave Project (JONSWAP) spectrum.

The instantaneous power P captured by the WEC equipment is directly proportional to the square of the float heave speed, as shown in Eq. 6.

$$P = |C_{PTO} \dot{z}^2| \tag{6}$$

For the case of an irregular wave, the average power $\bar{P}$ of a single float can be obtained from the instantaneous power P, as shown in Eq. 7.

$$\bar{P} = \frac{1}{T}\int_0^T P = \frac{1}{T}\sum_{i=1}^{NS} P_i T_s \tag{7}$$

here, T is the simulation duration, NS is the total number of time steps, and T_s is the time step.

Summing the average power $\bar{P}$ of all floats in the array, makes it possible to obtain the average power $\bar{P}_{array}$.

To establish the numerical model conveniently, the sea condition and WECs are measured by length scale $\lambda_L = 5$ to shrink based on the gravity similarity criterion. On the premise that the Froude numbers (Fr = $U/(gL)^{\frac{1}{2}}$) of the model and the prototype device are equal, the time scale $\lambda_L = \sqrt{5}$ and power scale $\lambda_P = 125\sqrt{5}$ are calculated. In this study, the simulation time T is set to 200 s; moreover, $T_s = 0.1$ s can not only ensure the accuracy of the model calculation, but also reduce the calculation cost.

▶ *Discretization of variables*

In Section "Optimisation Method", all variables defined by MAA-system can be used as optimisation variables. The draft, linear PTO damping, and plane location of the cylindrical float WECs are only considered as optimisation variables to test the optimisation performances of different metaheuristic algorithms. The optimisation problem is to find the appropriate draft, linear damping, and plane position of the cylindrical floats in the region of interest to maximize the average power generation. The float radius R and height H are set to a constant value, $R = 0.4$ m, $H = 0.4$ m. All metaheuristic algorithms adopt discrete coding and the parameter change interval is set, which can reduce the search space of the feasible solution. Specifically, the GA and PSO adopt binary coding, whereas DE, SA, IA, ACO, and CMA-ES adopt integer coding.

The generation method of the array layout affects the final optimisation result of the WEC array. There are two common methods. One is that the target sea area is divided into two-dimensional grids, and the devices are fixed on the grid points randomly. The other is that the WECs are placed on randomly generated arbitrary coordinates, and the minimum safety distance must be set in advance to avoid collision or overlapping arrangement of WECs. The former method greatly reduces the search space, but the final optimisation results depend on the grid size. The latter method could find the best layout of WECs theoretically, but the search space of possible solutions is excessively large. In this research, the former method is used, and the advantages of the two methods are achieved by setting a smaller mesh size. Here, N WECs are limited to a rectangular area ($l \times w$). A safe distance of at least L_{safety} m should be maintained between the WECs. The Euclidean distance L_{dist} between each WEC and the other WECs in the space is judged. If $L_{safety} < L_{dist}$, the WEC position is regenerated until it is satisfied. Fig. 8 shows the spatial arrangement method of the two WECs in Section "Results". Here, l = 12.8 m, w = 12.8 m. WEC 1 is fixed at (0, 0) coordinates, and the position of WEC 2 is changed. The plane arrangement of the two WECs is symmetrical, so all layout schemes can be obtained only by changing the position of WEC 2 in area Ⅱ. In Fig. 8, the mesh size L_{size} is 0.1 m, and the safety distance L_{safety} is 1 m.

▶ *Memory mechanism*

The optimisation of WECs requires a large amount of time, making the optimisation expensive. Most of the time consumed is used to calculate the power generation of the device. A metaheuristic algorithm must calculate the fitness value (WEC power) of multiple generations to obtain the optimal solution. The new individuals in each iteration process are the same as the previous generation of individuals with a certain probability. In particular, for evolutionary algorithms that keep excellent individuals to the next generation, such as the GA and DE, the probability of individual repetition is higher. Repeatedly calculating the fitness values of these individuals is meaningless and wastes a significant amount of computing resources. Avoiding repeated calculations is important for reducing the time cost of WEC optimisation. To solve this problem, a memory mechanism is proposed that creates a memory base to store individual coding information and fitness information. As shown in Fig. 9, the fitness value of the initial individual or initial population is calculated, and the coding information of each individual and its fitness value is saved in the memory information base. In the subsequent iteration process, whenever an individual fitness calculation is needed, it is first determined whether there is individual information in the memory information base. If individual information exists, it means that the fitness value of the individual has been calculated, and the fitness value information is directly assigned to the individual. Otherwise, the MAA-system is started to calculate the fitness value, and the individual information and fitness value information are saved in the memory information base. The memory mechanism is applicable to any metaheuristic algorithm and has no effect on the optimisation results.

Table 1 shows a comparison between the fitness calculation times of each metaheuristic algorithm with the memory mechanism and the original algorithm in the single float draft and damping optimisation test. The population number of all algorithms is 20, and each algorithm is tested 10 times. The results indicate that the memory mechanism can reduce the fitness calculation times of all algorithms, especially the GA, CMA-ES and DE.

▶ *Termination and evaluation*

As the metaheuristic algorithm proceeds, some criteria are needed to stop the search at some point. The number of iterations depends on whether an acceptable solution is reached or a set

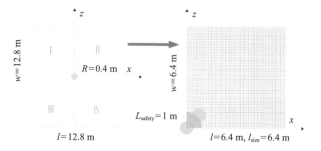

The target sea area is divided into two-dimensional grids, and the devices are fixed on the grid points randomly. Moreover, the minimum safety distance must be set in advance to avoid collision or overlapping arrangement of WECs.

Fig. 8　Two float array layout method settings

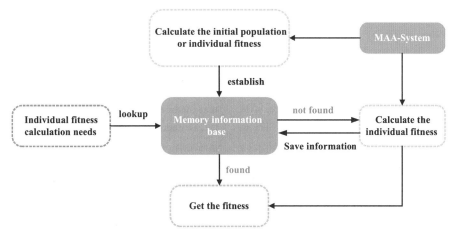

It can avoid calculating the fitness of repeated individuals by creating a memory base to store individual coding information and fitness information.

Fig. 9 Framework of memory mechanism

Table 1 Comparison of fitness calculation times between each metaheuristic algorithm with memory mechanism and the original algorithm

Method	Iterative times	Original fitness calculation times	Fitness calculation times with memory mechanism	Reduce the percentage of fitness calculation times
GA	10	200	57.9	71.05%
DE	9	180	145.2	19.33%
IA	9.7	194	175.6	9.48%
CMA-ES	11.5	230	150.8	34.43%
PSO	10.1	202	184.7	8.56%
ACO	10	200	194.1	2.95%
SA	8	160	151.2	5.50%

number of iterations is exceeded. The goal is to develop a metaheuristic algorithm that can obtain WEC optimisation results quickly and efficiently. Therefore, the following termination criteria are applied and checked at each iteration.

1. A maximum number of iterations (MaxIt) is reached;

2. The historical optimal solution ceases to improve after a certain number (I_{stop}) of iterations.

If one of these conditions is fulfilled, the algorithm stops, and the optimal solution of the last iteration is the final optimisation result. MaxIt and I_{stop} are normally determined by the number of potential solutions N_{ps}. In this study, MaxIt = 10 × lg (N_{ps}) and I_{stop} = lg (N_{ps}) + 2.

The final optimisation result and the consumption of time are two important indices for evaluating the performance of the metaheuristic algorithm. Previous metaheuristic algorithms implemented in WEC optimisation have no memory mechanism, so it is unfair to evaluate the performance of evolutionary algorithms with optimisation time, because they spend a lot of time on repeated computation. Furthermore, the power calculation time of WECs varies with the performance of computers. It is proposed that the times of fitness calculation of the metaheuristic algorithm be taken as the evaluation index of the algorithm time consumption. The information of the fitness calculation times is recorded by the memory information base. It represents the number of times different individuals are searched in the feasible solution space. Clearly, this index has the ability to evaluate accurately and objectively the optimisation time of each metaheuristic algorithm with a memory mechanism.

▶ *Computational resources*

The metaheuristic optimisation approaches studied were evaluated and compared for WECs for unidirectional irregular waves. Three high-performance desktop PCs with an Intel Xeon 2.40 GHz processor and 128 GB RAM completed all optimisation tests. To reduce the influence of randomness on the optimisation performance of the metaheuristic algorithm, each optimisation test run independently 10 times. Different tests can encounter repeated individuals. To avoid repeated calculations of individual fitness, a memory information base for all tests was built, as shown in Fig. 10. When there is no individual information in the current test memory information base, all test memory information bases are queried first. If there is, the individual fitness value is obtained, and the individual information is saved to the current test memory. If not, the MAA-system is started to calculate individual fitness, and the individual information is saved to the current test memory information base and the memory information base of all tests. Before each test, the current test memory base is cleared.

Results

The performances of seven metaheuristic algorithms in different search spaces were tested. All seven algorithms shared the same number of individuals (NP) and the same stopping conditions to make a fair comparison among them possible. The configurations of the optimisation tests are listed in Table 2, and the details of the different algorithms are summarised in Table 3. To reduce the influence of randomness and obtain more reliable results, each algorithm is run independently 10 times in the search space.

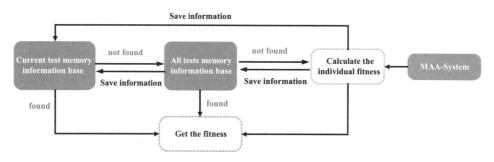

The memory mechanism is extended to cover all tests.

Fig. 10　Framework of all tests' memory mechanism

Table 2　The configurations of the optimisation tests

Search space	Variables		Value range	NP	MaxIT	I_{stop}	N_{sp}
1	WEC 1	D	0.06:0.02:0.36 m	20	30	5	1024
		C_{PTO}	0:20:1280 N/(m/s)				

to be continued

Search space	Variables		Value range	NP	MaxIT	I_{stop}	N_{sp}
2	WEC 1	D	0.06:0.02:0.36 m	40	60	8	1,048,576
		C_{PTO}	0:20:1280 N/(m/s)				
	WEC 2	D	0.06:0.02:0.36 m				
		C_{PTO}	0:20:1280 N/(m/s)				
3	WEC 1	D	0.20:0.02:0.36 m	50	80	10	268,435,456
		C_{PTO}	100:20:700 N/(m/s)				
	WEC 2	D	0.20:0.02:0.36 m				
		C_{PTO}	100:20:700 N/(m/s)				
		X	0:0.1:6.4 m				
		Y	0:0.1:6.4 m				

Table 3　The details of the different algorithms

Algorithms	Search space	Parameter setting
GA	1, 2, 3	$p_c = 0.6, p_m = 0.01$
DE	1, 2, 3	$F_0 = 0.4, F = e^{1-\frac{MaxIt}{MaxIt+1-I}}, Cr = 0.1$
IA	1, 2	$p_s = 50\%, p_r = 50\%, N_{cl} = 2$
	3	$p_s = 50\%, p_r = 50\%, N_{cl} = 3$
CMA-ES	1, 2, 3	The default settings
PSO	1, 2, 3	$c_1 = c_2 = 1.5, \omega_{max} = 0.8, \omega_{min} = 0.4,$ $\omega = \omega_{max} = MaxIt\frac{\omega_{max} - \omega_{min}}{T_{max}}$
ACO	1, 2, 3	$R_{ho} = 0.9, P_0 = 0.2$
SA	1, 2, 3	$T_0 = 100, K = 0.9$

▶ *Single wave energy converter draft and PTO damping optimisation*

The fitness values (generation power) of all solutions of a single WEC are obtained through an exhaustive algorithm, as shown in Fig. 11. The combined distribution of fitness values, draft, and PTO damping is unimodal for a single WEC. When D = 0.32 m and C_{PTO} = 360 N/(m/s), the average power generated by a single WEC reaches the maximum ($\bar{P}$ = 2.390 W). Table 4 shows the best solution of the different algorithms in the 10 independent tests. Each algorithm clearly can find the optimal solution in the feasible solution space, which also verifies that these metaheuristic algorithms have the ability to find the optimal solution in WEC optimisation. Fig. 12 shows box plot and point plot of the best fitness and the fitness calculation times of different metaheuristic algorithms, respectively. The upper and lower limits of the box represent the maximum and minimum values of the test data, respectively. The upper and lower edges of the box are the upper and lower quartiles of the data, respectively, which means that the box contains 50% of the data. The height of the box reflects the fluctuation of the data. The middle line in the box represents the median of the data, and the square point represents the average.

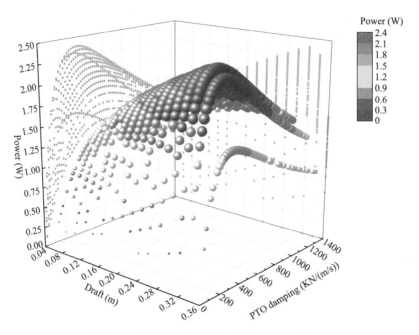

This is an obvious unimodal optimisation problem.

Fig. 11　The feasible solution space of single WEC draft and PTO damping optimisation test

Table 4　The best solution obtained by different algorithms in single WEC draft and PTO damping optimisation test

Method	Optimal solution in 10 tests		
	Draft (m)	PTO damping (N/(m/s))	Fitness (w)
GA	0.32	360	2.390
DE	0.32	360	2.390
IA	0.32	360	2.390
CMA-ES	0.32	360	2.390
PSO	0.32	360	2.390
ACO	0.32	360	2.390
SA	0.32	360	2.390

The data points obtained from all tests are drawn on the right side of the box. Fig. 12 shows that all metaheuristic algorithms can find or approach the optimal solution in most tests. These algorithms' convergence curves are presented in Fig. B.1. In particular, the IA, CMA-ES and PSO show better performance. In almost every test, the optimal solution could be found, and the optimisation result is extremely stable. However, the optimisation results of the SA algorithm fluctuate significantly in different tests. The number of fitness calculation times represents the search ability and convergence speed of the algorithm. The GA has fewest fitness calculations and the fastest convergence speed, but premature convergence and a large number of repeated individuals in the genetic process make the

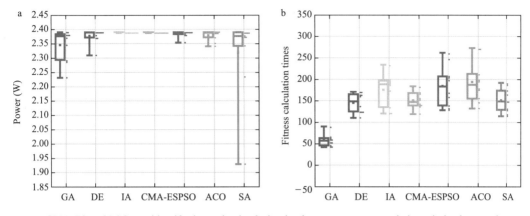

IA, CMA-ES and PSO can identify the optimal solution in almost every test, so their optimisation results are highly steady. However, the optimisation results of SA and GA fluctuate considerably in different tests.

Fig. 12 Optimal fitness of different algorithms in single WEC draft and PTO damping optimisation test (a) and fitness calculation times of different algorithms in single WEC draft and PTO damping optimisation test (b)

search ability of the GA algorithm poor. PSO and ACO possess strong search ability, which means that they require more computing time to reach convergence. Fig. 13 shows the average value of the optimal results and the fitness calculation times of each algorithm in the single WEC draft and PTO damping optimisation test. Owing to the small search space, each algorithm provides an excellent draft and PTO damping optimisation solution for a single WEC. In comparison, the performances of GA and SA are not as good as those of other five algorithms.

▶ *Double wave energy converter draft and PTO damping optimisation*

Two WECs are arranged along the wave direction, and the WEC spacing is half the average wavelength, as shown in Fig. 14. Table 5 shows the best solution of the different algorithms in the 10 independent tests. The distribution relations between the fitness values and each variable are shown in Fig. 15. All distributions have a clear peak in their shape, and the variable value corresponding to the peak value is consistent with the optimal configuration in Table 2, demonstrating that the double WECs draft and damping optimisation is also a typical unimodal optimisation problem. Because of the masking effect of WEC 1, the fitness value is less sensitive to the draft of WEC 2 than that of WEC 1, but the optimal draft of the two WECs is consistent with that of a single WEC. Furthermore, the optimal damping of the double WEC is slightly less than that of a single WEC. Fig. 16 shows the results of the best fitness and fitness calculation times in 10 tests of the double WEC draft and PTO damping optimisation with different metaheuristic algorithms. Fig. B.2 exhibits the convergence diagrams of the optimal fitness of seven compared algorithms. The accuracy and reliability of the IA and CMA-ES optimisation results are still better than those of the other algorithms. The maximum fitness in all tests is 5.104 w. It is believed that this is the best solution for the optimisation search space. CMA-ES can find this solution in nine tests, which fully demonstrates its high optimisation accuracy. DE can find the solution in six tests, but there is a "premature" phenomenon in other tests. Consequently, its global search ability is lower than that of CMA-ES. While the IA has the most fitness calculation times, it can always find a better solution. The standard deviation of the best solution obtained from the 10 tests is very small. This illustrates that the IA algorithm has a strong global search ability and does not easily fall into a local optimal value and be premature. PSO and ACO

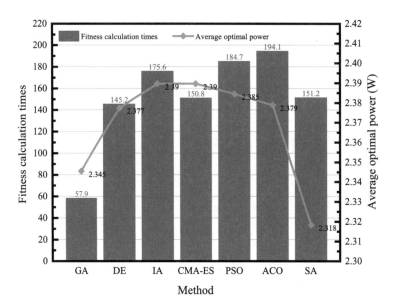

Owing to the small search space, each algorithm provides an excellent draft and PTO damping optimisation solution for a single WEC. In comparison, GA and SA perform slightly worse than other algorithms.

Fig. 13　Mean value of optimal fitness and mean value of fitness calculation times of different algorithms in single WEC draft and PTO damping optimisation test

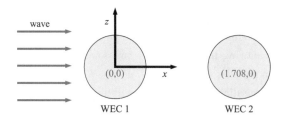

The spacing of double WECs is 0.5 times the average wavelength.

Fig. 14　WECs layout in double WEC draft and PTO damping optimisation test

Table 5　Best solutions in 10 tests of each algorithm in double WECs draft and PTO damping optimisation test

Method	WEC 1			WEC 2			Fitness (w)
	Draft (m)	PTO damping (N/(m/s))	Power (W)	Draft (m)	PTO damping (N/(m/s))	Power (W)	
GA	0.32	300	2.532	0.32	280	2.541	5.073
DE	0.32	340	2.562	0.32	320	2.542	5.104
IA	0.32	340	2.562	0.32	320	2.542	5.104
CMA-ES	0.32	340	2.562	0.32	320	2.542	5.104
PSO	0.32	340	2.552	0.32	360	2.546	5.098

to be continued

Method	WEC 1			WEC 2			Fitness (w)
	Draft (m)	PTO damping (N/(m/s))	Power (W)	Draft (m)	PTO damping (N/(m/s))	Power (W)	
ACO	0.32	340	2.562	0.32	320	2.542	5.104
SA	0.32	440	2.571	0.3	400	2.471	5.042

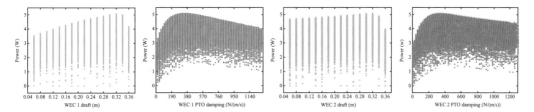

All distributions have a clear peak to their shape and the variable value corresponding to the peak value is consistent with the optimal configuration in Table 2, which demonstrates that the double float draft damping optimisation is also a typical unimodal optimisation problem.

Fig. 15 The distribution relation between fitness values and each variable

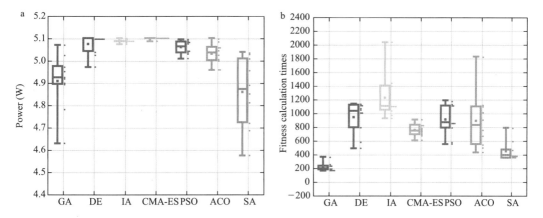

The accuracy and reliability of the IA and CMA-ES optimisation results are still better than other algorithms. DE has a premature in some tests. ACO performs worse than PSO. GA and SA are the two algorithms with the worst optimisation performance.

Fig. 16 Optimal fitness of different algorithms in double WEC draft and PTO damping optimisation test (a) and fitness calculation times of different algorithms in double WECs draft and PTO damping optimisation test (b)

are the two main swarm intelligence algorithms. In the draft and PTO damping optimisation tests of a single WEC and two WECs, the optimisation performance of PSO is higher than that of ACO. The GA and SA are the two algorithms with the worst optimisation performance, and both have an obvious precocity. The optimisation results of the different tests fluctuates greatly, and the reliability is poor. The reason for these phenomena is that the global search ability of the GA and SA is not strong enough, and the initial population or initial search point has an enormous impact on the final results. Increasing the number of populations is an effective method to avoid too fast a convergence of

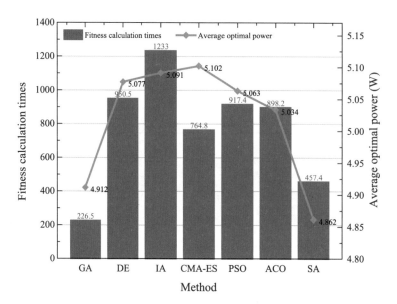

CMA-ES can realise excellent optimisation results only by calculating fitness values for a few times. While the IA requires the most fitness calculations, it can always find a better solution.

Fig. 17　Mean value of optimal fitness and mean value of fitness calculation times of different algorithms in double WEC draft and PTO damping optimisation test

the GA algorithm. The SA has a strong local search ability, but it is not suitable for solving optimisation problems with a large search space. Combining the SA with an algorithm with strong global search ability may achieve better optimisation. The optimisation performances of different metaheuristic algorithms are shown more intuitively in Fig. 17. CMA-ES, IA, and DE algorithms perform better in the double WEC draft and PTO damping optimisation tests. In particular, CMA-ES requires only a few fitness calculations to obtain a highly reliable optimisation solution.

▶ *Double wave energy converter draft, PTO damping, and layout optimisation*

The previous optimisation tests indicated that CMA-ES, IA, and DE are more suitable for the optimisation of WECs. To test the optimisation performances of these three algorithms in a larger search space, the draft and PTO damping of double WECs and the plane coordinates of WEC 2 are considered as optimisation variables. Table 6 shows the best solutions achieved by the three algorithms in the 10 tests. When WECs are positioned to benefit from the radiated and diffracted waves, the device can generate more power than it would in isolation[51]. This is proved by the optimisation results. The best layout of WEC 2 is mainly concentrated in two areas, as shown in Fig. 18. In region Ⅰ, the wave diffraction of two WECs arranged close to each other causes a near trapping[52]. The wave resonates with the structure, and the sharp increase in the local wave amplitude causes the double-WECs array to obtain more wave energy[53]. In region Ⅱ, the array energy acquisition mainly benefits from the positive hydrodynamic interaction generated by the two WEC radiation waves. In all tests, the generation power of the best optimized scheme increases by 21.2% compared with the power accumulation of two single WECs with optimal draft and PTO damping operating independently.

Table 6 Best solutions in 10 tests of each algorithm in double WECs draft, PTO damping and position optimisation test

Method	WEC 1			WEC 2					Fitness (w)
	Draft (m)	PTO damping (N/(m/s))	Power (W)	Draft (m)	PTO damping (N/(m/s))	X (m)	Y (m)	Power (W)	
DE	0.32	320	3.318	0.32	340	1.1	0	2.442	5.760
IA	0.32	260	3.285	0.32	380	1.1	0	2.507	5.792
CMA-ES	0.32	360	2.866	0.32	360	0	2.4	2.866	5.732

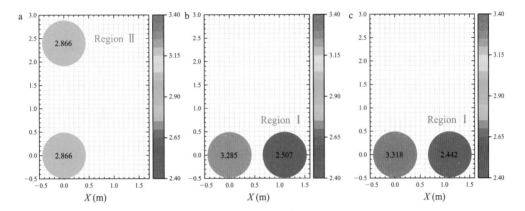

a. $\bar{P}$ = 5.760w, by DE; b. $\bar{P}$ = 5.792w, by IA; c. $\bar{P}$ = 5.732w, by CMA-ES. The best layout of WEC 2 is predominantly concentrated in two areas as shown in figure. This is because WEC devices could benefit from the radiated and diffracted waves.

Fig. 18 The optimal layout of double-WECs found by the three algorithms

Fig. 19 shows the distribution relation between fitness values and different variables. The distribution shape of draft and damping variables is still approximately unimodal, indicating that WEC draft and damping optimisation is generally unimodal optimisation. The change of array layout does not change the optimal draft, but the optimal damping clearly deviates from the optimal damping of a separated WEC, and the distribution shape is obviously different. The influence of layout on optimal fitness is significant. Two main peaks are clearly visible in the distribution between fitness and layout. With the increase in the number of WECs, the feasible solution space has a more local extremum. Consequently, finding the best WEC array layout is a complex multimodal optimisation.

Fig. 20 and Fig. B.3 show that DE has premature convergence in this large search space and cannot evolve to global optimisation. The optimisation performance of CMA-ES seems to be almost the same as that of DE. It can quickly converge to the local optimum, but the lack of global search ability makes it difficult to find the global optimal solution. Compared with these two algorithms, the IA has strong global search ability and is successful in converging near the global optimal solution in most tests. However, its convergence speed is slow, and more fitness values usually need to be calculated. The average optimisation time of the IA in this optimisation test is approximately twice that of DE and CMA-ES. Balancing the convergence speed and optimisation performance of the algorithm is

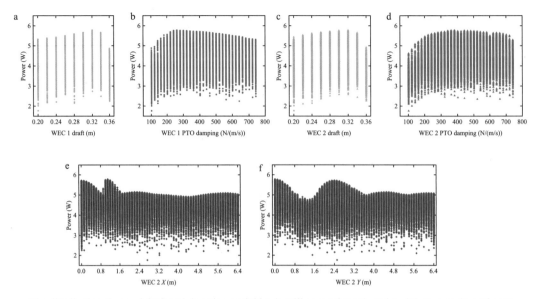

The distribution shape of draft and damping variables is still approximately unimodal. Two main peaks are clearly visible in the distribution between fitness and layout. Consequently, finding the best layout of WEC array is a complex multimodal optimisation.

Fig. 19　The distribution relation between fitness values and different variables

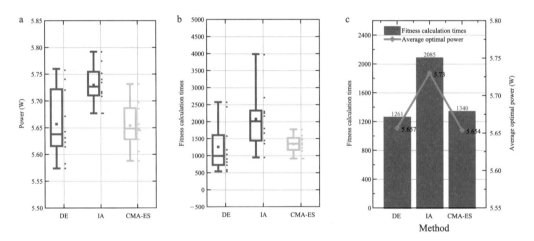

a, b, c respectively shows the optimal fitness, the fitness calculation times, the mean value of optimal fitness and fitness calculation times of three algorithms in double WEC draft, PTO damping and position optimisation test. IA has more powerful global search capabilities and is more appropriate in the application of solving the multimodal optimisation problem of WEC array layout.

Fig. 20　Three algorithms in double WEC draft, PTO damping and position optimisation test

a difficult problem. The final optimisation results reveal that the IA is more suitable for solving the multimodal optimisation problem of WEC array layout.

Conclusions

The optimisation of the WEC parameter configuration is a computationally costly, multimodal, large-scale, and intricate problem.

The key of solving this problem is to realise a fast and accurate simulation of the WEC and adopt an appropriate optimisation algorithm. The MAA-system proposed connects the modelling software APDL and the commercial BEM solver AQWA through MATLAB to realise the simulation process of WEC modelling, frequency-domain calculation, time-domain calculation, and data output. Taking the optimisation of the cylindrical float point absorbing WEC as an example, the optimisation performances of seven widely used metaheuristic algorithms (GA, DE, IA, CMA-ES, PSO, ACO, and SA) in different search spaces were compared and analysed. A memory mechanism was proposed to avoid the repeated calculation of fitness. This method effectively reduces the optimisation cost of metaheuristic algorithms, especially in evolutionary algorithms. Compared with the original algorithm, the optimisation time of the GA, CMA-ES and DE with the memory mechanism decreased by 71.05%, 34.33%, and 19.33%, respectively. In addition, using the fitness calculation times as the evaluation index of the algorithm time consumption is more accurate and objective. The test results show that, when the search space is small, the seven algorithms can find the optimal solution or come close to the optimal one in most tests, but the performances of the GA and SA are lower than those of other algorithms. With the increase in search space, the poor global search ability of the SA and GA leads to premature convergence of the algorithm. The optimisation results of CMA-ES, IA, and DE still have high reliability and stability. In particular the CMA-ES can converge quickly to the global optimal solution. The optimisation performances of PSO and ACO are better than those of the GA and SA, but they are obviously inferior to the above CMA-ES, IA, and DE. Then, the optimisation performances of CMA-ES, IA, and DE were tested further in a huge search space possessing 268,435,456 potential solutions. WEC array optimisation is often multimodal optimisation with multiple local extrema. The test results indicate that, although the IA consumes more optimisation time, the optimisation results do not easily fall into local optimisation. In contrast, DE and CMA-ES converge prematurely. In conclusion, the optimisation of the WEC draft and PTO damping is usually a unimodal optimisation problem; therefore, CMA-ES is the best choice to solve this type of optimisation problem. In addition, it is recommended that IA be utilised to solve multimodal optimisation problems, such as WEC array layout optimisation.

Future Work

This research has gone some way towards optimising WEC configuration using metaheuristic optimisation methods. It is recommended that further research be undertaken in the following areas.

1. The draft and damping optimisation of WECs is usually unimodal and the layout optimisation is multimodal. In addition, the array generation power is more susceptible to the layout. Therefore, using different optimisation methods to optimise the above variables in batches is an important direction for improving optimisation efficiency.

2. The metaheuristic algorithms selected in this research are sufficiently mature to be widely available. It would be intriguing to utilise more emerging combinatorial or single metaheuristic algorithms to optimise WECs and confirm their effectiveness.

3. The variable dimensions and computation cost increase exponentially with the number of WECs, so it is necessary to enhance the WEC array simulation method. Using graphics processing unit acceleration technology to improve the current hydrodynamic coupling calculation program based on central processing unit calculations should attract more attention from researchers.

4. Further research should consider more optimisation variables and pay particular attention to the influence of the wave frequency and wave direction on the optimal WEC configuration.

Acknowledgements

The authors would like to acknowledge the support of the National Key R&D Program of China (Grant No.2018YFB1501900), the National Natural Science Foundation of China (No. 52071303), the Shandong Provincial Key Research and Development Program (Grant No.2019JZZY010902), the Shandong Provincial Natural Science Foundation (Grant No. ZR2021ZD23), the National Natural Science Foundation of China (No. 41706100), the Joint Project of NSFC-SD (Grant No. U1906228), and the Taishan Scholars Program of Shandong Province (No. ts20190914).

Author Contributions

Feifei Cao: Conceptualization, methodology, funding acquisition, writing (review and editing). Meng Han: Software, investigation, data curation, writing (original draft). Hongda Shi: Conceptualization, formal analysis, resources, writing (review and editing), project administration. Ming Li: Writing (review and editing), supervision. Zhen Liu: Writing (review and editing), supervision.

Competing Interests

The authors declare that they have no known competing financial interests or personal relationships that could have appeared to influence the work reported in this paper.

References

1. Mørk G, Barstow S, Kabuth A, et al. Assessing the global wave energy potential[R]. Proceedings of the ASME 2010 29th International Conference on Ocean, Offshore and Arctic Engineering, 2010.
2. Ocean Energy Europe. Ocean Energy Key Trends and Statistics 2020[R]. Brussles: Ocean Energy Europe, 2021.
3. Ocean Energy Europe. 2030 Ocean Energy Vision: Industry analysis of future deployments, costs and supply chains. Supported by European Technology and Innovation Platform for Ocean Energy, Brussels.
4. Wu J, Shekh S, Sergiienko N Y, et al. Fast and effective optimisation of arrays of submerged wave energy converters[R]. Proceedings of the Genetic and Evolutionary Computation Conference, 2016.
5. McCabe A P, Aggidis G A, Widden M B. Optimizing the shape of a surge-and-pitch wave energy collector using a genetic algorithm[J]. Renew. Energy, 2010, 35: 2767-2775.
6. McCabe A P. Constrained optimisation of the shape of a wave energy collector by genetic algorithm[J]. Renew. Energy, 2013, 51: 274-284.
7. Sirigu S A, Foglietta L, Giorgi G, et al. Techno-Economic optimisation for a wave energy converter via genetic algorithm[J]. J. Mar. Sci. Eng., 2020, 8(7): 482.
8. Garcia-Teruel A, DuPont B, Forehand D I M. Hull geometry optimisation of wave energy converters: On the choice of the optimisation algorithm and the geometry definition[J]. Appl. Energy, 2020, 280: 115952.
9. Silva S, Gomes R P F, Falcão A F O. Hydrodynamic optimisation of the UGEN: Wave energy converter with U-shaped interior oscillating water column[J]. International Journal of Marine Energy, 2016, 15: 112-126.
10. Esmaeilzadeh S, Alam M R. Shape optimisation of wave energy converters for broadband directional incident waves[J]. Ocean Eng., 2019, 174: 186–200.
11. Capillo A, Luzi M, Pasc M, et al. Energy Transduction Optimisation of a Wave Energy Converter by Evolutionary Algorithms[R]. Proc. Int. Jt. Conf. Neural Networks 2018-July.
12. M'zoughi F, Bouallègue S, Garrido A J, et al. Water cycle algorithm–based airflow control for oscillating water column–based wave energy converters[J]. Proc. Inst. Mech. Eng. Part I J. Syst. Control Eng., 2020, 234: 118-133.
13. Jusoh M A, Ibrahim M Z, Daud M Z, et al. Parameters estimation of hydraulic power take-off system for wave energy conversion system using genetic algorithm[J]. IOP Conf. Ser. Earth Environ. Sci, 2020, 463: 12129.
14. Calvário M, Gaspar J F, Kamarlouei M, et al. Oil-hydraulic power take-off concept for an oscillating wave surge converter[J]. Renew. Energy, 2020, 159: 1297-1309.
15. Gao H, Xiao J. Effects of power take-off parameters and harvester shape on wave energy extraction and output of a hydraulic conversion system[J]. Appl. Energy, 2021, 299: 117278.
16. Jusoh M A, Ibrahim M Z, Daud M Z, et al. An estimation of hydraulic power take-off unit parameters for wave energy converter device using non-evolutionary nlpql and evolutionary ga approaches[J]. Energies, 2021, 14(1): 79.

17. Amini E, Golbaz D, Asadi R, et al. A comparative study of metaheuristic algorithms for wave energy converter power take-off optimisation: A case study for eastern australia[J]. J. Mar. Sci. Eng., 2021, 9: 490.
18. Neshat M, Sergiienko N Y, Mirjalili S, et al. Multi-mode wave energy converter design optimisation using an improved moth flame optimisation algorithm[J]. Energies, 2021, 14(13): 3737.
19. Sun P, Hu S, He H, et al. Structural optimisation on the oscillating-array-buoys for energy-capturing enhancement of a novel floating wave energy converter system[J]. Energy Convers. Manag., 2021, 228: 113693.
20. Zheng S, Zhang Y, Iglesias G. Power capture performance of hybrid wave farms combining different wave energy conversion technologies: The H-factor[J]. Energy, 2020, 204: 117920.
21. Budal K. Theory for absorption of wave power by a system of interacting bodies[J]. J. Sh. Res., 1977, 21: 248-253.
22. Child B F M, Venugopal V. Optimal configurations of wave energy device arrays[J]. Ocean Eng., 2010, 37: 1402-1417.
23. Sharp C, DuPont B. A multi-objective real-coded genetic algorithm method for wave energy converter array optimisation[R]. Proceedings of the ASME 2016 35th International Conference on Ocean, Offshore and Arctic Engineering.
24. Sharp C, DuPont B. Wave energy converter array optimisation: A genetic algorithm approach and minimum separation distance study[J]. Ocean Eng., 2018, 163: 148-156.
25. Sarkar D, Contal E, Vayatis N, et al. Prediction and optimisation of wave energy converter arrays using a machine learning approach[J]. Renew. Energy, 2016, 97: 504-517.
26. Tay Z Y, Venugopal V. Optimisation of spacing for oscillating wave surge converter arrays using genetic algorithm[J]. J. Waterw. Port, Coastal, Ocean Eng., 2017, 143: 04016019.
27. Giassi M, Göteman M. Parameter optimisation in wave energy design by agenetic algorithm[R]//Proc. Of the 32nd International Workshop on Water Waves and Floating Bodies (IWWWFB), 23–26th April,. Dalian, China, April 23–26, 2017.
28. Giassi M, Göteman M. Layout design of wave energy parks by a genetic algorithm[J]. Ocean Eng., 2018, 154: 252-261.
29. Ruiz P, Nava V, Topper M, et al. Layout optimisation of wave energy converter arrays[J]. Energies, 2017, 10(9): 1-17.
30. Fang H W, Feng Y Z, Li G P. Optimisation of wave energy converter arrays by an improved differential evolution algorithm[J]. Energies, 2018, 11(12): 1-19.
31. Faraggiana E, Masters I, Chapman J. Design of an optimisation scheme for the wavesub array[R].// Proceedings of 3rd International Conference on. Renewable Energies Offshore, Lisbon, Portugal, 2018.
32. Penalba M, Kelly T, Ringwood J. Using NEMOH for modelling wave energy converters: A comparative study with WAMIT[R]. 12th European Wave and Tidal Energy Conference, Cork, 2017.
33. Lyu J, Abdelkhalik O, Gauchia L. Optimisation of dimensions and layout of an array of wave energy converters[J]. Ocean Eng., 2019, 192: 106543.
34. Neshat M, Alexander B, Wagner M, et al. A detailed comparison of meta-heuristic methods for optimising wave energy converter placements[R]. Proceedings of the Genetic and Evolutionary Computation Conference, 2018: 1318-1325.
35. Neshat M, Abbasnejad E, Shi Q, et al. Adaptive neuro-surrogate-based optimisation method for wave energy converters placement optimisation[R]. International Conference on Neural Information Processing, 2019: 353-366.
36. Neshat M, Alexander B, Sergiienko N Y, et al. A hybrid evolutionary algorithm framework for optimising power take off and placements of wave energy converters[R]. Proceedings of the Genetic Evolutionary Computation Conference, 2019: 1293-1301.
37. Neshat M, Alexander B, Sergiienko N Y, et al. New insights into position optimisation of wave energy converters using hybrid local search[J]. Swarm Evol. Comput., 2020, 59: 100744.
38. Holland J. Adaptation in natural and artificial systems[J]. SIAM Rev., 1975, 18(3): 529-530.
39. Storn R, Price K. Differential evolution—a simple and efficient heuristic for global optimisation over continuous spaces[J]. Journal of Global Optimisation, 1997. 11: 341-359.
40. Tasoulis D K, Pavlidis N, Plagianakos V, et al. Parallel differential evolution[R]. Proceedings of the 2004 Congress on Evolutionary Computation (IEEE Cat. No.04TH8753), 2023-2029 Vol.2.
41. Castro L, Zuben F V. Learning and optimisation using the clonal selection principle[J]. IEEE Trans. Evol. Comput., 2002, 6(3): 239-251.
42. Kennedy J, Eberhart R. Particle swarm optimisation[R]. Proceedings of ICNN'95 -International Conference on Neural Networks, 1995, 4: 1942-1948.
43. Poli R, Kennedy J, Blackwell T. Particle swarm optimisation[J]. Swarm Intelligence, 2007, 1: 33-57.

44. Kennedy J, Eberhart R C. A discrete binary version of the particle swarm algorithm[R]. 1997 IEEE International Conference on Systems, Man, and Cybernetics. Computational Cybernetics and Simulation, 5: 4104-4108.
45. Wang Z. A modified particle swarm optimisation[J]. Journal of Harbin University of Commerce, 2009.
46. Dorigo M, Maniezzo V, Colorni A. Ant system: Optimisation by a colony of cooperating agents[J]. IEEE Transactions on Systems, Man, and Cybernetics, Part B, 1996, 26(1): 29-41.
47. Mishra R, Jaiswal A. Ant colony Optimisation: A Solution of Load balancing in Cloud[J]. International Journal of Web & Semantic Technology, 2012, 3: 33-50.
48. Kirkpatrick S, Gelatt D, Vecchi M. Optimisation by Simulated Annealing[J]. Science, 1983, 220: 671-680.
49. Hansen N J A. The CMA Evolution Strategy: A Tutorial[J/OL]. 2016. https://arxiv.org/abs/1604.00772v1.
50. Mostapha Kalami Heris. CMA-ES in MATLAB[EB/OL]. 2015. https://www.mathworks.com/matlabcentral/fileexchange/52898-cma-es-in-matlab.
51. McNatt J C, Venugopal V, Forehand D. A novel method for deriving the diffraction transfer matrix and its application to multi-body interactions in water waves[J]. Ocean Eng., 2015, 94: 173-185.
52. Evans D V, Porter R. Near-trapping of waves by circular arrays of vertical cylinders[J]. Appl. Ocean Res., 1997, 19: 83-99.
53. Ning D, He Z, Gou Y, et al. Near trapping effect on wave-power extraction by linear periodic arrays[J]. Sustainability, 2019, 12(1): 29.
54. Faraggiana E, Chapman J C, Williams A J. Genetic based optimisation of the design parameters for an array-on-device orbital motion wave energy converter[J]. Ocean Eng., 2020, 218: 108251.

Appendix A. Related Works

Table A.1 A brief survey of the recent investigation on WECs optimisation with metaheuristic algorithms

Variables	WECs mode	Algorithms	Simulation methods	Reference
Shape	Surge-and-Pitch WEC	GA	Frequency-domain	[33]
Shape	Surge WEC	GA	Frequency-domain	[6]
Shape	Surge, Surge-Heave-and-Pitch WEC	GA, PSO	Frequency-domain	[8]
Shape	Roll-Sway UGEN	GA, COBYLA	Frequency-domain	[9]
Shape, PTOs	Pressure-Difference WEC	GA	Frequency-domain	[10]
Shape	Pitch-ISWEC	GA, PSO, HG-PSO	N/A	[11]
PTOs	OWSC	GA	Frequency-domain	[14]
PTOs	Swing-arm WEC	GA	N/A	[13]
PTOs	OWC	WCA	Frequency-domain	[12]
PTOs	Swing-arm WEC	NLPQL\GA	Frequency-domain	[16]
PTOs	Heave-Cylinders	GA	Time-domain	[15]
PTOs	Mooring-Submerged Balls	CMA-ES, GWO, HHO, MVO, GOA	Time-domain	[17]
PTOs, Shape, Mooring	Mooring-submerged Cylinders	CMA-ES, PSO, GWO, WOA, MFO, IMFO	Frequency-domain	[18]

to be continued

Variables	WECs mode	Algorithms	Simulation methods	Reference
Layouts	Heave Cylinders	PI, GA	Frequency-domain	[22]
Layouts	Heave-Cylinders	GA	Frequency-domain	[23,24]
Layouts	OWSCs	GA	Frequency-domain and Machine learning prediction	[25]
Layouts	OWSCs	GA	Frequency-domain	[26]
Layouts	Heave Cylinders	GA	Time-domain	[27,28]
Layouts	Mooring submerged balls	(1+1)-EA, CMA-ES	Frequency-domain	[4]
Layouts	Surge-Barges	CMA-ES, GA, GSO	Frequency-domain	[29]
Layouts	Heave-Cylinders	DE	Frequency-domain	[30]
Layouts, PTOs	Mooring WaveSub	GA, PSO	Time-domain	[31,54]

Appendix B. Convergence Curves

To intuitively compare the effectiveness and convergence rate of different algorithms, this paper shows the convergence curves of each algorithm in the 10 independent tests with different search spaces.

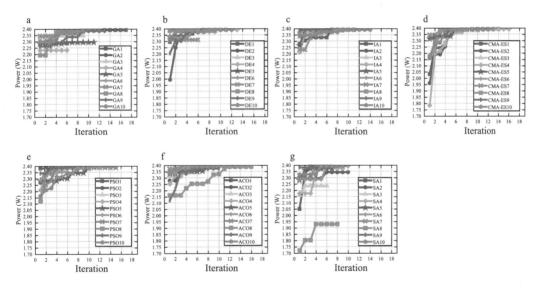

a. GA; b. DE; c. IA; d. CMA-ES; e. PSO; f. ACO; g. SA. CMA, PSO and IA could converge near the optimal fitness value in almost every test, and the optimisation performance of GA and SA is not satisfactory.

Fig. B.1　Evolution and convergence rate of the optimal fitness of 7 algorithms in single WEC draft and PTO damping optimisation test

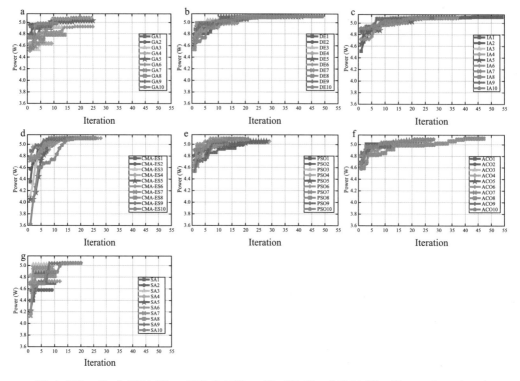

a. GA; b. DE; c. IA; d. CMA-ES; e. PSO; f. ACO; g. SA. DE, IA and CMA-ES still outperform the other methods. In particular, CMA-ES succeeds in attaining higher fitness value as well as faster convergence speed.

Fig. B.2 Evolution and convergence rate of the optimal fitness of seven algorithms in double WEC draft and PTO damping optimisation test

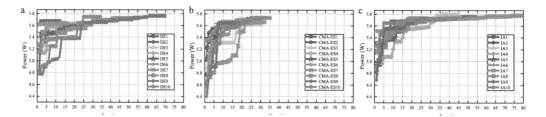

a. DE; b. CMA-ES; c. IA. CMA-ES and DE are easily trapped into local optimal solutions. Although the IA converges slowly, it has stronger global search ability and is suitable for optimizing WEC multimodal optimisation problems.

Fig. B.3 Evolution and convergence rate of the optimal fitness of 3 algorithms in double WEC draft, PTO damping and layout optimisation test

第三篇
海洋碳增汇应用典型示范与推广

Part III
Typical Demonstrations and Promotion of Implemented Ocean Carbon Sequestration

Optimization of aquaculture sustainability through ecological intensification in China①

Shuang-lin Dong[1]*, Yun-wei Dong[1], Ling Cao[2], Johan Verreth[3], Yngvar Olsen[4], Wen-jing Liu[5], Qi-zhi Fang[5], Yan-gen Zhou[1], Li Li[1], Jing-yu Li[1], Yong-tong Mu[1], Patrick Sorgeloos[6]

1 Key Laboratory of Mariculture (Ocean University of China), Ministry of Education, Ocean University of China, Qingdao 266003, China
2 School of Oceanography, Shanghai Jiao Tong University, Shanghai, China
3 Aquaculture & Fisheries Group, Wageningen University and Research, Wageningen, the Netherlands
4 Department of Biology, Norwegian University of Science & Technology, Trondheim, Norway
5 School of Mathematical Science, Ocean University of China, Qingdao 266100, China
6 Laboratory of Aquaculture & Artemia Reference Center, Ghent University, Gent, Belgium
* Corresponding author: Shuang-lin Dong (dongsl@ouc.edu.cn); Patrick Sorgeloos (patrick.sorgeloos@UGent.be)
Shuang-lin Dong and Yun-wei Dong contributed equally to this work.

Abstract

As the world's leading producer of farmed aquatic food products, China faces great uncertainty concerning the further sustainable development of its aquaculture industry due to high and increasing pressure from environmental and resource constraints. To realize truly sustainable development, it is imperative to establish an evaluation system that integrates social, economic, environmental, and resource criteria, and optimizes aquaculture systems through a holistic approach. Here, we used analytical hierarchy processes combined with expert judgment and objective data to assess the sustainability of 10 major aquaculture production systems (APSs) in China. Based on the evaluation results, we propose the ecological intensification of aquaculture systems (ELIAS), rationally integrating anthropogenic inputs with aquaculture ecosystem services, to improve the sustainability of APSs. The holistic evaluation system and ELIAS are fundamental and critical for the future development of aquaculture in China and globally.

Keywords: production systems; sustainability; ecological intensification; optimization; China

Introduction

Aquaculture is the fastest-growing food production industry in terms of annual production growth and has become a significant contributor of essential macro- and micronutrients to the diets of the global population[1-3]. Increasing the consumption of farmed aquatic food products over land-raised animal meat could potentially reduce the amount of land required for growing feed crops for a global population expected to reach nine billion people by 2050[4]. However, there is a great deal of uncertainty in the further development of aquaculture due to the high and increasing pressure of environmental challenges and resource constraints[2,5].

China is a dominant aquatic food producer that contributed 58% of the world's farmed aquatic food products in 2018[2]. World aquaculture production grew by an average rate of 5.3% per year from 2001 to 2018, compared to only 4% in 2017 and

① 本文于2022年1月发表在Reviews in Aquaculture第14卷, https://doi.org/10.1111/raq.12648。

3.2% in 2018 due to a dramatic slowdown in the growth of China's aquaculture industry. The Food and Agriculture Organization (FAO) estimated that China's aquaculture production would continue to grow by 36.5% and 24.7% in 2030 compared with 2016 under the no-plan (without the implementation of a stringent environmental protection policy) and full-plan (full implementation of a stringent environmental protection policy) scenarios, respectively; accordingly, world aquaculture production will increase by 38.3% and 34.8%, respectively[6].

Aquaculture takes place in a wide range of production systems in which aquatic plants, herbivores (e.g., grass carp), omnivores (e.g., common carp), carnivores (e.g., trout), filter-feeders (e.g., silver carp), and deposit-feeders (e.g., sea cucumber) are raised in fresh, brackish, and marine waters. In the past decade, the sustainability of some aquaculture production systems (APSs) has been studied from many perspectives, such as environmental impacts[7], resource constraints[8,9], disease prevention[3,10], improvement of genetic stocks[11], feeds[3], markets[12], finance[13], and social responsibility[14]. However, comparative assessments of the sustainability of different APSs involving economic, social, environmental, and resource criteria are extremely limited due to the lack of adequate objective measures of all these criteria for major APSs[15].

Like other food production systems, aquaculture has undergone a process of intensification in recent decades, achieving high and predictable yields in the short term but potentially facing numerous challenges in the long term[5,16,17]. Reducing the use of energy, water, land, feed, and fertilizer inputs is key to providing healthy aquatic food products for a growing population in an environmentally responsible manner. Hereby, ecological intensification is proposed to achieve the sustainable goals of food production systems by integrating anthropogenic inputs with ecosystem services.

China has the highest diversity of aquaculture species and production systems[3,18]; however, aquaculture development is unbalanced among different regions in China. Some intensive production systems originated in developed countries have been implemented in developed areas, while many extensive systems exist in developing regions[19]. Like the debates on the role of aquatic food products in the Global South and Global North[20], the farms in developing regions and small-scale farms in China tend to emphasize low value species and food security, whereas the farms in developed regions and larger farms tend to give more weight to luxury species and food safety[19].

This paper provides a comparative assessment of the sustainability of 10 major APSs in mainland China through two analytical hierarchy processes (AHPs)[21], covering 11 criteria of society, economy, environment, and resources (see "Materials and Methods" for details). Based on the evaluation results, a holistic approach to ecological intensification of aquaculture systems (ELIAS) is proposed, and feasible recommendations are suggested to improve the sustainability of aquaculture in the future.

Results and Discussion

▶ *Footprint of highly diversified aquaculture systems in China*

In the last 30 years, China's aquaculture production (excluding seaweed) has increased by 9-fold from 5.45×10^6 t in 1989 to 47.56×10^6 t in 2018 (Fig. S1A). A total of 84.9% of its edible production (see "Materials and Methods" for further details) was derived from inland APSs, and 63.4% was derived from fed (feed-based) APSs in 2018. The share of fed aquaculture in total farmed aquatic animal production rose by about 11% from 2000 to 2018[2]. The diversity of China's APSs is among the highest in the world[3]. Generally, the APSs in China can be divided into 10 major types based on feeding strategy, location, and environment (Fig. S1B). The major farmed species are low trophic level carps in inland APSs and filter-feeding mollusks in marine APSs[3] (Fig. S1C).

The ecological footprints of APSs in China are

very different (Table 1). In general, non-fed APSs, such as non-fed nearshore aquaculture and non-fed aquaculture in ponds, have smaller ecological footprints in terms of fossil energy consumption, aquaculture pollution, and fishmeal consumption. In contrast, some intensively fed APSs, such as recirculating aquaculture systems (RAS) and fed nearshore aquaculture, have larger ecological footprints. Fed aquaculture in ponds is the most important APS in terms of volume, value, and edible production and is also the largest consumer or user of fishmeal, freshwater, and arable land. In terms of value non-fed APSs account for 36.4% of the total aquaculture production in China. In terms of volume and value production non-fed nearshore aquaculture is the second-largest APS in China, and is regarded as a carbon sequestration aquaculture system[22,23]. Fed aquaculture nearshore is one of the leading opportunities for Chinese fishermen whose employment has undergone a transition from capture fisheries to aquaculture. However, crowded net cages nearshore have caused eutrophication, diseases, and other problems[5,24]. An alternative strategy for managing land and water scarcity and waste accumulation is to move aquaculture activities offshore[5,25].

Table 1 Main characteristics and ecological footprints of the 10 major aquaculture systems in China

Systems	nFEN	nFAL	nFAP	PFA	FOA	RAS	SALA	FAL	FAP	FEN
Production ($\times 10^6$ t, live weight)	10.37	3.25	5.02	2.33	0.15	0.47	1.5	0.68	19.60	0.59
Production value (mRMB)	242.87	31.95	51.76	43.57	11.88	35.20	20.61	8.64	402.98	46.73
Food security ($\times 10^6$ t, edible production, FSe)	1.73	2.83	4.52	2.03	0.13	0.41	1.23	0.59	17.04	0.51
Economic growth (%, EG)[a]	5.8	2.8	4.3	7.1	15.7	8	3.8	−7.0	4.4	5.7
Pollution (COD[b] g/kg, PO)	0	0	0	2.18	85.9 (59.9–154)	60.4 (2.28–227)	37.1 (6.35–126)	38.8 (2.54–196)	34.1 (2.18–276)	97.3 (59.9–154)
Fishmeal consumption (portion, FC)	0	0	0	0.01	0.03	0.11	0.05	0.01	0.68	0.12
Freshwater footprint (m^3/kg, FW)	0	0	2.83 (1.66–5.04)	0.01	1.6 (1.1–2.2)	1.50 (0.71–8.73)	5.22 (3.89–11.5)	2.49 (1.40–11.8)	4.91 (0.95–47.15)	1.6 (1.10–2.20)
Land use (m^2/kg, LU)	0	0	0.178	0	0	0.116	0	0	0.132	0
Energy consumption (kW·h·kg^{-1}, EC)	0.017	0.017	0.37	0.017	3.16	8.66	0.37	3.16	0.37	3.16

Note: nFEN: non-fed nearshore aquaculture; nFAL: non-fed aquaculture in large inland waters; nFAP: non-fed aquaculture in ponds; PFA: paddy field aquaculture; FOA: fed offshore aquaculture; RAS: recirculating aquaculture systems; SALA: waterlogged salt-alkali land aquaculture; FAL: fed aquaculture in large inland waters; FAP: fed aquaculture in ponds; FEN: fed nearshore aquaculture.
a. Average annual growth rates of production from 2009 to 2018. b. COD: chemical oxygen demand.

Driven by economic interests and encouraged or constrained by government policies, paddy field aquaculture has grown fast in the past decade, while fed aquaculture in large inland waters, such as reservoirs and lakes, is being outlawed in China. Paddy field aquaculture is a conventional household integrated aquaculture system in many countries, which allows fish or other farmed aquatic animals, such as crayfish, crabs and turtles, and rice shoots to share water and land space; it can increase both aquaculture and agriculture yields and reduce environmental burdens through mutually beneficial effects between fish and rice shoots[16,22,26]. However, disproportionate fish farming and rice planting within a paddy field may lead to waste accumulation and discharge.

China consists of 35 million hectares of salt-alkali lands, and more than 3 million hectares of them can be reclaimed to pond-terrace systems (Fig. S2). Waterlogged salt-alkali land aquaculture has not only exploited the vast unused land resources but also formed a mutually beneficial development pattern of ponds and terraces[27]. However, attention should be paid to the re-salinization of the reclaimed land with the development of the APS due to freshwater consumption.

▶ *Sustainability assessment of aquaculture production systems*

The sustainability of APSs is described as long-term development integrating four domains, namely society, economy, environment, and resources, and involving 11 criteria, namely aquaculture pollution, economic growth, food safety, freshwater consumption, arable land use, employment, food security, policy, fossil energy consumption, fishmeal consumption, and ecological risks[28] (also see "Materials and Methods" for criterion selection). The overall sustainability of aquaculture depends on the sustainability of each APS. Assessing the sustainability of each APS is a multi-criteria decision-making problem involving various conflicting criteria. Therefore, an analytical hierarchy process (AHP) was used to assess the sustainability of the 10 APSs (Fig. S3). According to expert judgment, the top five criteria with respect to aquaculture sustainability among the 11 criteria were aquaculture pollution, economic growth, food safety, freshwater consumption, and arable land use (Fig. 1), implying the importance of environmental degradation and the scarcity of fresh water and arable land in China[24].

The multidimensional characteristics or each criterion weight of each APS in China are shown in Fig. 2a–2c. Although the criteria of non-fed nearshore aquaculture (nFEN) and non-fed aquaculture in large inland waters (nFAL) are in the best condition overall compared with the other eight systems, each APS has its obvious advantages and disadvantages. For example, RAS (Fig. 2c) is strong in ecological risks, arable land use, and freshwater consumption compared with other land-based aquaculture systems but weak in criteria such as fishmeal consumption, economic growth, and employment.

The overall sustainability weight of each APS calculated using the expert AHP method is shown in Fig. 2d. The nFEN and nFAL have significantly higher sustainability weights, while FAL, FEN, and FAP rank as the lowest three among APSs in China. The objective AHP used seven objective criteria data (Table 1), involving aspects of society and economy (food security and economic growth), environment (aquaculture pollution and fossil energy consumption), and resources (fishmeal consumption, freshwater consumption, and arable land use). The results of the objective AHP showed that the 10 systems could be divided into two distinct groups (Fig. 2e), and the overall sustainability weights of non-fed APSs (nFEN, nFAL, and nFAP) were higher than those of fed APSs.

Both expert and objective AHPs ranked non-fed systems (nFEN, nFAL, and nFAP) and supplement fed systems (PFA) as the top four in terms of sustainability weights. These four APSs account for 36.6% of the total aquaculture edible production in China, which is the reason why China's aquaculture is one of the most ecologically efficient industries in the world[2,23,24,29]. The other six fed APSs had lower sustainability weights due to several deficiencies such as aquaculture pollution, food safety, fossil energy consumption, and fishmeal consumption. Compared with the expert AHP, the objective AHP can better quantify the overall sustainability weights

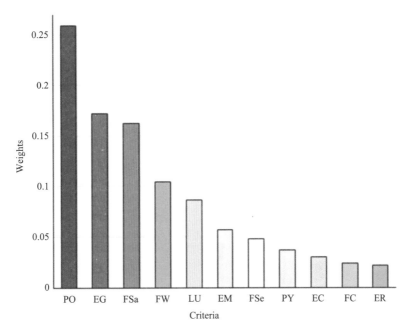

PO: aquaculture pollution; EG: economic growth; FSa: food safety; FW: freshwater consumption; LU: land use; EM: employment; FSe: food security; PY: policy; EC: energy consumption; FC: fishmeal consumption; ER: ecological risks (farmed fish escape and pathophoresis).

Fig. 1　Criteria rating with respect to aquaculture sustainability

of APSs if sufficient criteria are available, rather than simply ranking them (Fig. 2d and 2e).

Over the past decades, the growth of aquaculture production in China has been achieved mainly through aquacultural area expansion and intensification. By increasing external inputs of energy and pelleted feeds, the intensification of food production systems has replaced ecological functions that would otherwise be provided by diverse biomes, as in the case of fed APSs. Therefore, this strategy can lead to environmental degradation, resource constraints (Fig. 2a–2c), and poor sustainability (Fig. 2d, 2e) [2,4,5]. Meanwhile, the intensification of non-fed APSs has also encountered the challenge of overstocking[24]. Therefore, an ecological intensification approach should be adopted to promote and improve the development and sustainability of China's APSs.

▶ *System optimization through ecological intensification*

Ecological intensification of aquaculture systems (ELIAS) is an approach to improve the overall efficiency and productivity of agriculture or aquaculture systems through the rational integration of anthropogenic inputs and ecosystem services[30,31]. Ecosystem services are processes or conditions that lead to benefits for humans[32] and can be classified into four categories: provision, regulation, support, and culture. In food production ecosystems, including aquaculture, the target subjects are farmed organisms[33]. The boundary of an APS is the extent of a farm or water area, including aquaculture waters, ancillary land, and the available space above them. Due to the high diversity of APSs, various ecosystem services, such as trophic synergies, mutualism, and physical purification, can be integrated into aquaculture ecosystems to improve their outputs, comprehensive benefit, and system sustainability[23,30] (Fig. S4). Additionally, the integrated use of these ecosystem services, such as land space and water for irrigation, is another way to improve the economic benefits of aquaculture

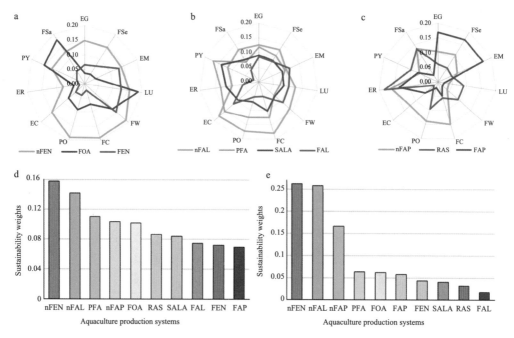

a. Local weights of nFEN, FOA, and FEN with respect to each criterion. b. Local weights of nFAL, PFA, SALA, and FAL to each criterion. c. Local weights of nFAP, RAS, and FAP to each criterion. d. Sustainability weights of each APS evaluated using the expert AHP. e. Sustainability weights of each APS evaluated using the objective AHP. FSe: food security; EC: fossil energy consumption; EG: economic growth; PO: aquaculture pollution; FC: fishmeal consumption; FW: freshwater consumption; LU: arable land use. nFEN: non-fed nearshore aquaculture; nFAL: non-fed aquaculture in large inland waters; nFAP: non-fed aquaculture in ponds; PFA: paddy field aquaculture; FOA: fed offshore aquaculture; RAS: recirculating aquaculture systems; SALA: waterlogged salt-alkali land aquaculture; FAL: fed aquaculture in large inland waters; FAP: fed aquaculture in ponds; FEN: fed nearshore aquaculture.

Fig. 2   Multidimensional characteristics and sustainability rating of each aquaculture production system in China

ecosystems.

To realize ELIAS for a specific APS, firstly, the boundary of the system should be determined. Secondly, its footprints and main constraints should be calculated or identified using the holistic evaluation system as mentioned above. Thirdly, approaches to ecological intensification for the APS are identified through integrating anthropogenic aquaculture inputs with ecosystem services. Finally, a holistic solution is proposed and implemented in consideration of multiple stakeholders (Fig. S5).

Different aquaculture systems have different concerns, so different systems should take different pathways for improving sustainability under the framework of ELIAS (Fig. 3). APSs in large and public waters (reservoirs, lakes, nearshore areas etc.) are generally sensitive to water quality, and organic matter discharge from these systems is prohibited or severely restrained. APSs in small private or collectively owned waters (ponds, RAS etc.) are generally profit sensitive and prioritize high economic benefits. Paddy field aquaculture, waterlogged salt-alkali land aquaculture and offshore aquaculture are policy encouraged and are currently promoted by the government and researchers[2,5,24,25]. For these systems, it is imperative to make full use of the mutually relationship between aquaculture and agriculture (i.e., fish-rice, ponds-terraces) and the physical self-purification function of the ocean. If social and economic profits are the main goals for future aquaculture, APS expansion should be encouraged under a relatively loose environmental

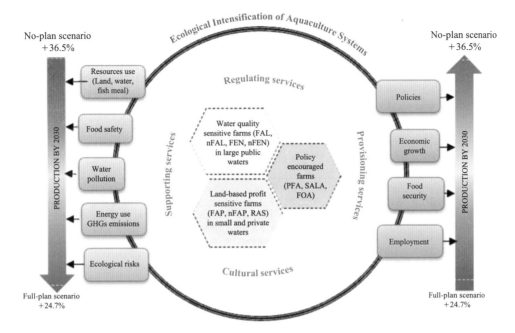

No-plan scenario: without implementation of a stringent environmental protection policy; full-plan scenario: full implementation of a stringent environmental protection policy. 36.5% and 24.7% are the increments of China's aquaculture production in 2030 compared with 2016 under the no-plan and full-plan (6). See Figure 1 for the acronyms of each APS.

Fig. 3 Conceptual framework of ecological intensification of aquaculture systems (ELIAS)

protection policy, while considering environmental and resource criteria, aquaculture should be constrained in some APSs. ELIAS emphasizes the trade-offs among criteria and aims to produce aquatic food efficiently while protecting the environment, conserving natural resources, ensuring food security and safety, and promoting social and economic development.

The comprehensive benefit of each APS, with its own advantages and disadvantages, can be maximized through specific strategies of ecological intensification. In general, for non-fed APSs, the strategy is to make full use of their natural resources, especially natural food organisms for farmed animals, through rational stocking; while for fed or intensive APSs, the strategy is to reduce their negative impacts on the environment through ecological intensification (Table 2).

Table 2 Approaches to ecological intensification for major APSs in China within the frame of ELIAS

APSs	Undesirable criteria or constraints	Major ecosystem services that need to be integrated	Explanation
RAS	Pollution, fishmeal consumption, energy consumption, economic feasibility	Regulating services (RS) of photosynthesis, Supporting services (SS) of alternative energy	Integrating with the photosynthesis of aquatic plants in situ or ex situ[35,36]; leveraging alternative sources of energy, such as thermal drainage from power plants or geothermal water or underground cool water[27]; adopting wind and/or solar power generation, and integrating with tourism and education activity[23,33].

to be continued

APSs	Undesirable criteria or constraints	Major ecosystem services that need to be integrated	Explanation
Fed aquaculture in ponds	Pollution, fishmeal consumption, freshwater consumption, land use, food safety	RS of trophic synergism, SS for power generation, Cultural services (CS) of tourism	Integrating ca 20% of filter-feeders in freshwater ponds, and integrated multi-trophic aquaculture (IMTA) in seawater ponds, such as 1:1 of shrimp: clams in the standing crop[27,37,38]; adopting the models of raceways in ponds and partitioned aquaculture systems[34]; codesigning the integration with other sectors of wind and/or solar power generation, tourism, and education activity.
Fed nearshore aquaculture	Pollution, fishmeal consumption, energy consumption, food safety, ecological risks, conflicts with other users	RS of trophic synergism and self-purification, SS for power generation, CS of tourism	IMTA[40,41]; moving to offshore areas to take advantage of the physical self-purification function of the ocean[5]; informed planning based on ecological and social carrying capacities[42]; codesigning the integration with other sectors of wind power generation, tourism, and education activity.
Fed offshore aquaculture	Pollution, energy consumption, ecological risks, economic feasibility	Provisioning services (PS) of trophic synergism and self-purification, SS for power generation	IMTA; informed planning based on ecological carrying capacity of the ocean[43]; codesigning the integration with wind and/or solar power generation; stocking triploid varieties.
Non-fed nearshore aquaculture	Over stocking, conflicts with other users	PS of natural productivity, SS for power generation, CS of tourism	Stocking suitable species and optimal quantities of seed according to the compositions and productivity of natural feed organisms[27]; codesigning the integration with other sectors of wind and/or solar power generation, tourism, and education activity; informed planning based on ecological and social carrying capacities.
Non-fed aquaculture in large inland waters	Disproportionality of fish stocking	PS of natural productivity, CS of tourism	Stocking suitable multi-species and optimal quantities of seed according to the compositions and productivity of natural feed organisms, such as stocking ca 70% silver carp and ca 30% bighead carp in smaller lakes, but stocking ca 70% bighead carp and ca 30% silver carp in larger reservoirs[27]; codesigning the integration with other sectors of wind and/or solar power generation, tourism, and education activity.
Non-fed aquaculture in ponds	freshwater consumption, land use	PS of natural productivity, SS for power generation, CS of tourism	Stocking suitable species and optimal quantities of seed according to the compositions and productivity of natural feed organisms; integrated aquaculture with agriculture and animal husbandry[27]; codesigning the integration with other sectors of wind and/or solar power generation, tourism, and education activity.
Paddy field aquaculture	food safety, disproportionality of fish and rice	PS of trophic synergism and natural productivity	Farming fish and planting rice in proportion, commercialized production, banned pesticides[16,24]. The area for animal culture is less than ca 10% of the total paddy field area[27].

to be continued

APSs	Undesirable criteria or constraints	Major ecosystem services that need to be integrated	Explanation
Waterlogged salt-alkali land aquaculture	freshwater consumption, re-salinization	PS of trophic synergism, SS for power generation	Area ratio of terraces + roads to pond is about 6:4, and polyculture in ponds[27]; farming salt-tolerant aquatic species, such as whiteleg shrimp and tilapia[44]; application of freshwater conservation techniques, such as fishnet cages in ponds[27]; codesigning the integration with wind and/or solar power generation.
Fed aquaculture in large inland waters	Pollution, freshwater consumption, conflicts with other users	Regulating services (RS) of photosynthesis, Supporting services (SS) of alternative energy	Gradually banning conventional aquaculture models and develop in the water-based recirculating aquaculture systems with zero discharge.

Recirculating aquaculture systems (RASs) are highly intensive APSs that offer several advantages over conventional aquaculture systems, such as reduced water use, land conservation, increased feed efficiency, and improved biosecurity[5,34]. Currently, however, most RASs are highly artificial, farming fed-species and do not take advantage of potential ecosystem services, such as photosynthesis. These systems are constrained by high carbon footprints, high production costs, waste disposal challenges, and the risk of catastrophic disease failures[3,5] (Table 1). Therefore, current RASs are considered less sustainable systems in China (Fig. 2c–2e). Comparing with developed countries, China's higher energy cost and relatively lower labor cost make the products from RAS uncompetitive in international and even in local markets. To overcome the barriers of higher operation costs of dissolved waste removal, RAS should be integrated with ecosystem services, like nutrient recovery through the photosynthesis of aquatic plants, in situ or ex situ[35,36]. Aquatic plants integrated in RAS can absorb inorganic nitrogen, release dissolved oxygen, and may serve as a commercial product (such as edible seaweed, vegetables, or fruits). To make RAS more profitable and sustainable, RAS can also run with ecosystem services from nature or another industry: leveraging alternative sources of energy, such as thermal drainage from power plants, geothermal water, or underground cool water[27]. Moreover, non-fossil energy can be adopted to reduce the carbon footprint of RAS goods.

Fed aquaculture in pond (FAP) involves a vast number of smallholders or family farms from inland to coastal areas. In the future, the development of commercialized integrated aquaculture with an allowable amount of waste discharge should be encouraged. Integrating ecosystem services (including trophic synergism, mutualism, and commensalism) into FAP systems (Fig. S4) can improve pond yields, reduce the discharge of organic matter, and improve the food safety, essential micro-nutrient deficiency, and sustainability of pond systems[37,38]. For example, by exploiting synergies between species in polyculture ponds, production can be greatly improved without increasing waste discharge[39]. In addition, some novel intensified integrated models such as "raceways in ponds" and "partitioned aquaculture systems" may resolve the issue of operational complexity[34].

To meet the regulations for environmental protection, a large portion of the current fed nearshore aquaculture (FEN) should be moved to more exposed areas or offshore areas. Meanwhile, as a stopgap measure, FEN should be integrated with seaweed and/or mollusk farming, known as integrated multi-trophic aquaculture (IMTA), to reduce the negative effects of FEN to the maximum extent possible[38,40,41]. Trade-offs among various stakeholders should be considered in informed planning based on ecological and social carrying capacities[42].

Offshore aquaculture can make good use of the physical and biological self-purification functions of the ocean through dilution and microbial degradation, which are important marine ecosystem services. Fed offshore aquaculture (FOA) should be designed carefully to address environmental concerns associated with conventional FEN, including, among others, the risks of gene contamination from escaped fish and disease transmission[5]. Furthermore, the carrying capacity of FOA within a certain area and the suitability of the local habitat should be carefully investigated during the planning phase to reduce potential long-term negative environmental effects[43]. Moreover, IMTA and non-fossil energy should be adopted to reduce the environmental influences and carbon footprint from FOA[25,38,40,41].

Considering the crucial roles of ecosystem services in non-fed APSs, stocking suitable species and optimal quantities of seed (spores, spat, post larvae, fry, or fingerlings) according to the compositions and productivity of natural feed organisms are the most economically feasible ways for improving their yields[27]. Informed planning is needed based not only on physical and production carrying capacities but also on ecological and social carrying capacities[27,42]. Conventional household paddy field aquaculture (PFA) needs to be scaled up and commercialized, and farmed fish and planted rice should be in proportion[16,24,27]. In addition, pesticides should be banned to ensure the food safety of PFA. Waterlogged salt-alkali land aquaculture (SALA) has formed a mutually beneficial development pattern between aquaculture and agriculture; however, attention should be paid to the re-salinization of the reclaimed land with the development of SALA due to freshwater consumption. Therefore, farming euryhaline aquatic species, such as penaeid shrimp and tilapia[44], and the application of freshwater conservation techniques should be implemented to reduce freshwater consumption[27]. The ecosystem services such as cultural service of aquaculture ecosystems can also be used to magnify the economic benefits of APSs. Integrating aquaculture with tourism, education activity, game fishing, solar power generation (Fig. S6), or wind turbines can improve the economic efficiency and employment level of aquaculture systems[22,34].

▶ *Policy recommendations*

China's aquaculture production has grown at an annual rate of 7.5% over the past 30 years, mainly by the intensification of APSs, leading to higher farming costs and potentially greater environmental risks. The government should formulate and implement detailed targeted development plans according to the sustainability of various farming systems. Initiating measures can include:

1. Encouraging and providing subsidies for the certification of farmed aquatic food products from non-fed APSs that comply with the rules of organic aquaculture45 as organic products to promote the market competitiveness of their products and the development of these farming systems.

2. Strengthening research on the physical, production, ecological and social carrying capacity of aquaculture waters; encouraging ecosystem approach to aquaculture management to achieve Good Environmental Status; confirming the production potential of various aquaculture waters to coordinate the interests of various stakeholders and realize the sustainable development of aquaculture.

3. Providing preferential loans for farmers to adopt ELIAS approaches, while implementing stringent environmental policies to promote the ecological intensification of fed aquaculture in ponds, fed nearshore aquaculture, and recirculating aquaculture systems.

4. Providing preferential policies in terms of land or sea use, as well as loans, to promote the development of waterlogged salt-alkali land aquaculture, paddy field aquaculture, and fed offshore aquaculture. For example, providing

subsidies for construction equipment and fuel oil for fed offshore aquaculture.

5. Encouraging the integration of aquaculture activities with tourism, education, and wind and/or solar power generation in order to realize the economic and social benefits of ecological intensification.

6. Phasing out fed aquaculture in large inland waters to protect the water quality of large inland waters, unless some closed aquaculture facilities without pollution are developed and adopted.

The realization of ELIAS will require close collaboration among policymakers, scientists, farmers, and the supporting industry. Further advances in breeding[11], feed innovation (decoupling aquafeeds from wild fish and terrestrial plant ingredients)[12], and disease prevention[10] will ensure a solid underpinning for achieving the sustainable development of the aquaculture industry. The history of aquaculture development is an evolutionary process of ELIAS, in which intensification and ecological intensification processes have alternated with the increasing demand for aquatic food products, technology advances, and increasing environmental concerns[46,47]. With the increasing proportion of non-fossil energy in China's total energy consumption and the realization of ELIAS, China's aquaculture will achieve the goals of high productivity, zero waste discharge, and a smaller carbon footprint simultaneously in near future.

Materials and Methods

▶ *Criterion selection*

In this study, we used keyword co-occurrence analysis to identify the most popular issues in the topic of aquaculture sustainability. A total of 1611 articles were obtained from the Web of Science Core Collection with the time interval setting as 1985–2021. The keywords provided by the author were extracted and co-occurrence analysis was conducted using VOS viewer (version 1.6.16). Based on the keyword cluster analysis (Fig. S7) and literature28, 11 criteria were selected in the study (Fig. 2).

▶ *Building the metric data of APSs related to system sustainability*

The data on aquaculture production and areas of aquaculture waters in mainland China were collected from the China Fishery Statistical Yearbook (http://data.cnki.net/Trade/yearbook/single/N2018120050?z=Z009).

Total production value of each APS was its production in volume multiplied by the weighted price of the top three or six major species farmed in that APS. The annual average market prices (RMB/kg) of the major farmed species in China in 2018 were provided by the China Aquatic Products Processing and Marketing Alliance (Table S1). The average conversion factors of the major aquatic species group used for converting live weight to edible meat were as follows: fish 1.15 (gutted, head-on), crustaceans 2.80 (tails/meat, peeled), and mollusks 6.0 (meat, without shells)[48]. The data of discharge coefficients of pollution sources in China's aquaculture were taken from the Handbook of the First National Census of Pollution Sources[49]. The discharge coefficient of chemical oxygen demand (COD) was the weighted average of the top three or six major farmed species in each system (Table 1). The feed conversion ratios of species groups (total feed fed/total species-group biomass increase) were collected from Tang et al.[50]. The proportions of fishmeal and oil in feeds of a specific species or a species group were taken from a previous publication[51]. The share of fishmeal and oil consumption of each system was calculated from the weighted value of the major species (Table 1). The total water footprint of freshwater species and the water footprint of the feed of mariculture species were taken from previous publications[52-54]. The total freshwater footprint of each system was calculated from the weighted value of the major species in each aquaculture system. For calculating land use for pond farming and RAS, the ratios of 1:1.5 (water area of ponds to total land use for the farm) and 1:10 (water area of tanks to land use for the RAS farm) were applied, respectively[55]. The energy consumption (kW·h·kg^{-1}) of aquatic products from different aquaculture systems in China was collected or calculated based on Xu et al.[56].

▶ *Calculation of sustainability weights*

The sustainability evaluation of various APSs is a multi-criteria decision-making problem involving various conflicting criteria. Eleven criteria contribute to the sustainability of APSs, but only seven of them have available objective data in China (Table 1). Therefore, two methods were comparatively used to deal with the problem: an analytical hierarchy process (AHP)[21] based on experts' judgments, or expert AHP; and AHP partially based on objective data or objective AHP.

Expert AHP

This method involved all 11 criteria, however, all local weights used were based on pairwise comparison judgments provided by experts. The decision problem was firstly structured as a three-level hierarchy (Fig. S3). The top level was the overall goal of sustainability of APSs. All 11 criteria contributing to the overall goal were represented in the intermediate level. The lowest level comprised the 10 APSs, which were evaluated in terms of the criteria in the intermediate level.

As some criteria (such as food safety and policy) were unavailable or immeasurable, the relative importance of various elements in the same level with respect to the elements in their upper level was compared pairwise by 30 invited senior well-trained Chinese experts from related fields. The invited experts included seven members of the Fisheries Group of Discipline Assessment Organization of the State Council of China, 10 members of the Preparatory Committee of Chinese Society of Aquaculture Ecology, seven experienced experts from disciplines of fisheries management, fisheries economics, aquatic food product safety, and aquaculture engineering, one from administration, three representatives of NGOs, and two from freshwater and seawater aquaculture industry.

In the intermediate level, a pairwise comparison matrix of 11 criteria with respect to the overall goal was created by aggregating the 30 experts' judgments through the geometric mean. The consistency ratio of the matrix was 0.0147. Therefore, the estimation of the eigenvector $v = (v_1, \cdots, v_{11})$ associated with the principal eigenvalue was accepted. Then the local weight (Fig. 2B) of the ith criteria with respect to the overall goal was given by

$$w_i = \frac{v_i}{\sum_{k=1}^{11} v_k}, i = 1, \cdots, 11 \quad (1)$$

In the lowest level, each APS had a numerical value for each criterion, which was the geometric mean of the 30 experts' judgments. The local weights of 10 systems with respect to each criterion were calculated by normalizing their values under the same criterion. Formally, the local weight of the jth APS with respect to the ith criterion was given by

$$p_{ij} = \frac{b_{ij}}{\sum_{k=1}^{10} b_{ik}}, j = 1, \cdots, 10, i = 1, \cdots, 11 \quad (2)$$

where b_{ij} is the numerical value of the jth APS under the ith criterion.

The sustainability weight of the jth aquaculture system was given by

$$sp_j = \sum_{i=1}^{11} w_i p_{ij}, j = 1, \cdots, 10 \quad (3)$$

Based on the sustainability weights, the sustainability ranking of 10 APSs was obtained.

Objective AHP

This method took full advantage of existing information, including the relative importance of criteria based on experts' judgments and seven objective criterion data of 10 APSs. The decision problem was also structured as a three-level hierarchy. However, the intermediate level only comprised seven criteria having objective data. In the intermediate level, the weights of the seven criteria with respect to the sustainable goal were obtained through a pairwise comparison matrix of the seven criteria given by 30 experts' judgments. The consistency ratio of the matrix is 0.0223 and the estimation of the eigenvector $v = (v_1, \cdots, v_7)$ associated with the principal eigenvalue was accepted. Then the weight of the th criterion with respect to the overall goal was given by

$$w_i = \frac{v_i}{\sum_{k=1}^{7} v_k}, i = 1, \cdots, 7 \quad (4)$$

In the lowest level, there are actual data from APSs under the seven criteria. Thus, the local weights of 10 APSs under each criterion can be calculated using the ratio method as described in the above-classified aggregation method.

The sustainability ranking of 10 APSs was obtained based on the sustainability weights of the systems, which were given by

$$sp_j = \sum_{i=1}^{7} w_i p_{ij}, j = 1, \cdots, 10 \quad (5)$$

where p_{ij} is the local weight of the jth APS under the th criterion.

Acknowledgments

This study was funded by the National Blue Granary S & T Innovation Program (2019YFD0901000), Natural Science Foundation of China grants (31572634 and U1906206). We thank George N Somero and Rosamond L. Naylor for helpful comments on the paper, and thank Bin-lun Yan, Da-peng Li, Guo-xing Nie, Jia-shou Liu, Ji-ting Sun, Jin-long Yang, Jia-song Zhang, Hong Lin, Hui Liu, Pao Xu, Qing Fang, Song-lin Wang, Wei-dong Li, Xiang-li Tian, Xiao-dong Li, Xiao-juan Cao, Xiao-jun Yan, Yao-guang Zhang, Ying Liu, Yong Liang, Yue Wang, Yu-ze Mao, Wei-min Wang, and Wen Zhao for making their pairwise comparison judgements on relative importance of 11 criteria and sustainability of 11 aquaculture systems.

References

1. Hicks C C, Cohen P J, Graham N A J, et al. Harnessing global fisheries to tackle micronutrient deficiencies[J]. Nature, 2019, 574(7776): 95-98.
2. FAO. The state of world fisheries and aquaculture 2020: Sustainability in action[R]. Rome, Italy: Food and Agriculture Organization of the United Nations, 2020.
3. Naylor R L, Hardy R W, Buschmann A H, et al. A 20-year retrospective review of global aquaculture[J]. Nature, 2021, 591(7851): 551-563.
4. Froehlich H E, Runge C A, Gentry R R, et al. Comparative terrestrial feed and land use of an aquaculture-dominant world[J]. Proc. Natl. Acad. Sci. U. S. A., 2018, 115(20): 5295-5300.
5. Klinger D, Naylor R. Searching for solutions in aquaculture: Charting a sustainable course[J]. Annu. Rev. Environ. Resour., 2012, 37(1): 247-276.
6. FAO. The state of world fisheries and aquaculture 2018: Meeting the sustainable development goals[R]. Rome, Italy: Food and Agriculture Organization of the United Nations, 2018.
7. Henriksson P J G, Belton B, Murshed-e-Jahan K, et al. Measuring the potential for sustainable intensification of aquaculture in Bangladesh using life cycle assessment[J]. Proc. Natl. Acad. Sci. U. S. A., 2018, 115(12): 2958-2963.
8. Cao L, Naylor R, Henriksson P, et al. China's aquaculture and the world's wild fisheries[J]. Science, 2015, 347(6218): 133-135.
9. Damerau K, Waha K, Herrero M. The impact of nutrient-rich food choices on agricultural water-use efficiency[J]. Nat. Sustain., 2019, 2(3): 233-241.
10. Reverter M, Sarter S, Caruso D, et al. Aquaculture at the crossroads of global warming and antimicrobial resistance[J]. Nat. Commun., 2020, 11(1): 1870.
11. Houston R D, Bean T P, Macqueen D J, et al. Harnessing genomics to fast-track genetic improvement in aquaculture[J]. Nat. Rev. Genet., 2020, 21(7): 389-409.
12. Roheim C A, Bush S R, Asche F, et al. Evolution and future of the sustainable seafood market[J]. Nat. Sustain., 2018, 1(8): 392-398.
13. Jouffray J B, Crona B, Wassénius E, et al. Leverage points in the financial sector for seafood sustainability[J]. Sci. Adv., 2019, 5(10): eaax3324.
14. Kittinger J N, Teh L C L, Allison E H, et al. Committing to socially responsible seafood[J]. Science, 2017, 356(6341): 912-913.
15. National Academies of Sciences, Engineering, and Medicine. Science Breakthroughs to Advance Food and Agricultural Research by 2030[M]. Washington, DC: The National Academies Press, 2019.
16. Little D C, Newton R W, Beveridge M C. Aquaculture: A rapidly growing and significant source of sustainable food? Status, transitions and potential[J]. Proc. Nutr. Soc., 2016, 75(3): 274-286.
17. Boyd C E, D'Abramo L R, Glencross B D, et al. Achieving sustainable aquaculture: Historical and current perspectives and future needs and challenges[J]. J. World Aquac. Soc., 2020, 51(3): 578-633.
18. Metian M, Troell M, Christensen V, et al. Mapping

diversity of species in global aquaculture[J]. Rev. Aquac., 2020, 12(2): 1090-1100.
19. Newton R, Zhang W, Xian Z, et al. Intensification, regulation and diversification: The changing face of inland aquaculture in China[J]. Ambio, 2021, 50(9): 1739-1756.
20. Belton B, Reardon T, Zilberman D. Sustainable commoditization of seafood[J]. Nat. Sustain., 2020, 3(9): 677-684.
21. Saaty T L, Vargas L G. Models, Methods, Concepts & Applications of the Analytic Hierarchy Process[M]. Boston, MA, USA: Springer, 2012.
22. Tang Q S, Zhang J H, Fang J G. Shellfish and seaweed mariculture increase atmospheric CO2 absorption by coastal ecosystems[J]. Mar. Ecol. Prog. Ser., 2011, 424: 97-104.
23. Alleway H K, Gillies C L, Bishop M J, et al. The ecosystem services of marine aquaculture: Valuing benefits to people and nature[J]. Bioscience, 2019, 69(1): 59-68.
24. Gui J F, Tan Q S, Li Z J, et al. Aquaculture in China: Success Stories and Modern Trends[M]. Oxford, UK: John Wiley & Sons Ltd, 2018.
25. Lester S E, Gentry R R, Kappel C V, et al. Offshore aquaculture in the United States: Untapped potential in need of smart policy[J]. Proc. Natl. Acad. Sci. U. S. A., 2018, 115(28): 7162-7165.
26. Shepon A, Gephart J A, Henriksson P J G, et al. Reorientation of aquaculture production systems can reduce environmental impacts and improve nutrition security in Bangladesh[J]. Nature Food, 2020, 1(10): 640-647.
27. Dong S L, Tian X L, Gao Q F. Aquaculture Ecology[M]. Bejing, China: Science Press, 2017.
28. Stentiford G D, Bateman I J, Hinchliffe S J, et al. Sustainable aquaculture through the One Health lens[J]. Nature Food, 2020, 1(8): 468-474.
29. Waite R, Beveridge M, Brummet R, et al. Improving Productivity and Environmental Performance of Aquaculture[M]. Washington, DC: World Resources Institute, 2014.
30. Bommarco R, Kleijn D, Potts S G. Ecological intensification: Harnessing ecosystem services for food security[J]. Trends Ecol. Evol., 2013, 28(4): 230-238.
31. Dong S L. On ecological intensification of aquaculture systems in China[J]. Chinese Fisheries Economics, 2015, 33(5): 3-9.
32. Daily G C. Nature's Services: Societal Dependence on Natural Ecosystems[M]. Washington, DC.: Island Press, 1997:392.
33. Willot P A, Aubin J, Salles J M, et al. Ecosystem service framework and typology for an ecosystem approach to aquaculture[J]. Aquaculture, 2019, 512: 734260.
34. Tidwell J H. Aquaculture Production Systems[M]. Oxford, UK: Wiley-Blackwell, 2012.
35. Corey P, Kim J K, Duston J, et al. Growth and nutrient uptake by Palmaria palmata integrated with Atlantic halibut in a land-based aquaculture system[J]. Algae, 2014, 29(1): 35-45.
36. Chang B V, Liao C S, Chang Y T, et al. Investigation of a farm-scale multitrophic recirculating aquaculture system with the addition of rhodovulum sulfidophilum for milkfish (Chanos chanos) coastal aquaculture[J]. Sustainability-Basel, 2019, 11(7): 1880.
37. Alexander K A, Freeman S, Potts T. Navigating uncertain waters: European public perceptions of integrated multi trophic aquaculture (IMTA) [J]. Environ. Sci. Policy, 2016, 61: 230-237.
38. Knowler D, Chopin T, Martínez-Espiñeira R, et al. The economics of integrated multi-trophic aquaculture: Where are we now and where do we need to go?[J]. Rev. Aquac., 2020, 12: 1579-1594.
39. Bosma R H, Verdegem M C J. Sustainable aquaculture in ponds: Principles, practices and limits[J]. Livest Sci., 2011, 139(1-2): 58-68.
40. Chopin T. Progression of the integrated multi-trophic aquaculture (IMTA) concept and upscaling of IMTA systems towards commercialization[J]. Aquac. Eur., 2011, 36: 5-12.
41. Sorgeloos P. Aquaculture: The blue biotechnology of the future[J]. World Aquac., 2013, 35: 16-25.
42. McKindsey C W, Thetmeyer H, Landry T, et al. Review of recent carrying capacity models for bivalve culture and recommendations for research and management[J]. Aquaculture, 2006, 261(2): 451-462.
43. Stigebrandt A. Carrying capacity: General principles of model construction[J]. Aquac. Res., 2011, 42(s1): 41-50.
44. Sun W M, Dong S L, Jie Z L, et al. The impact of net-isolated polyculture of tilapia (Oreochromis niloticus) on plankton community in saline–alkaline pond of shrimp (Penaeus vannamei)[J]. Aquac. Int., 2011, 19(4): 779-788.
45. Lembo G, Mente E. Organic Aquaculture Impacts and Future Developments: Impacts and Future Developments[M]. Springer, 2019.
46. Costa-Pierce B A. Sustainable ecological aquaculture systems: The need for a new social contract for aquaculture development[J]. Mar. Technol. Soc. J., 2010, 44(3): 88-112.
47. Nash C E. The History of Aquaculture[M]. Wiley

Blackwell, 2011.
48. Tacon A G J, Metian M. Fish matters: Importance of aquatic foods in human nutrition and global food supply[J]. Rev. Fish. Sci., 2013, 21(1): 22-38.
49. Census C P S. Handbook of the First National Census of Pollution Sources - Discharge Coefficient of Pollution Sources in Aquaculture In Chinese 2009[Z/OL]. https://www.doc88.com/p-037431165661.html.
50. Tang Q S, Han D, Mao Y Z, et al. Species composition, non-fed rate and trophic level of Chinese aquaculture[J]. Journal of Fishery Sciences of China, 2016, 23: 729-758.
51. Tacon A G J, Metian M. Feed matters: Satisfying the feed demand of aquaculture[J]. Rev. Fish. Sci. Aquac., 2015, 23(1): 1-10.
52. Pahlow M, van Oel P R, Mekonnen M M, et al. Increasing pressure on freshwater resources due to terrestrial feed ingredients for aquaculture production[J]. Sci. Total Environ., 2015, 536: 847-857.
53. Mohanty R K, Ambast S K, Panigrahi P, et al. Water quality suitability and water use indices: Useful management tools in coastal aquaculture of *Litopenaeus vannamei*[J]. Aquaculture, 2018, 485: 210-219.
54. Ouyang Y T. Study on water footprint of freshwater cultured fish and spatial optimization of cultured policy in China[D]. Dalian: Dalian University of Technology, 2018.
55. Boyd C E, McNevin A A. Aquaculture, Resource Use, and the Environment[M]. Hoboken, New Jersery, USA: John Wiley & Sons, Inc., 2014.
56. Xu H, Zhang Z I, Zhang J H, et al. The research and development proposals on fishery energy saving and emission reduction in China[J]. Journal of Fisheries of China, 2011, 35(3): 472-480.

Weakened fertilization impact of anthropogenic aerosols on marine phytoplankton—a comparative analysis of dust and haze particles[①]

Chao Zhang[1,2], Qiang Chu[3], Yingchun, Mu[4], Xiaohong Yao[1,2], and Huiwang Gao[1,2]*

1 Frontiers Science Center for Deep Ocean Multispheres and Earth System, and Key Laboratory of Marine Environment and Ecology, Ministry of Education, Ocean University of China, Qingdao 266100, China
2 Laboratory for Marine Ecology and Environmental Sciences, Pilot National Laboratory for Marine Science and Technology, Qingdao 266071, China
3 Laboratory of Environmental Protection in Water Transport Engineering, Tianjin Research Institute for Water Transport Engineering, Ministry of Transport, Tianjin 300000, China
4 Estuarine and Coastal Environment Research Center, Chinese Research Academy of Environmental Sciences, Beijing 100012, China
* Corresponding author: Huiwang Gao (hwgao@ouc.edu.cn)

Abstract

Although increases in air pollutants are changing chemical compositions of atmosphere, the resultant impacts on marine biogeochemistry remains elusive. We performed a collective analysis of 12 microcosm experimental data concerning treatments of dust particles (DPs, typically mineral aerosols), haze particles (HPs, typically anthropogenic aerosols), and various nutrients in varying trophic seawaters of the Northwest Pacific Ocean. The addition of DPs and HPs generally stimulated phytoplankton growth, as indicated by total chlorophyll a (Chl a), and shifted the phytoplankton size structure towards larger cells (> 2 μm in cell size), as indicated by size-fractionated Chl a. We further found that DP/HP-derived Chl a increase relative to the control ($RC_{Chl\ a}$) was proportional to the proportion of nitrogen (N) supplied by DPs/HPs relative to the baseline N concentration in seawater (P_{SN}) and was higher than that in the N alone treatment when the P_{SN} exceeded about 480%. The enhanced utilization of dissolved organic P potentially contributed to the stimulation of DPs/HPs. The slope of fitted line based on $RC_{Chl\ a}$ and P_{SN} in the DP treatments (0.14) was higher than that in the HP treatments (0.11). When the particle loading was extremely high (2 mg·L^{-1}), the addition of HPs exhibited an obvious inhibition impact on phytoplankton and was adverse to the shift of the size structure towards larger cells. These results suggest that the impact of HPs on phytoplankton is a composite result of stimulation by nutrients and inhibition by toxic matter, which may affect carbon sequestration efficiency in the ocean by regulating phytoplankton biomass and size structure.

Keywords: atmosphere deposition; dust; haze; phytoplankton; nutrient; incubation experiment

---

① 本文于2022年1月发表在 *Ecotoxicology and Environmental Safety* 第230卷，https://doi.org/10.1016/j.ecoenv.2022.113162。

Introduction

Atmospheric deposition is regarded as an important source of nutrients to the upper ocean[1,2]. On a global scale, the contribution of atmospheric inputs with respect to nutrients is generally comparable to or greater than that of riverine inputs[3]. In contrast to the impact of riverine inputs trapped in coastal waters, however, atmospheric deposition can be transported over an extensive distance, thereby affecting almost the whole ocean[4,5]. It has been reported that sustained atmospheric nitrogen (N) deposition has the potential to alter the nutrient limitation from N to phosphorus (P) in the North Pacific Ocean[2,6] and that atmospheric iron (Fe) deposition has been regarded as the primary way to alleviate Fe limitation in high-nutrient low-chlorophyll (HNLC) regions[7]. The considerable nutrient input by atmospheric deposition exerts an important impact on biogeochemical cycles of biogenic elements such as carbon (C), N, and P in the ocean[2,8,9].

Dust deposition, as a natural event, is generally regarded as an important carrier of nutrients and noticeably affects marine ecosystems and primary productivity. Statistically, over 80% of atmospheric Fe and P depositions originate from dust on a global scale[1,10]. Field observations and remote sensing have confirmed that dust events can cause the rapid growth of phytoplankton and even cause outbreaks in the upper ocean[11,12]. Onboard incubation experiments further explored the response process, and illustrated the impact of dust-derived alterations in the concentration and structure of nutrients on phytoplankton size and community structure[13-15]. However, most field studies are based on a one-off cruise and are inadequate to unearth a generalized conclusion. On the other hand, during the long-range transport of dust from desert sources to the ocean, anthropogenic air pollutants can mix with dust and affect its chemical characteristics, such as the N content and solubility of P and Fe[16,17]. At present, therefore, the impact of dust deposition on marine ecosystems and primary productivity tends to be a composite result of natural and anthropogenic sources of matter[17,18]. With the increasing influence of anthropogenic activities on natural processes, such composite results may change continuously for a period of time in the future[6,19].

In contrast to dust, the impact of anthropogenic aerosols on the ocean is more complicated. On the one hand, anthropogenic aerosols contain more bioavailable nutrients such as N, P, and Fe, and theoretically exert a stronger fertilization impact on marine primary productivity[1,10,20]. In fact, anthropogenic emissions have contributed over 80% of the atmospheric N deposition over the ocean, and this contribution likely to increase in the future[19]. Isotope technology was used to illustrate that anthropogenic Fe deposition accounted for 21% to 59% of the soluble Fe in the surface seawater of the North Pacific Ocean[9]. On the other hand, the higher content of heavy metals, such as copper (Cu) and organic matter, such as polycyclic aromatic hydrocarbons (PAHs), has the potential to inhibit phytoplankton growth[21,22]. Paytan et al.[23] combined experimental and modeling measures to illustrate that anthropogenic Cu deposition was likely to cause a toxic impact in the area downwind of the East Asian continent. However, another study in the East China Sea (ECS) revealed that Cu tends to stimulate marine primary productivity[18]. Such paradoxical results are closely related to trophic status, phytoplankton biomass and community structure in seawater[24]. In addition, the concentration of heavy metals in seawater due to atmospheric deposition is much lower than the traditionally toxic threshold that causes obvious cell death[23,25], which also likely increases the uncertainty related to interpreting the so-called toxic impact. Collectively, it is necessary to understand the comprehensive impact, i.e., stimulation and inhibition, of anthropogenic aerosols on marine primary productivity and thus more readily distinguish the difference in their impact from that of dust.

The Northwest Pacific Ocean (NWPO) is located in the area downwind of the Asian continent, and as such, it is a typical area that has been noticeably influenced by dust events and anthropogenic activities. The Asian continent, as the second largest dust source in the world, is responsible for contributing tens to hundreds of tons of dust per year, thereby significantly affecting primary productivity therein[4,11]. With the burgeoning human and industrial activities in East Asia, the acute influence of anthropogenic air pollutants on the ocean has gradually become obvious[2,9]. Our previous studies determined the key role of N (e.g., relief of N limitation, changes in the N:P ratio) supplied by dust particles (DPs, typically mineral aerosols) and/or haze particles (HPs, typically anthropogenic aerosols) in affecting primary productivity in the NWPO[15,26,27], while the potentially supplementary effects of P and/or trace metals remain more complicated, which may lead to varying fertilization effects of DPs and HPs on phytoplankton. In this study, we performed a collective analysis of the data from a series of onboard incubation experiments enriched with DPs and/or HPs through five cruises from 2014 to 2016 in the NWPO. The study region includes China coastal seas, the subtropical gyre (SG), and the Kuroshio Extension (KE), where seawater ranges from oligotrophic to eutrophic (Fig. 1). Based on the induction and analysis of these experimental data, we aimed to distinguish the impacts of DPs and HPs on phytoplankton growth and size structure, and attempt to establish the general pattern of phytoplankton response to atmospheric deposition.

Methods

▶ Collection and analysis of dust and haze particle samples

Soil samples used for preparing DPs were collected in three deserts of China, i.e., the Gobi Desert (42.37°N, 112.97°E, used for DP addition experiments conducted in 2014), Mu Us Desert (37.92°N, 107.11°E, used for DP addition experiments conducted in 2015), and Tengger Desert (38.79°N, 105.51°E, used for DP addition experiments conducted in 2016). The collected soil samples were subsequently crushed and sieved to obtain particles with a diameter less than 20 μm[28]. To simulate the mixing process between dust and air pollutants during long-range transport from the desert to the ocean, i.e., the dust aging process, we followed the simulation method of Guieu et al.[17], which mixed the soil samples and synthetic cloud water (including HNO_3 and H_2SO_4) and obtained DP samples after the evaporation of the aqueous phase[26]. During a strong Asian dust event, the loadings of dissolved inorganic N (DIN), PO_4^{3-} and soluble Fe in atmospheric particles collected from the Yellow Sea (YS) were 90–1264 $\mu mol·g^{-1}$, 1.0–4.8 $\mu mol·g^{-1}$ and 67–1624 $\mu g·g^{-1}$, respectively[11], which generally cover the DP values used in this study (i.e., 577–1007 $\mu mol·g^{-1}$ for DIN, 0.8–4.5 $\mu mol·g^{-1}$ for PO_4^{3-} and 69–766 $\mu g·g^{-1}$ for soluble Fe, Table 1). The contents of nutrients and trace metals in DPs were comparable to those of aerosol samples during a dust storm that occurred over the YS[11]. HP samples were collected directly using the KC-1000 high-volume TSP sampler (Laoshan Electron Ltd., China) during pollution weather in Qingdao, which is adjacent to the YS. In total, we collected four groups of HP samples in 2014, 2015, and 2016. Details of the weather conditions during the sampling period are displayed in Table S1. The contents of nutrients and trace metals in HPs were close to the median in aerosols reported in other cities (Fig. S1); thus, HPs in our study can be regarded as a typical type of anthropogenic aerosol.

Collected DP and HP samples were stored at −20 °C prior to chemical analysis, and nutrients and soluble trace metals in DPs and HPs were determined following the protocols used in previous studies[26,27]. Briefly, the ultrasonic bath method was used for extracting nutrients and soluble trace metals into a certain volume of aqueous phase medium, and the concentrations of nutrients and soluble trace metals were determined using continuous flow colorimetric autoanalyzer (SEAL Analytical) and inductively coupled plasma-mass spectrometry

(ICP–MS, Agilent 7500c), respectively. Details of aerosol treatment are provided in Zhang et al.[26,27].

▶ *Cruises and environmental design*

We conducted 12 incubation experiments on five cruises covering the China coastal seas and open oceans of the NWPO from 2014 to 2016, i.e., Cruise Ⅰ from March to April 2014, including stations Ar4 and G7 in the SG of the NWPO; Cruise Ⅱ from April to May 2014, including stations H10 and B7 in the YS; Cruise Ⅲ from April to May 2015, including station A1-b in the KE of the NWPO and station PN3 in the ECS; Cruise Ⅳ from March to April 2016, including station YS1 in the YS and stations M1, M1B in the KE of the NWPO; and Cruise Ⅴ from May to June 2016, including stations Seats, E4, and D5 in the South China Sea (SCS, Fig. 1). Details of seawater sampling and assemblage have been presented previously[26]. Briefly, filtered surface seawater (200 μm nylon mesh) and added materials (DPs, HPs, and various nutrient reference materials mainly referring to N, P, Fe, and their combinations) were mixed uniformly in 20 L acid-washed incubation bottle that were placed in three plastic vessels filled with cycled seawater continuously pumped from the ocean surface. During the incubation periods (4–10 days), a certain amount of seawater was sampled on a certain date to determine various parameters. It is noted that we used two volumes of incubation bottles in the 2016 experiments, i.e., 20 L incubation bottles primarily spiked with DPs and/or HPs, and 2 L incubation bottles primarily spiked with various nutrients. For full details of amendments and samplings during the incubations, see Fig. 2, Tables 1, S2 and S3.

▶ *Chemical and biological parameters*

Chemical and biological parameters, including nutrients and chlorophyll a (Chl a) in collected seawater, were measured as previously described[26]. Briefly, seawater filtered by acid-washed cellulose acetate membranes (about 200 mL) was stored onboard at −20 °C, and used to measure nutrients by continuous flow colorimetric autoanalyzer, i.e.,

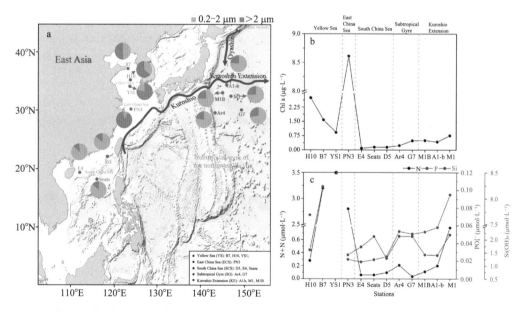

a. The contribution of size-fractionated Chl a (0.2–2 μm and > 2 μm) to total Chl a; b, c. Total Chl a concentration, and nutrient concentration in the baseline seawater. The concentrations of N+N, PO_4^{3-}, and $Si(OH)_4$ at station YS1 labelled on the upper X axis were 10.70 μmol·L^{-1}, 0.67 μmol·L^{-1}, and 15.00 μmol·L^{-1}, respectively.

Fig. 1 Baseline conditions of sampling stations used for microcosm incubation experiments

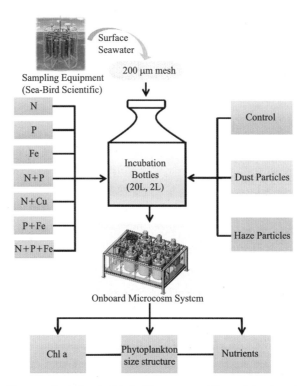

Fig. 2 Experimental design of onboard microcosm incubations amended with dust/haze particles and various nutrients

QuAAtro analyzer for experiments in 2014 and AA3 analyzer for experiments 2015 and 2016, in the laboratory on land.

The measurement of Chl a is divided into two groups based on the volume of incubation bottles (20 L and 2 L). For 20 L incubation bottles, sized-fractionated Chl a pigments were obtained by filtering 100–200 mL of seawater through 20- (Millipore NY2004700), 2- (Whatman 111111), and 0.2- (Whatman 111106) μm pore size membranes. For 2 L incubation bottles (only for 2016 experiments), total Chl a pigments were obtained by filtering 100–200 mL seawater through a 0.7 μm pore size membrane (Whatman GF/F 1825-047). All Chl a pigments were extracted onboard in 90% acetone under the following conditions: darkness, –20 °C, and 18–22 h duration[29]. The fluorescence of extracted pigments was calibrated as the Chl a concentration on a Turner Trilogy fluorometer[30].

The sum of size-fractionated Chl a was regarded as the total Chl a for 20 L incubation experiments.

▶ *Protocol of data analysis*

To illustrate the phytoplankton response to various additions, the relative change in Chl a ($RC_{Chl\,a}$, %) was used[13] and is defined as:

$$RC_{Chl\,a} = \frac{Chl\,a_T - Chl\,a_C}{Chl\,a_C} \times 100\%$$

where $Chl\,a_C$ and $Chl\,a_T$ are the average concentrations of Chl a during the incubation experiments in the control and various treatments, respectively. The calculation of $RC_{Chl\,a}$ was based on the whole incubation process and reflects the overall impact of various additions on phytoplankton growth.

In general, the contributions of size-fractionated Chl a to total Chl a were used to depict the phytoplankton size structure[31,32]. As the high biomass of large phytoplankton induced by added materials can be sustained only over a short duration, the phenomenon that reflects the change in phytoplankton size structure is likely covered if the average contribution during the incubation experiments is used[26]. In addition, considering that nutrient input generally promotes the shift in phytoplankton size structure towards larger cells[26,31,32] as well as the dominance of pico-sized cells in the baseline seawater, the maximum contribution of large cells (> 2 μm) to total Chl a ($C_{L\text{-Max}}$, corresponding to the minimum contribution of pico-sized cells to total Chl a), rather than the average contribution, is introduced during the incubation experiments to better illustrate the shift of phytoplankton size structure in this study.

The significance of the difference in the mean values between varying treatments is evaluated using a one-way analysis of variance (one-way ANOVA), and a comparison between the control and other treatments is then applied by Dunnett's test using Statistical Product and Service Solutions software (SPSS v.19.0).

Results

▶ *Chemical and biological conditions of the baseline seawater*

Seawater in the SCS and SG generally suggests a state of oligotrophy (Fig. 1). Baseline Chl a concentrations range between 0.09 μg·L^{-1} and 0.50 μg·L^{-1}. The concentrations of N+N, PO_4^{3-}, and $Si(OH)_4$ are no greater than 0.21, 0.06, and 2.12 μmol·L^{-1}, respectively. In the KE, the range of Chl a and N+N concentrations increases to 0.43−0.76 μg·L^{-1} and 0.11−0.79 μmol·L^{-1}, respectively, revealing oligo-mesotrophic characteristics. Furthermore, the trophic statuses in the YS and ECS are generally greater than those in the SCS and open oceans (i.e., SG and KE) of the NWPO, with the highest N+N concentration reaching 10.70 μmol·L^{-1} at station YS1. The dominant contributor to total Chl a is pico-sized phytoplankton at all stations except H10, B7, and PN3, where the contribution of large phytoplankton is not less than half (Fig. 1).

The N:P ratios in the baseline seawater at most stations (except YS1, PN3, and B7) ranged between 0.7 and 9, which is lower than the Redfield ratio (N:P=16:1). Correspondingly, the phytoplankton response in Chl a showed positive responses to N-related additions (Fig. S2), indicative of N-dominant limiting conditions. In contrast, phytoplankton in the oligotrophic seawater (i.e., SG and SCS) tended to be colimited by N and P, while those in the KE and YS tended to be primarily limited by N alone[26]. In addition to N limitation, we also found the occurrence of P limitation in the inshore waters of the YS at station B7 (N:P ≈ 31), which was severely influenced by continental inputs such as riverine runoff and atmospheric deposition with high N:P ratios[6,33]. At stations YS1 in the YS and PN3 in the ECS, phytoplankton showed insignificant responses to nutrient inputs due to the extreme physiological statuses of the phytoplankton (Fig. S3), i.e., an exponential growth phase (Chl a concentration increased from 0.92 μg·L^{-1} to 12.97 μg·L^{-1} during the five days of the incubations) at station YS1 and a decline phase (Chl a concentration decreased from 8.61 μg·L^{-1} to 2.48 μg·L^{-1} during the five days of the incubations) at station PN3[27].

▶ *Distinct phytoplankton responses to the addition of dust and haze particles*

The addition of DPs and HPs can supply a considerable amount of DIN and a slight amount of P relative to the baseline seawater (Table 1, Fig. 1). In terms of trace metals, DP additions supplied substantial crustal metals such as Fe and manganese (Mn). In contrast, anthropogenic metals such as Cu and cadmium (Cd) supplied by HPs were generally higher than those supplied by DPs at the same particle loadings (Table 1). In addition, considering the high N content in particles and the frequent occurrence of N shortage in the baseline seawater, we expressed the particle loading in the unit of μmol·L^{-1} to replace that of mg·L^{-1} (Table 1, Fig. 3), thereby improving the comparison between DP and HP treatments.

Table 1 Concentrations of added inorganic nutrients and soluble trace metals from dust and haze particles during incubation

Particle	Added amount (mg·L^{-1})	Stations	Nutrients (nmol·L^{-1})			Soluble trace metals (ng·L^{-1})						
			NO$_3^-$+NO$_2^-$	NH$_4^+$	PO$_4^{3-}$	Pb	Cu	Cd	As	Co	Fe	Mn
DPs-A	2	Ar4, G7, H10, B7	1134	20.6	8.6	0.48	0.46	0.08	—	5.30	946.4	826.8
DPs-B	0.2	A1-b	197.8	3.55	0.9	—	—	—	6.24E-2	2.18E-2	153.3	128.7
	1	A1-b, PN3	989.2	17.8	4.5	—	—	—	0.31	0.11	766.4	643.6
DPs-C	0.3	Seats, E4, D5	290.2	0.88	0.3	0.12	0.57	—	0.63	0.69	22.8	57.2
	0.5	M1, M1B	439.7	1.33	0.4	0.18	0.86	—	0.96	1.05	34.6	86.7
	1	M1B, Seats, E4, D5	879.4	2.66	0.8	0.36	1.7	—	1.9	2.10	69.2	173.4
	2	YS1	1759	5.31	1.7	0.71	3.4	—	3.8	4.20	138.4	346.9
HPs-A	0.4	A1-b, PN3	1026	1536	0.8	62.8	14.9	0.5	2.4	0.32	55.8	62.6
	2	Ar4, G7, H10, B7	5130	7680	4.0	313.8	74.4	2.6	12.2	1.60	279.2	313.0
HPs-B	0.1	M1	240.9	617.3	0.5	3.8	5.2	0.4	2.2	0.10	9.8	13.7
	0.3	M1	722.7	1852	1.5	11.3	15.7	1.1	6.7	0.30	29.5	41.2
	0.6	M1	1445	3704	2.9	22.7	31.3	2.3	13.3	1.60	59.0	82.3
HPs-C	0.03	M1B	145.0	228.7	0.3	3.1	2.8	0.4	1.6	0.07	5.8	9.7
	0.06	M1B	290.1	457.4	0.6	6.2	5.5	0.8	3.2	0.13	11.6	19.4
HPs-D	0.05	YS1, Seats	153.9	234.4	0.3	2.2	3.4	0.2	2.3	0.09	8.4	17.7
	0.1	YS1, Seats, E4, D5	307.7	468.8	0.7	4.3	6.8	0.5	4.6	0.18	16.7	35.3

Note: DPs indicates dust particles; HPs indicates haze particles.

Dust particles

In general, DP additions stimulated phytoplankton growth in Chl a at all stations except stations YS1 and PN3 (no significant difference relative to the control at both stations, Fig. 3). However, the stimulation effect was generally more obvious in lower trophic seawater than it was in higher trophic seawater, as follows.

1. Phytoplankton in lower trophic seawaters relative to higher trophic seawaters exhibited more obvious responses to the same added amounts of DPs (Fig. 3). At stations Ar4 and G7 in the SG, RC$_{Chl\,a}$ after DP addition (1.16 μmol·L^{-1}) was about 67% and about 122%, respectively. These changes were higher than those at the same DP loadings at stations H10 (about 19%) and B7 (about 11%) in the YS. In the KE, RC$_{Chl\,a}$ at similar DP loadings showed higher values at station M1B than at station M1. The

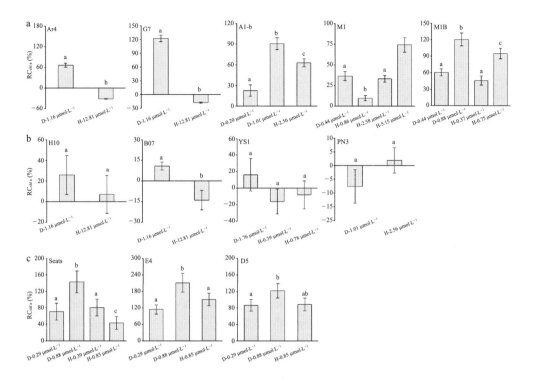

a. Open oceans; b. YS and ECS; c. SCS. The added amounts of dust and haze particles are expressed as the N loading ($\mu mol \cdot L^{-1}$). D- indicates dust particle treatments, H- indicates haze particle treatments. The $RC_{Chl\,a}$ labelled with the same letter indicates no significant difference between these treatments ($P > 0.05$), with the different letters indicating significant differences between these treatments ($P < 0.05$).

Fig. 3 Relative change in Chl a ($RC_{Chl\,a}$) in dust and haze particle treatments

maximum $RC_{Chl\,a}$ of about 212% was observed in the higher DP treatments (0.88 $\mu mol \cdot L^{-1}$) at station E4, where seawater exhibited the most oligotrophic characteristics (Fig. 1). This maximum value was 48% and 74% higher than those in the higher DP treatments (0.88 $\mu mol \cdot L^{-1}$) at stations Seats and D5, respectively. Similar responses in the lower DP treatments were observed among the three stations.

2. The phytoplankton response with an increasing DP loading in higher trophic seawater was more obvious than that in lower trophic seawater (Fig. 3). In the KE, the relative change in Chl a ($RC_{Chl\,a}$) increased by 200%–300% with increasing DP loadings (from 0.20 $\mu mol \cdot L^{-1}$ to 1.01 $\mu mol \cdot N \cdot L^{-1}$). In contrast, $RC_{Chl\,a}$ increased by less than 100% with increasing DP loadings (from 0.29 $\mu mol \cdot L^{-1}$ to 0.88 $\mu mol \cdot L^{-1}$) in the SCS, as characterized by lower trophic statues relative to the KE.

Haze particles

In contrast to the DP treatments, the addition of HPs generally stimulated phytoplankton growth at low particle loadings and gradually showed an inhibition effect with increasing particle loadings (Fig. 3). Such effects exhibited the following characteristics.

1. The phytoplankton response with increasing HP loading showed an opposite trend in the KE and SCS (Fig. 3). Similar to DP treatments, the stimulation effect of HP additions became increasingly pronounced with increasing HP loading at stations M1 and M1B in the KE (Fig. 3). In contrast, at station Seats in the SCS, the higher additions of HPs (0.85 $\mu mol \cdot L^{-1}$) had a weaker stimulation effect on the increase in Chl a relative to the lower additions (0.39 $\mu mol \cdot L^{-1}$).

2. The increases in Chl a due to HP additions

were generally smaller than those due to DP additions at the same N loadings (Fig. 3). In the SCS, $RC_{Chl\,a}$ in the DP treatments at 0.88 μmol·L^{-1} was 38% to 230% higher than that in the HP treatments at 0.85 μmol·L^{-1}. In addition, although HP additions supplied 2.56 μmol·L^{-1} at station A1-b, $RC_{Chl\,a}$ was only 70% of that in the DP treatment at 1.01 μmol·L^{-1}. At station M1, the stimulation effect of HPs at 2.58 μmol·L^{-1} was comparable to that of the DP treatment at 0.44 μmol·L^{-1}.

3. The phytoplankton response was primarily dominated by the inhibition effect at extremely high particle loadings (12.81 μmol·L^{-1}, Fig. 3). At stations Ar4 and G7 in the SG, and B7 in the YS, $RC_{Chl\,a}$ in the HP treatments ranged between −31% and −13%. At station H10, the addition of HPs had an insignificant impact on phytoplankton growth relative to the control treatment.

Phytoplankton size structure

Along with the fertilization effect of DP and HP additions (excluding HP additions at a loading of 12.81 μmol·L^{-1}), the phytoplankton size structure generally shifted towards larger cells (> 2 μm in cell size) at stations in the KE and SCS (Fig. 4). However, the extent of this enhanced shift induced by DP addition was more pronounced than that induced by HP addition. At station A1-b, $C_{L\text{-Max}}$ in the DP treatments was about 51% at 0.20 μmol·L^{-1} and about 57% at 1.01 μmol·L^{-1}; this contribution was higher than that in the control treatment (about 46%). In contrast, HP addition at 2.56 μmol·L^{-1} induced a $C_{L\text{-Max}}$ value of only about 54%. At stations M1 and M1B, $C_{L\text{-Max}}$ values were comparable for DP treatments at 0.44 μmol·L^{-1} and HP treatments at 2.58 μmol·L^{-1} and 0.75 μmol·L^{-1}. In the SCS, $C_{L\text{-Max}}$ in DP treatments at 0.29−0.88 μmol·L^{-1} was 25% to 163% higher than those in the control treatment, while HP additions at 0.85 μmol·L^{-1} induced only a 0−55% increase relative to the control. Hence, at the same N loadings, HP additions decelerated the shift of phytoplankton size structure towards larger size cells to some extent in contrast to DP additions.

When the inhibition effect clearly occurred at the 12.81 μmol·L^{-1} HP loading, the shift in phytoplankton size structure varied at different stations (Fig. 4). DP and HP additions at stations Ar4, H10, and B7 increased $C_{L\text{-Max}}$ in contrast to the control treatments. On the other hand, at station G7, $C_{L\text{-Max}}$ in the HP treatments was only about 70% of that in the control treatments, while the addition of DPs at 1.16 μmol·L^{-1} increased $C_{L\text{-Max}}$ relative to the control. At station YS1 in the YS, although there was no significant difference in total Chl a between the DP/HP and control treatments, HP additions generally increased the contribution of pico-sized cells to total Chl a. The impact of DP/HP additions on phytoplankton size structure at station PN3 in the ECS was insignificant relative to the control, consistent with the phytoplankton response in total Chl a.

Discussion

▶ *Promotion effects of dust and haze particles on phytoplankton*

In our study, the primary N limitation at most of the incubation stations qualitatively determined the important role of N supplied by DP/HP additions in stimulating phytoplankton growth. However, the phytoplankton response in $RC_{Chl\,a}$ varied at the same N loadings across different stations (Fig. 3). Indeed, the occurrence of primarily N limitation can be observed across oligotrophic to eutrophic seawater in this study and various other studies, which leads to the varying roles of external N input in affecting phytoplankton[15,34]. For instance, dust addition has the potential to relieve, change, or even intensify N limitation in seawater, depending on the contribution of the added amount of N to the N standing stock in seawater[15]. Given the important role of trophic status in affecting the phytoplankton response to DP/HP additions and the dominantly N fertilization impact, we introduced the P_{SN} ((the added N amounts/N stocks in the baseline seawater) ×100) to illustrate the extent of change in external N input relative to N stocks in the baseline seawater, and found a significant correlation between $RC_{Chl\,a}$

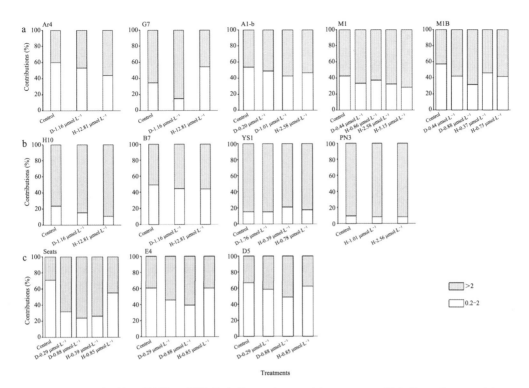

a. Open oceans; b. YS and ECS; c. SCS. D- indicates dust particle treatments, H- indicates haze particle treatments.

Fig. 4　Contributions of size-fractionated Chl a (> 2 μm and 0.2–2 μm) to total Chl a in dust and haze particle treatments

and P_{SN} ($P < 0.01$, Fig. 5). This quantitative relationship allows us to better evaluate and predict the impact of atmospheric deposition on primary productivity in the upper ocean across varying trophic statuses, where N is usually regarded as a limiting nutrient[3,34].

Furthermore, when the nutrient treatments (N and N+P) were considered, we found that $RC_{Chl\ a}$ in the N treatments was much lower than those fitted by DP treatments when the P_{SN} exceeded about 480% in this study. In contrast, $RC_{Chl\ a}$ in the N+P treatments was generally closer to the fitted line of DPs at the same N loading (Fig. 5). This suggests that the promotion effect of DP addition tends to be a combined result induced by N and P rather than N alone. However, the direct supply of P by DP additions was slight relative to the baseline seawater (Fig. 1, Table 1). This phenomenon was likely related to the enhanced utilization of dissolved organic phosphorous (DOP) due to DP additions[35,36]. In our previous study, the calculation of the bioavailable P budget for the incubation system illustrated the possibility of the utilization of DOP in the YS and SG of the NWPO[26]. Indeed, on the one hand, the substantial supply of N relative to P by DP additions strengthened the consumption of bioavailable P in seawater (Fig. S4), and facilitated the initiation of DOP utilization[34]; on the other hand, DP additions also supplied trace metals such as Fe as a cofactor of an alkaline phosphatase enzyme, which can convert DOP into dissolved inorganic phosphorous for phytoplankton uptake[36]. The enhanced utilization of DOP by DP additions has also been widely reported by various studies in the SCS[37], Red Sea[35], vast areas of the Atlantic Ocean, etc[36]. Our recent study conducted in the China coastal seas also revealed that anthropogenic aerosol additions significantly increased alkaline phosphatase activity (data not

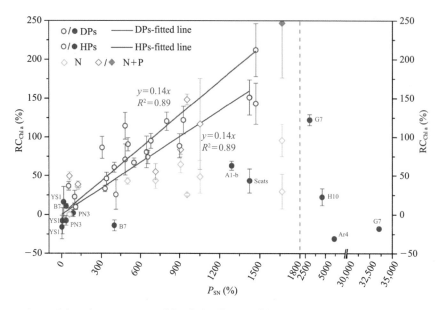

The experimental data that were not used for fitting lines (solid circle) includes: (1) data in all treatments at stations YS1 and PN3, where additions of DPs and HPs showed no significant impact on phytoplankton due to their extreme physiological statuses, i.e., exponential growth phase at station YS1 and decline phase at station PN3; (2) data in the HP treatments at stations Ar4, G7, H10, B7, and Seats (only referring to HP treatments at high N loadings of 0.85 μmol·L^{-1}) and A1-b, of which there was an obvious inhibition impact of HP additions on phytoplankton; (3) data in the DP treatments at stations B7 and G7, where the P limitation was more obvious than the N limitation (Fig. S2).

Fig. 5 The relationship between the relative change in Chl a ($RC_{Chl\,a}$) and the proportion of N input relative to N stocks in the baseline seawater (P_{SN}, (the added N amounts/N stocks in the baseline seawater)×100) in the dust particle (DP), haze particle (HP), N, and N+P treatments (on the left of the dotted grey line)

shown), which reflects the utilization of DOP. In addition, the Fe, Mn, etc. supplied by DP additions also have the potential to exert a direct stimulation effect on phytoplankton[38]. While these hypotheses must be further substantiated, it is certain that DP addition stimulated phytoplankton growth more efficiently than N alone.

The phytoplankton response in Chl a to HP addition was different from that to DP addition. When the P_{SN} was less than about 900%, $RC_{Chl\,a}$ in the HP treatment increased linearly with P_{SN} and was generally higher than that in the N treatment at the same N loading. Interestingly, the slope of the fitted line obtained from the HP treatments (0.11) was slightly lower than that from the DP treatments (0.14, Fig. 5). This is consistent with the result that the promotion impact of HPs on phytoplankton growth was generally lower than that of DPs at the same N loading (Fig. 3). Moreover, with the increase in P_{SN}, $RC_{Chl\,a}$ in the HP treatment decreased gradually and reached negative values when the P_{SN} was high enough (i.e., at a loading of 12.81 μmol·L^{-1}, Fig. 3 and 5). This suggests the existence of an inhibition effect on phytoplankton in the HP treatments. The inhibition effect will be further discussed in the following sections.

In terms of the promotion effect, our study demonstrated the combined effect of N and P in different types of atmospherically deposited matter (i.e., DPs and HPs) on phytoplankton growth. While various studies have reported the dominant N fertilization effect on phytoplankton, they have ignored the effect of P due to the slight P supply relative to N in atmospherically deposited matter[3,20]. Although the addition of DPs and HPs had different promotion efficiencies in phytoplankton response at the same N loadings, their impacts were generally higher than those induced by N alone (Fig. 5). Hence, if we considered only the effect of N deposition, the overall effect of atmospheric deposition on phytoplankton would be underestimated. On the other hand, the input of atmospherically deposited matter will increase the N:P ratios in seawater and the N:P uptake ratios of phytoplankton, which would significantly affect the phytoplankton community[14,15]. Our study suggests that the lack of consideration of the indirect supply of P induced by atmospheric deposition would cause the deviation of the quantitative relationship between N:P ratios and phytoplankton communities, which will affect the prediction of models for both regional and global oceans[14].

▶ *Distinct effects of dust and haze particles on phytoplankton size structure*

As shown in Fig. 6, the phytoplankton size structure generally shifted towards larger cells under the impact of DP addition. This is consistent with the results in our previous study[26] as well as other studies[31,32]. Indeed, the nutrients supplied by DP addition increased the trophic status in the incubated seawater and were favorable for the growth of large phytoplankton, which were characterized by a stronger nutrient storage ability and a higher biomass-specific growth rate[32]. In contrast to DP additions, some toxic matter in HPs weakened, to some extent, the fertilization impact of nutrients (primarily related to N) supplied by HPs on the phytoplankton size structure, i.e., shifting towards larger cells (Fig. 6). In addition, when the particle loadings were large enough (i.e., P_{SN} ⩾ about 900%), HP additions tended to shift the phytoplankton size structure towards small cells. In contrast to large cells, small cells are generally more sensitive to toxic matter because of their larger ratios of surface area to volume[32]. However, the short generation time allows small cells to recover more quickly from toxic impacts[39]. Note that HP additions at stations Ar4, H10, and B7 increased C_{L-Max} compared with the control treatments

(Fig. 4). This is because HP additions inhibited phytoplankton of all sizes first and then gradually increased the contribution of pico-sized cells to total Chl a due to their short generation period (Fig. S5 and S6). Overall, the toxic matter in HPs was not conducive to the growth of large phytoplankton relative to that of small phytoplankton.

The complicated impact of HPs on phytoplankton size structure has the potential to affect carbon sequestration of the ocean. In contrast to small phytoplankton, large phytoplankton characterized by higher biomasses and sinking rates more easily transport fixed carbon to the deep ocean[40]. Hence, the toxic impact of HP addition on the phytoplankton size structure likely reduces the carbon export efficiency induced by purely HP fertilization. With the increasing influence of anthropogenic activities, the impact of atmospheric deposition on phytoplankton may not be exhibited as only fertilization or only inhibition; Rather, it may tend to be the net effect of both. On the other hand, in the context of excess anthropogenic N input, the gradual change in nutrient limitation from N to P in the upper ocean has the potential to reduce the fertilization effect of atmospheric deposition characterized by high ratios of N:P[6,33], and consequently the toxic effect likely becomes obvious over time in the future[23,24].

▶ *Complex effects of HPs on phytoplankton*

There are many types of matter in HPs that potentially lead to toxic impacts, the most common being heavy metals, such as Cu, Cd, and lead (Pb) [24,25]. The toxic impact of Cu deposition has been widely reported in previous studies[22-24]. In our study, we found that the phytoplankton response in $RC_{Chl\,a}$ decreased with increasing added amounts of Cu in the HP treatments, while a positive correlation between $RC_{Chl\,a}$ and Cu was observed in the DP treatments (Fig. S7). However, we found no significant difference in Chl a between the N (1 μmol·L^{-1}) and Cu+N (0.1 μg·L^{-1}+1 μmol·L^{-1}) treatments at station A1-b (Fig. S8[27]). A seemingly paradoxical phenomenon concerning Cu toxicity on

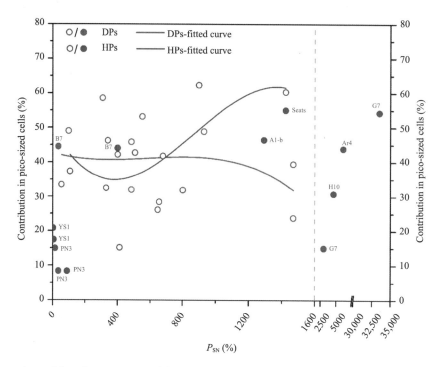

The experimental data that were not used for fitting curves (solid circle) includes: (1) data in all treatments at stations YS1 and PN3, where additions of DPs and HPs showed no significant impact on phytoplankton due to their extreme physiological statuses, i.e., exponential growth phase at station YS1 and decline phase at station PN3; (2) data in the HP treatments at stations Ar4, G7, H10, B7, and Seats (only referring to HP treatments at high N loadings of 0.85 μmol·L^{-1}) and A1-b, of which there was an obvious inhibition impact of HP additions on phytoplankton; (3) data in the DP treatments at stations B7 and G7, where the P limitation was more obvious than the N limitation (Fig. S2).

Fig. 6 The relationship between the maximum contribution of pico-sized cells to total Chl a (open circle) and the proportion of N input relative to N stocks in the baseline seawater (P_{SN}, (the added N amounts/N stocks in the baseline seawater)×100) in dust particle (DP) and haze particle (HP) treatments (on the left of the dotted grey line)

phytoplankton was also observed in the area of the NWPO adjacent to the East Asian continent[18,23]. Indeed, various factors, such as the varying tolerance of different phytoplankton and in/organic matter in seawater, can affect Cu toxicity[22,41]. Apart from Cu, the considerable supply of Cd and Pb also likely inhibited phytoplankton growth by disrupting the photosynthetic system, reproduction rate, etc.[25,41] (Table 1). Hence, the toxicity of HPs on phytoplankton is likely a combined impact of various types of matter. Although much work is still needed to ascertain the types of toxic matter in HPs, the complex impacts of HPs on phytoplankton growth and size structure appear to affect marine biogeochemical cycles in a unique way, rather than in the traditional way, i.e., fertilization alone or inhibition alone.

Given the rapid uptake by biomes and the vertical/horizontal water mixing in the ocean[26,42], the instantaneous increase in the concentration of anthropogenic aerosols in the seawater was generally too small to cause an obvious toxic impact. The estimated maximum cumulative mass loadings (ten days, maximum duration of a typical haze event) of anthropogenic aerosols in the surface seawater (10 m) were no larger than 0.06 mg·L^{-1} in the NWPO[27]. Moreover, the N loadings in the HP treatments that caused an obvious toxic impact (12.81 μmol·L^{-1}) in this study were much higher than the maximum accumulated N concentration induced by atmospheric deposition in the surface seawater of the NWPO (about 0.41 μmol·L^{-1}, which was calculated as (410 μmol·m^{-2}·d^{-1}/10 m)×10 d[33]). Note that an extreme dust storm event accompanied by strong wet deposition has the potential to create an obvious accumulation of aerosols (over 2 mg·L^{-1}) instantaneously in surface seawater, which causes a bloom of phytoplankton rather than a toxic impact[11]. At the present time, although various enrichment incubation experiments have illustrated that atmospherically deposited matter has the ability to inhibit phytoplankton growth[22,23,43], there is still a lack of adequate evidence proving the existence of overwhelmingly toxic impacts under realistic conditions. Therefore, our study argues that the dominant impact of anthropogenic deposition on marine primary productivity is still stimulated at the present time[27], while the potentially inhibition impact is not neglected given the extreme weather events and the possible accumulation of toxic matter in seawater[23,34].

Conclusion

In summary, we performed a collective analysis of a series of incubation experimental data from five cruises across the China coastal seas including the YS, ECS, SCS and open oceans of the NWPO including the SG and KE. By comparing phytoplankton responses to DPs, HPs, and various nutrient additions, we found that additions of DPs and HPs (referring to only low HP additions between 0.39 μmol·L^{-1} and 2.56 μmol·L^{-1}) generally stimulated phytoplankton growth and shifted the size structure towards larger cells. Such stimulation impact due to DP additions was more obvious than those due to HP additions at the same N loadings. In addition, the inhibition impact of HPs became dominant when the added amount of HPs was large enough, i.e., 12.81 μmol·L^{-1}. This suggests that the phytoplankton response to HP addition is a net effect of nutrient-derived stimulation and toxic matter-derived inhibition. In contrast to the pure fertilization effect of DPs, the net fertilization effect of HPs tends to reduce the efficiency of carbon fixation due to the lower stimulation efficiency and that of carbon export to the deep ocean due to the weaker shift in size structure towards larger cells[40,44].

As our study highlights the distinct impacts of DPs and HPs on phytoplankton growth and shifts in size structure, it has the following two implications. First, it emphasizes the varying roles of atmospheric deposition under different weather conditions in marine primary productivity. While the conventional estimate of monthly or annual nutrient deposition flux reflects the overall input of nutrients to the ocean, it cannot distinguish the varying stimulation efficiencies of the same nutrient, e.g., at the same N loading in our study, originating from different aerosols on marine primary productivity[3,33]. Second, with the enhanced emissions of anthropogenic pollutants, such as oxidized N in the near future, the chemical compositions of DPs and HPs are likely to change continuously (e.g., N content, solubility of P and trace metals)[45,46]. Moreover, the strongest dust storm that occurred in the last decade made inroads on the East Asian continent in 2021 (https://www.chinadailyhk.com/article/160558), and accordingly, it also potentially enhanced the influence of extreme dust storm events on the marine environment. Hence, the alternate occurrence of heavy haze weather and strong dust storms will likely become a trend, and the resultant change in DP and HP-derived stimulation efficiency in primary productivity and phytoplankton community structure deserves constant attention[2,6]. Note that the DPs used in this study are artificially modified dust particles, which are simulated primarily based on the mixing of soil samples and air pollutants including HNO_3 and H_2SO_4[26]. Considering the complexity of the dust aging process during long-range transport[16,17], more studies using realistic dust samples are necessary to verify and/or improve the relationship between the phytoplankton response and DPs concluded in this study. On the other hand, a recent study found that the enhanced grazing pressure on phytoplankton induced by atmospheric deposition exhibited an important regulatory effect on carbon export to the deep ocean[44]. The impact of different atmospherically deposited matter on zooplankton calls for further investigation.

Acknowledgments

This work was financially supported by National Natural Science Foundation of China (NSFC, 41906119 and 41876125), NSFC-Shandong Joint Fund (U1906215), and Major State Basic Research Development Program of China (973 Program, 2014CB953701). We thank the open research cruise NORC2015-05 supported by NSFC Shiptime Sharing Project (project number: 41449905) for supporting microcosm incubation experiments conducted in the SCS.

Author Contributions

Chao Zhang: Conceptualization, formal analysis, investigation, methodology, visualization, writing (original draft), funding acquisition. Qiang Chu: Data curation, investigation, visualization. Yingchun Mu: Formal analysis, investigation, methodology. Xiaohong Yao: Data curation, writing (review and editing), supervision. Huiwang Gao: Conceptualization, methodology, resources, writing (review and editing), supervision, funding acquisition.

Competing Interests

The authors declare that they have no known competing financial interests or personal relationships that could have appeared to influence the work reported in this paper.

References

1. Mahowald N, Jickells T D, Baker A R, et al. Global distribution of atmospheric phosphorus sources, concentrations and deposition rates, and anthropogenic impacts[J]. Global Biogeochem. Cycles, 2008, 22(4): GB4026.
2. Kim I-N, Lee K, Gruber N, et al. Increasing anthropogenic nitrogen in the North Pacific Ocean[J]. Science, 2014, 346(6213): 1102-1106.
3. Duce R A, Liss P S, Merrill J T, et al. The atmospheric input of trace species to the world ocean[J]. Global Biogeochem. Cycles, 1991, 5(3): 193-259.
4. Shao Y, Wyrwoll K H, Chappell A, et al. Dust cycle: An emerging core theme in Earth system science[J]. Aeolian Res., 2011, 2(4): 181-204.
5. Schulz M, Prospero J, Baker A, et al. Atmospheric transport and deposition of mineral dust to the ocean: implications for research needs[J]. Environ. Sci. Technol., 2012, 46(19): 10390-10404.
6. Moon J Y, Lee K, Lim W A, et al. Anthropogenic nitrogen is changing the East China and Yellow seas from being N deficient to being P deficient[J]. Limnol. Oceanogr., 2020, 66(3): 914-924.
7. Boyd P W, Jickells T, Law C, et al. Mesoscale iron enrichment experiments 1993–2005: Synthesis and future directions[J]. Science, 2007, 315(5812): 612-617.
8. Mahowald N M, Hamilton D S, Mackey K R, et al. Aerosol trace metal leaching and impacts on marine microorganisms[J]. Nat. Commun., 2018, 9(1): 1-15.
9. Pinedo-Gonzalez P, Hawco N J, Bundy R M, et al. Anthropogenic Asian aerosols provide Fe to the North Pacific Ocean[J]. Proc. Natl. Acad. Sci. U. S. A., 2020, 117(45): 27862-27868.
10. Luo C, Mahowald N, Bond T, et al. Combustion iron distribution and deposition[J]. Global Biogeochem. Cycles, 2008, 22(1): GB1012.
11. Shi J H, Gao H W, Zhang J, et al. Examination of causative link between a spring bloom and dry/wet deposition of Asian dust in the Yellow Sea, China[J]. J. Geophys. Res.:Atmos., 2012, 117: D17.
12. Yoon J E, Kim K, Macdonald A M, et al. Spatial and temporal variabilities of spring Asian dust events and their impacts on chlorophyll- a concentrations in the western North Pacific Ocean[J]. Geophys. Res. Lett., 2017, 44(3): 1474-1482.
13. Maranon E, Fernandez A, Mourino-Carballido B, et al. Degree of oligotrophy controls the response of microbial plankton to Saharan dust[J]. Limnol. Oceanogr., 2010, 55(6): 2339-2352.
14. Chien C T, Mackey K R, Dutkiewicz S, et al. Effects of African dust deposition on phytoplankton in the western tropical Atlantic Ocean off Barbados[J]. Global Biogeochem. Cycles, 2016, 30(5): 716-734.
15. Zhang C, Yao X H, Chen Y, et al. Variations in the phytoplankton community due to dust additions in eutrophication, LNLC and HNLC oceanic zones[J]. Sci. Total Environ., 2019, 669: 282-293.
16. Meskhidze N, Chameides W L, Nenes A, et al. Iron mobilization in mineral dust: Can anthropogenic SO_2 emissions affect ocean productivity?[J]. Geophys. Res. Lett., 2003, 30(21): 267-283.

17. Guieu C, Dulac F, Desboeufs K, et al. Large clean mesocosms and simulated dust deposition: A new methodology to investigate responses of marine oligotrophic ecosystems to atmospheric inputs[J]. Biogeosciences, 2010, 7(9): 2765-2784.
18. Meng X, Chen Y, Wang B, et al. Responses of phytoplankton community to the input of different aerosols in the East China Sea[J]. Geophys. Res. Lett., 2016. 43(13): 7081-7088.
19. Duce R A, LaRoche J, Altieri K, et al. Impacts of atmospheric anthropogenic nitrogen on the open ocean[J]. Science, 2008, 320(5878): 893–897.
20. Martino M, Hamilton D, Baker A R, et al. Western Pacific atmospheric nutrient deposition fluxes, their impact on surface ocean productivity[J]. Global Biogeochem. Cycles, 2014, 28(7): 712-728.
21. Echeveste P, Dachs J, Berrojalbiz N, et al. Decrease in the abundance and viability of oceanic phytoplankton due to trace levels of complex mixtures of organic pollutants[J]. Chemosphere, 2010, 81(2): 161–168.
22. Wang F J, Chen Y, Guo Z Z, et al. Combined effects of iron and copper from atmospheric dry deposition on ocean productivity[J]. Geophys. Res. Lett., 2017, 44(5): 2546-2555.
23. Paytan A, Mackey K R, Chen Y, et al. Toxicity of atmospheric aerosols on marine phytoplankton[J]. Proc. Natl. Acad. Sci. U. S. A., 2009, 106(12): 4601-4605.
24. Yang T J, Chen Y, Zhou S Q, et al. Impacts of aerosol copper on marine phytoplankton: A review[J]. Atmosphere, 2019, 10(7): 414.
25. Kumar K S, Dahms H U, Lee J S, et al. Algal photosynthetic responses to toxic metals and herbicides assessed by chlorophyll a fluorescence[J]. Ecotoxicol. Environ. Saf., 2014, 104: 51-71.
26. Zhang C, Gao H W, Yao X H, et al. Phytoplankton growth response to Asian dust addition in the northwest Pacific Ocean versus the Yellow Sea[J]. Biogeosciences, 2018, 15(3): 749-765.
27. Zhang C, Ito A, Shi Z B, et al. Fertilization of the Northwest Pacific Ocean by East Asia air pollutants[J]. Global Biogeochem. Cycles, 2019, 33(6): 690-702.
28. Maring H, Savoie D L, Izaguirre M A, et al. Mineral dust aerosol size distribution change during atmospheric transport[J]. J. Geophys. Res.: Atmos., 2003, 108(D19): 8592.
29. Strickland J, Parsons T. A Practical Handbook of Seawater Analysis[M]. 2nd. Ottawa: Fisheries Research Board of Canada Bulletin 167, 1972.
30. Welschmeyer N A. Fluorometric analysis of chlorophyll a in the presence of chlorophyll b and pheopigments[J]. Limnol. Oceanogr., 1994, 39(8): 1985-1992.
31. Guo C, Yu J, Ho T-Y, et al. Dynamics of phytoplankton community structure in the South China Sea in response to the East Asian aerosol input[J]. Biogeosciences, 2012, 9(4): 1519-1536.
32. Maranon E. Cell size as a key determinant of phytoplankton metabolism and community structure[J]. Annu. Rev. Mar. Sci., 2015, 7: 241-264.
33. Qi J H, Yu Y, Yao X H, et al. Dry deposition fluxes of inorganic nitrogen and phosphorus in atmospheric aerosols over the Marginal Seas and Northwest Pacific[J]. Atmos. Res., 2020, 245: 105076.
34. Moore C M, Mills M M, Arrigo K R, et al. Processes and patterns of oceanic nutrient limitation[J]. Nat. Geosci., 2013, 6(9): 701-710.
35. Mackey K R, Roberts K, Lomas M W, et al. Enhanced solubility and ecological impact of atmospheric phosphorus deposition upon extended seawater exposure[J]. Environ. Sci. Technol., 2012, 46(19): 10438-10446.
36. Browning T J, Achterberg E P, Yong J C, et al. Iron limitation of microbial phosphorus acquisition in the tropical North Atlantic[J]. Nat. Commun., 2017, 8(1): 15465.
37. Chu Q, Liu Y, Shi J, et al. Promotion effect of Asian dust on phytoplankton growth and potential dissolved organic phosphorus utilization in the South China Sea[J]. J. Geophys. Res. Biogeosci., 2018, 123(3): 1101-1116.
38. Sunda W. Feedback interactions between trace metal nutrients and phytoplankton in the ocean[J]. Front. Microbiol., 2012, 3: 204.
39. Levy J L, Stauber J L, Jolley D F. Sensitivity of marine microalgae to copper: the effect of biotic factors on copper adsorption and toxicity[J]. Sci. Total Environ., 2007, 387(1-3): 141-154.
40. Bach L T, Riebesell U, Sett S, et al. An approach for particle sinking velocity measurements in the 3–400 μm size range and considerations on the effect of temperature on sinking rates[J]. Marine Biol., 2012, 159(8): 1853-1864.
41. Debelius B, Forja J M, DelValls Á, et al. Toxicity and bioaccumulation of copper and lead in five marine microalgae[J]. Ecotoxicol. Environ. Saf., 2009, 72(5): 1503-1513.
42. Zhang J, Guo X Y, Zhao L. Tracing external sources of nutrients in the East China Sea and evaluating their contributions to primary production[J]. Prog. Oceanogr., 2019, 176: 102122.
43. Zhou W, Li Q P, Wu Z. Coastal phytoplankton responses to atmospheric deposition during summer[J]. Limnol. Oceanogr., 2020, 66(4): 1298-1315.

44. Xiu P, Chai F. Impact of atmospheric deposition on carbon export to the deep ocean in the subtropical Northwest Pacific[J]. Geophys. Res. Lett., 2021, 48(6): e2020GL089640.
45. Li W J, Xu L, Liu X H, et al. Air pollution–aerosol interactions produce more bioavailable iron for ocean ecosystems[J]. Sci. Adv., 2017, 3(3): e1601749.
46. Zhang J X, Gao Y, Leung L R, et al. Impacts of climate change and emissions on atmospheric oxidized nitrogen deposition over East Asia[J]. Atmos. Chem. Phys., 2019, 19(2): 887-900.

Effect of anthropogenic aerosol addition on phytoplankton growth in coastal waters: Role of enhanced phosphorus bioavailability[①]

Qin Wang[1,2], Chao Zhang[1,2], Haoyu Jin[1,2], Ying Chen[3], Xiaohong Yao[1,2], Huiwang Gao[1,2*]

1 Frontiers Science Center for Deep Ocean Multispheres and Earth System, Key Laboratory of Marine Environment and Ecology, Ocean University of China, Ministry of Education of China, Qingdao, China.
2 Laboratory for Marine Ecology and Environmental Sciences, Pilot National Laboratory for Marine Science and Technology, Qingdao, China.
3 Shanghai Key Laboratory of Atmospheric Particle Pollution Prevention, Department of Environmental Science and Engineering, Fudan University, Ministry of Education of China, Shanghai, China.
* Corresponding author: Huiwang Gao (hwgao@ouc.edu.cn)

Abstract

Atmospheric deposition can supply nutrients to induce varying responses of phytoplankton of different sizes in the upper ocean. Here we collected surface and SCM (subsurface chlorophyll a maximum) seawaters from the Yellow Sea and East China Sea to conduct a series of onboard incubation experiments, aiming to explore the impact of anthropogenic aerosol (AR, sampled in Qingdao, a coastal city in Northern China) addition on phytoplankton growth using schemes with (unfiltered seawater, UFS) and without (filtered seawater, FS) microsized (20–200 μm) cells. We found that AR addition stimulated phytoplankton growth obviously, as indicated by chlorophyll a (Chl a) in surface incubations, and had stimulatory or no effects in SCM incubations, which was related to nutrient statuses in seawater. The high ratio of nitrogen (N) to phosphorus (P) in the AR treatments demonstrated that P became the primary limiting nutrient. The alkaline phosphatase activity (APA), which can reflect the rate at which dissolved organic P (DOP) is converted into dissolved inorganic P, was 1.3–75.5 times higher in the AR treatments than in the control, suggesting that AR addition increased P bioavailability in the incubated seawater. Dinoflagellates with the capacity to utilize DOP showed the dominant growth in the AR treatments, corresponding to the shift in phytoplankton size structure towards larger cells. Surprisingly, we found that nanosized (2–20 μm) and picosized (0.2–2 μm) Chl a concentrations in UFS were generally higher than those in FS. The APA in UFS was at least 1.6 times higher than in FS, and was proportional to the contribution of microsized cells to the total Chl a, suggesting that microsized cells play an important role in the increase in APA, which contributes to the growth of nanosized and picosized phytoplankton. Current work provides new insight into the increase of P bioavailability induced by atmospheric deposition and resultant ecological effect in coastal waters.

Keywords: atmospheric deposition; nutrients; phytoplankton; size structure; China coastal waters; alkaline phosphatase

---

① 本文于2022年6月发表在 *Frontiers in Microbiology* 第13卷，https://doi.org/10.3389/fmicb.2022.915255。

Introduction

Atmospheric deposition can supply a considerable amount of nutrients, including macronutrients such as nitrogen (N) and phosphorus (P), and micronutrients such as iron (Fe) and zinc (Zn), to the ocean[1,2], and affect the phytoplankton size structure and community composition[3-5]. The deposition of a large amount of N promoted the growth of diatoms and inhibited the growth of diazotrophs in the Bay of Bengal and the Arabian Sea[6]. Dust additions increased the N:P ratios in the seawater and induced the dominant growth of nanosized (2–20 μm) phytoplankton in the East China Sea (ECS)[3]. The shift in phytoplankton size structure is always accompanied by the competition for nutrients among phytoplankton of different sizes[7,8]. Large phytoplankton (≥2 μm in cell size) have a greater capacity for biomass accumulation and anti-predator defense[9,10], leading to an advantageous growth in eutrophic seawater. With minimal diffusion boundary layer thickness and a larger specific surface area[9,11-13], picosized (0.2–2 μm in cell size) phytoplankton have a competitive advantage in oligotrophic seawater[9,14]. However, this consensus is roughly defined and the nutrient competition mechanism between phytoplankton of different sizes is rather complicated in realistic conditions, where the trophic status is not ideally eutrophic or oligotrophic[9,14]. For example, in contrast to diatoms, dinoflagellates have a growth advantage in high-nitrate and low-phosphate seawaters due to their acclimatization to high ratios of N:P[3,15], even if there is an overlap in the size structure of diatoms and dinoflagellates. In HNLC (high nutrient low chlorophyll) and coastal seawaters, dust additions can induce the rapid growth of different kinds of diatoms covering nanosized and microsized cells[16,17]. Our quantitative knowledge of the relationship between nutrient uptake and the growth of different sized phytoplankton is still inadequate.

The impact of atmospheric deposition on primary productivity is generally associated with the substantial supply of N and/or Fe nutrients, while few studies focus on P due to its negligible supply relative to N and Fe[5,18,19]. Although some studies pointed out that atmospheric deposition can promote the utilization of DOP to relieve P limitation by providing cofactors such as Fe and Zn in open oceans primarily characterized by oligotrophy[20,21], there are few studies reported in coastal waters characterized by mesotrophy and even eutrophy. In the context of the overwhelming input of N relative to P through various ways such as riverine input and atmospheric deposition, the phenomenon of P limitation becomes increasingly prevailing in coastal waters[22]. The impact of atmospheric deposition on marine phytoplankton is not only confined to the traditional relationship between supply (e.g., N and Fe supply) and demand (e.g., N and Fe limitation), but also considering the acclimatization mechanism to copy with the potential P deficiency. A few studies have deduced that atmospheric deposition might enhance the utilization of DOP in P-deficient environments by calculating the P budget in the system and setting up model parameters[16,23]. However, there is still a lack of direct evidence to verify this hypothesis in coastal waters, and the resultant ecological effect is still poorly understood.

The Yellow Sea (YS) and ECS adjacent to the East Asian continent are marginal seas of the northwestern Pacific Ocean and are obviously influenced by anthropogenic air pollutants from the surrounding continent[24,25]. A series of studies reported that anthropogenic aerosols can transport a long distance to reach coastal seas and even open oceans[26-28]. The source apportionment results also showed that particles collected in the YS were full of secondary, biomass burning and soot-like particles, indicating that marine aerosols are strongly affected by anthropogenic activities[28-30]. The N:P ratio in anthropogenic aerosol (AR) is generally much higher than the phytoplankton stoichiometry (i.e., Redfield ratio: N:P=16:1). In the Jiaozhou Bay of the YS, the N:P ratio of atmospheric dry deposition is higher than 100, and even exceeds 1000 in some specific conditions[19,31].

It has been reported that anthropogenic N deposition has the potential to change nutrient structure in the seawater[18]. Atmospheric N deposition is regarded as an important factor that induces phytoplankton blooms[32,33]. On the other hand, under the impact of vertical water mixing, the nutrients in atmospheric deposition can be transferred to the subsurface layers. Model studies have shown that the supplementation of N in surface waters to the lower layer is an important reason for the formation of subsurface chlorophyll a maximum (SCM)[34,35]. In contrast to the surface layer, few studies focused on the impact of atmospheric deposition on phytoplankton in the SCM layer.

In this study, we carried out three onboard incubation experiments enriched with AR using surface and SCM seawaters in the YS and ECS. The unfiltered and filtered (through 20 μm membrane) seawaters were used to illustrate the effects of microsized (20–200 μm) phytoplankton on the growth and nutrient uptake of nanosized (2–20 μm) and picosized ones. On this basis, our study intended to (1) reveal the difference in phytoplankton response to AR addition in surface and SCM seawaters; (2) identify the main factor of AR addition that affects the growth and community structure succession of phytoplankton; and (3) explore the interaction between different sized phytoplankton under the effects of AR addition.

Materials and Methods

▶ *Incubation experiments*

The AR samples used for incubation experiments were collected with a cellulose acetate filter membrane (Whatman 41) on the Laoshan campus of Ocean University of China (36°9′39″N, 120°29′29″E) on 30 June 2019. During the sampling period, the AQI was 57–83 μg·m^{-3}, indicating the air quality is moderate. The detailed air quality conditions were shown in Table 1.

Table 1 Air quality conditions during the AR sampling period

Parameter	Concentration
Humidity (%)	62–91
AQI (μg·m^{-3})	57–83
PM2.5 (μg·m^{-3})	29–61
PM10 (μg·m^{-3})	63–94
NH_4^+ (μmol·m^{-3})	0.17
$NO_3^- + NO_2^-$ (μmol·m^{-3})	0.43
PO_4^{3-} (nmol·m^{-3})	7.37
Fe (nmol·m^{-3})	31.88
Zn (nmol·m^{-3})	2.31
Al (nmol·m^{-3})	72.24
Mn (nmol·m^{-3})	0.98
Cu (nmol·m^{-3})	0.19
Cd (nmol·m^{-3})	0.01
Ni (nmol·m^{-3})	0.13
Pb (nmol·m^{-3})	0.13
Co (nmol·m^{-3})	0.01

Note: Humidity data during sampling were obtained from National Meteorological Information Center (NMIC, https://data.cma.cn/). AQI, PM$_{2.5}$ and PM$_{10}$, were obtained from China National Environmental Monitoring Centre (CNEMC, http://www.cnemc.cn/). The soluble nutrients and total trace metal concentrations in aerosols were sampled and determined in the laboratory[72].

In the summer of 2019, three onboard microcosm experiments were conducted during the cruise of R/V Beidou in the YS and the northern part of the ECS (Fig. 1A). The initial seawater at U1, U2 and U3 was collected from surface layers (about 3–5 m below the water surface) and SCM layers (captured by the CTD profile data) using Niskin bottles with Sea Bird CTD-General Oceanic Rosette assembly (Table 2). The site name is abbreviated as Ui$_{Sur}$ and Ui$_{SCM}$, respectively, i refers to the site number, i.e., 1/2/3.

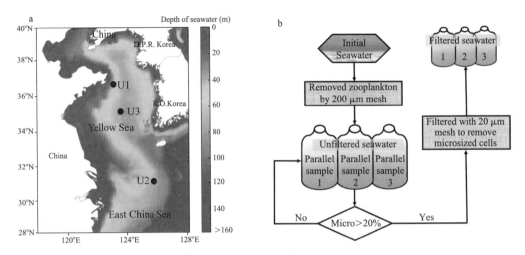

Where micro refers to the contribution of microsized Chl a to total Chl a.

Fig. 1 Water sampling stations used for the microcosm incubation experiments in the Yellow Sea and East China Sea (a) and the treatment procedure of the incubations (b)

Table 2 Background conditions of seawater at the experimental sites

Site	U1		U2		U3	
Incubation dates (2019)	Aug 16–21		Sep 3–8		Aug 22–28	
Water layer	Surface	SCM	Surface	SCM	Surface	SCM
Water depth (m)	3	19	3	34	3	29
Temperature (°C)	24.2	18.0	28.6	26.7	27.7	20.3
Salinity	31.5	31.8	31.4	32.8	30.2	32.6
$NO_3^- + NO_2^-$ (μmol·L^{-1})	0.19	1.70	0.08	2.59	0.10	0.11
PO_4^{3-} (μmol·L^{-1})	0.01	0.17	ND	0.11	0.01	0.01
Si(OH)$_4$ (μmol·L^{-1})	2.85	4.78	1.34	4.90	1.71	0.98
N:P (μmol:μmol)	15:1	10:1	20:1	24:1	12:1	11:1
APA (nmol·L^{-1}·h^{-1})	35.3	3.4	2.2	3.7	18.4	4.7
Chl a (μg·L^{-1})	0.38	1.19	0.55	0.95	0.62	0.58
Micro Chl a (%)	9	26	8	8	9	15
Nano Chl a (%)	18	22	27	30	36	38
Pico Chl a (%)	73	52	65	62	55	47
Dinophyceae (%)	60	95	85	89	92	96

Note: ND refer to Not Detectable.

Based on the content of N in AR aerosols, three treatments were conducted in triplicate for the incubation experiments: (1) control, no AR addition; (2) low AR addition, the added amount of AR was expressed in the unit of N (i.e., 1 μmol·L^{-1}) at U2, indicating the added amount of AR contains 1 μmol·L^{-1}; (3) high AR addition, 1.7 μmol·L^{-1} at U3 and 2 μmol·L^{-1} at U1. AR sample was firstly ultrasonically extracted in deionized water at 0 °C for 1 h, and the leaching solution including particles was added to the incubation bottles directly[36]. Apart from inorganic nutrients, the aerosols addition could stimulate mixotrophic dinoflagellates by promoting the utilization of organic matter[37-39]. With N addition of 1 μmol·L^{-1}, 1.7 μmol·L^{-1} and 2 μmol·L^{-1} by AR particles, the P concentrations added by AR particles were only 17.98 nmol·L^{-1}, 30.56 nmol·L^{-1} and 35.95 nmol·L^{-1}. The added amount of AR was determined according to the deposition flux of N (574.0–970.0 mg·m^{-2} per event) divided by an averaged mixed layer of 30 m[40,41]. Nutrient enrichment experiments were set up to illustrate nutrient limitation in original seawaters and interpret phytoplankton response to AR additions (Table S1).

The sampled seawater was passed through 200 μm sieves, mixed well in a clean 120 L high-density polyethylene barrel, and then dispensed into clean (acid-washed) 20 L polycarbonate incubation bottles. These incubation bottles were placed into three microcosm devices filled with continually updated surface seawater to keep the incubation system temperature relatively stable[3]. To explore the impact of microsized cells on the phytoplankton community, seawater from incubation bottles of each treatment was filtered through 20 μm sieves when the contribution of microsized Chl a to total Chl a in the control exceeded 20%, and then the filtered seawater (hereafter FS) was transferred uniformly into three 2 L bottles to continue the cultivation. The incubations unfiltered with 20 μm sieves were defined as the unfiltered seawater (hereafter UFS, Fig. 1B). Bottles were shaded to have approximately 40% light attenuation, matching light levels at depths on 3–5 m as previously used[3,42]. All experiments ran for 5–6 d.

▶ *Measurements of Chl a, nutrients, and the phytoplankton community structure*

Chl a

Approximately 150 mL seawater from incubated bottles was sampled at about 07:00 a.m. every day during the incubations. The sampled seawater was subsequently filtered through 20 μm, 2 μm, and 0.2 μm filters, to obtain microszied, nanoszied and picosized cells. After 20–24 h of extraction by 90% acetone in darkness at –20 °C, the pigments collected by different filters were measured using a Trilogy fluorometer (Turner Designs). The total Chl a concentration was obtained by summing three size-fractionated Chl a concentrations.

Nutrients

An ultrasonic method was used to leach nutrients in AR samples. Briefly, AR samples were ultrasonically extracted in deionized water at 0 °C for 1 h. The leaching solution was then filtered through a 0.45 μm polyethersulfone syringe filter[43]. The filtrates were used for the determination of soluble nutrients from aerosols, including NO_3^-, NO_2^-, NH_4^+, $Si(OH)_4$ and PO_4^{3-}. In addition, about 200 mL of incubated seawater (sampled every day) was filtered through acid-washed cellulose acetate membranes into 125 mL acid-washed high-density polyethylene bottles (prerinsed with the filtrates three times). The water samples were frozen at –20 °C immediately prior to the determination of $NO_3^- + NO_2^-$, PO_4^{3-} and $Si(OH)_4$ in the university laboratory. All nutrient samples were measured with a QuAAtro continuous-flow analyser (SEAL Analytical). The detection limits for NH_4^+, NO_3^-, NO_2^-, PO_4^{3-} and $Si(OH)_4$ were 0.04, 0.02, 0.005, 0.01 and 0.03 μmol·L^{-1}, respectively. For convenience, $NO_3^- + NO_2^-$ is abbreviated to N+N.

The concentrations of total trace metals were analyzed by inductively coupled plasma mass spectrometry (ICP-MS)[40]. The 8 cm^2 cellulose acetate filter was put into the Teflon high pressure vial, with 2 mL 69% HNO_3 and 0.5 mL 40% HF. After digestion at 180 °C for 48 h and evaporation at 160 °C, the residue was dissolved with 2% HNO_3 and diluted to 50 mL for determination.

Alkaline phosphatase activity (APA)

45 mL seawater was sampled from the incubated bottles and mixed with 0.5 mL fluorogenic substrate 4-methylumbelliferone phosphate (MUF-P) as the mixed substrate. After the addition of the mixed

borax-sodium carbonate buffer solution (pH ≈ 11) and the mixed substrate to the sample tube, the fluorogenic substrate MUF-P hydrolyzed by AP was converted into equimolar phosphate group and 4-methylumbelliferone (MUF). The fluorescence value of MUF was measured and recorded by a Trilogy fluorometer (Turner Designs) at 0, 0.5, and 1 h[44]. The slope was calculated as the hydrolysis rate, which reflect the alkaline phosphatase activity (APA).

High-throughput sequencing

Approximately 1 L seawater from incubated bottles was filtered using 0.22 μm Whatman polycarbonate filters (Shanghai Mosh Science Equipment Co., Ltd.) under gentle vacuum pressure (≤0.02 MPa). Filters were stored immediately in liquid nitrogen until DNA extraction and high-throughput sequencing (Shanghai Personal Biotechnology Co., Ltd.). The V4 hypervariable region was selected as the target region of the 18S rDNA[45]. The primers for polymerase chain reaction (PCR) were forward primer 582F, 5′-CCAGCASCYGCGGTAATTCC-3′ and reverse primer V4R, 5′-ACTTTCGTTCTTGATYRA-3′[46]. The Illumina NovaSeqPE250 platform was used for paired-end sequencing of community DNA fragments. First, we demultiplexed the raw sequence data and then invoked qiime cutadapt trim-paired to cut the primers[47]. Quality control of these sequences was performed using the DADA2 plugin with qiime dada2 denoise-paired[48]. Then, we merged amplicon sequence variants (ASVs) and removed singleton ASVs. A pretrained naive Bayes classifier plugin was used to annotate the species for each ASV using QIIME2 software (2019.4)[49]. The Silva database (Release132, http://www.arb-silva.de)[50] was used for species annotation. The microbiome bioinformatics of communities was analyzed using QIIME2 (2019.4). The accession number in NCBI Sequence Read Archive was PRJNA835313.

▶ *Data analysis*

We used one-way ANOVA to assess whether there was a significant difference in the Chl a concentration between the control and treatments[51], and evaluated the nutrient limitation in surface seawater at the three sites. Statistical analysis was performed using IBM SPSS Statistics 20 (SPSS 20.0). CANOCO software (version 5.0) was used to analyze the relationships between environmental factors and phytoplankton. The detrended correspondence analysis (DCA) used species-sample data showed that the first axis of gradient was less than 3. Therefore, redundancy analysis (RDA) was better choice.

Results

▶ *Overview of original seawater in surface and SCM layers*

In general, trophic statuses in surface seawater at $U1_{Sur}$, $U2_{Sur}$, and $U3_{Sur}$ and in SCM seawater at $U3_{SCM}$ were lower than those in SCM seawater at $U1_{SCM}$ and $U2_{SCM}$. Phytoplankton at $U1$–3_{Sur} were colimited by N and P based on the significant increase in Chl a after N+P addition relative to the control treatment (on days 4–6, $P < 0.05$, Fig. S1).

At $U1_{Sur}$, the concentrations of N+N, PO_4^{3-}, and $Si(OH)_4$ were 0.19 μmol·L^{-1}, 0.01 μmol·L^{-1}, and 2.85 μmol·L^{-1}, respectively. APA was 35.3 nmol·L^{-1}·h^{-1} in the original seawater, consistent with the reported value of about 35 nmol·L^{-1}·h^{-1} in this region[52]. The total Chl a concentration was as low as 0.38 μg·L^{-1}, of which picosized cells (73%) contributed the most of the total Chl a, followed by nanosized cells (18%) and microsized cells (9%) (Table 2). Chloropicophyceae and Dinophyceae codominated the community, with a relative abundance of 33% and 60%, respectively.

At $U2_{Sur}$, $U3_{Sur}$ and $U3_{SCM}$, low concentrations of N+N and PO_4^{3-} were observed, i.e., 0.08–0.11 μmol·L^{-1} and about 0.01 μmol·L^{-1}, respectively. The APA was less than 18.4 nmol·L^{-1}·h^{-1}, and the total Chl a concentration (0.55–0.62 μg·L^{-1}) was more than 1.4 times higher than that at $U1_{Sur}$. Picosized Chl a contributed the most of the total Chl a (47%–65%), and the contribution of microsized Chl a was less than 15%. Dinophyceae dominated the communities with a relative abundance of ≥85% (Table 2).

Seawaters at $U1_{SCM}$ and $U2_{SCM}$ contained abundant nutrients, with 1.70–2.59 $\mu mol \cdot L^{-1}$ of N+N, 0.11–0.17 $\mu mol \cdot L^{-1}$ of PO_4^{3-}, and 4.78–4.90 $\mu mol \cdot L^{-1}$ of $Si(OH)_4$. The APA at $U1_{SCM}$ and $U2_{SCM}$ (3.4–3.7 $nmol \cdot L^{-1} \cdot h^{-1}$) were of the same orders of magnitude as those of $U2_{Sur}$ and $U3_{SCM}$. The total Chl a concentration (≥0.95 $\mu g \cdot L^{-1}$) was more than 1.7 times higher than that of $U1_{Sur}$ and $U2_{Sur}$. Picosized phytoplankton (52%–62%) were the primary contributors to the total Chl a. Dinophyceae dominated the communities with a relative abundance of ≥89% (Table 2).

▶ *Changes in inorganic nutrients*

The N:P ratio after AR addition increased from 11:1–20:1 to 44:1–54:1 at all sites in the surface incubated seawater and at $U3_{SCM}$ in the SCM incubated seawater. Due to the sufficient nutrient stock in the original seawater at $U1_{SCM}$ and $U2_{SCM}$, the N supplied by AR addition only increased the N:P ratio from 10:1–24:1 to 16:1–28:1. Because of the low contents of PO_4^{3-} and $Si(OH)_4$ in the AR, the changes in the concentrations of PPO_4^{3-} and $Si(OH)_4$ in AR-amended seawater were slight at all sites.

During the incubations at $U1–3_{Sur}$ and $U3_{SCM}$, the concentrations of N+N did not change significantly in the control and AR treatments (Fig. 2). In the AR treatments, the N+N concentrations remained relatively stable and were significantly higher than those in the control treatments. In contrast, the concentrations of PO_4^{3-} were close to the detection limit in all treatments. The maximum consumption of $Si(OH)_4$ in the control and AR treatments was less than 13% at $U1–3_{Sur}$ and $U3_{SCM}$. At $U1_{SCM}$ and $U2_{SCM}$, the concentrations of N+N and PO_4^{3-} decreased gradually by more than 85% in the control and AR treatments relative to the original values. The concentrations of $Si(OH)_4$ decreased sharply in the control (>28%) and AR treatments (>69%) on days 3–5 at both sites (Figs. 2&S2).

There were no obvious differences in the concentrations of N+N, PO_4^{3-}, and $Si(OH)_4$ between FS and UFS in the control and AR treatments at $U1–3_{Sur}$ and $U3_{SCM}$. In the FS at $U1_{SCM}$ and $U2_{SCM}$, the consumption of N+N and $Si(OH)_4$ in AR treatments was lower than that in the UFS at the end of the incubations ($P < 0.05$, Figs. 2, S2).

▶ *Changes in alkaline phosphatase activity*

In the UFS, the APA in the AR treatments was 1.3–75.5 times higher than that in the control at the end of the incubations ($P < 0.05$, Fig. 3), and this phenomenon could be observed in both surface and SCM incubations. In contrast, there was almost no difference in APA between control and AR treatments on the last day of FS incubations at all sites except $U1_{SCM}$. For the AR treatments, the APA in the UFS at all sites was generally higher (1.6–7.3 times) than that in the FS (Fig. 3).

▶ *Changes in total and size-fractionated Chl a*

At $U1–3_{Sur}$, the concentration of the total Chl a in the AR treatments was generally higher than that in the control. At $U1_{SCM}$, the total Chl a concentration in the AR treatments was more than 1.2 times higher than that in the control on days 2–5. At $U2_{SCM}$ and $U3_{SCM}$, there was no significant difference in Chl a between the control and AR treatments (Fig. 4). The responses of phytoplankton of different sizes varied with AR addition. At $U1_{Sur}$, the dominant size of phytoplankton changed from picosize to nanosize (46% contribution to total Chl a). A similar pattern in the size shift towards larger cells also occurred at $U2_{Sur}$ and $U3_{Sur}$, although picosized cells always dominated the contribution to total Chl a (Fig. 5). At $U1_{SCM}$ and $U2_{SCM}$, the dominant contributor of phytoplankton in AR treatments were picosized cells on days 1–2, and shifted to microsized cells (≥49% contribution to total Chl a) on day 5. At $U3_{SCM}$, the dominant contributor was always picosized phytoplankton during the incubations (Fig. 5).

For AR treatments, the concentrations of nanosized and picosized Chl a in UFS were generally higher than those in FS at $U1_{Sur}$ and $U3_{Sur}$. Specifically, at $U1_{Sur}$, nanosized and picosized Chl a concentrations in the UFS enriched with AR were 2.4 and 1.8 times higher than those in the FS on day 5.

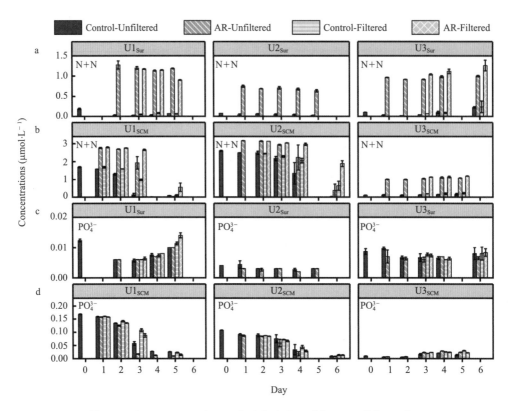

The error bar represents the standard deviation of three parallel samples.

Fig. 2 Changes in (A, B) N+N and (C, D) PO_4^{3-} concentrations in the control and AR treatments incubated with surface and SCM (subsurface chlorophyll a maximum) seawaters

At $U3_{Sur}$, the picosized Chl a concentration in UFS enriched with AR was 2.7 times higher than that in FS on day 6. Similar to the incubations with surface seawater, the picosized Chl a concentration enriched with AR in FS (0.10 μg·L^{-1}) at $U3_{SCM}$ was lower than that in UFS (0.25 μg·L^{-1}) on day 6. In contrast, at $U1_{SCM}$, nanosized and picosized Chl a concentrations in FS (4.72 μg·L^{-1} and 0.36 μg·L^{-1}) were higher than those in UFS (2.89 μg·L^{-1} and 0.22 μg·L^{-1}) on day 5. At $U2_{SCM}$, there was no significant difference in the concentrations of size-fractionated Chl a between UFS and FS (Fig. 4).

▶ *Changes in the phytoplankton community*

The ASVs at all sites assigned to phytoplankton could be classified into 25 groups of eukaryotic microalgae at class level (level 3). The ASV richness of Dinophyceae (dinoflagellates) accounted for ⩾60% of phytoplankton in the original seawater at each site. In terms of UFS, Dinophyceae dominated the phytoplankton community in the AR treatments at all sites (Fig. 6). The dominant class changed from Dinophyceae to Chloropicophyceae (71%) in the control on day 5 of the incubations (corresponding to the maximum Chl a concentration) at $U1_{Sur}$. *Chloropicon* spp. was the main component of Chloropicophyceae (Fig. S3). With AR addition, the relative abundance of Dinophyceae increased to 46% being the dominant phytoplankton (Fig. 6). The succession of phytoplankton communities at $U3_{SCM}$ was similar to that at $U1_{Sur}$. Dinophyceae maintained the dominant status in the control and AR treatments at the rest of the incubation sites (Fig. 6).

Discussion

On the basis of the distinct nutrient concentrations (N, P and Si) in the original seawaters, the sites

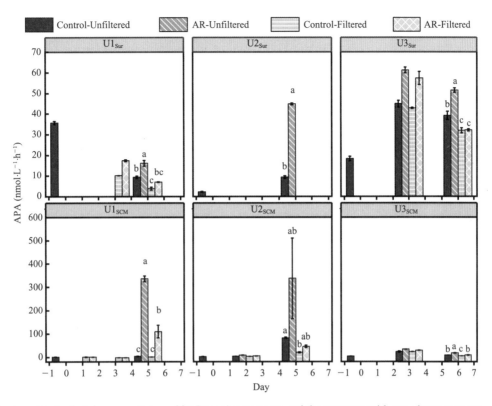

Treatment in the last day were compared by Ducon's range test, and the treatment with same letter were not significantly different ($\alpha = 0.05$). The error bar represents the standard deviation of three parallel samples.

Fig. 3   Changes in APA in the control and AR treatments incubated with surface and SCM (subsurface chlorophyll a maximum) seawaters

used for incubations were classified into two types: U1–3$_{Sur}$ and U3$_{SCM}$ with lower trophic status, where the concentrations of N+N, PO_4^{3-}, and $Si(OH)_4$ did not exceed 0.50 µmol·L^{-1}, 0.02 µmol·L^{-1}, or 3.00 µmol·L^{-1}, respectively; U1$_{SCM}$ and U2$_{SCM}$ with higher trophic status, where the concentrations of N+N, PO_4^{3-}, and $Si(OH)_4$ exceeded 1.50 µmol·L^{-1}, 0.10 µmol·L^{-1}, and 4.50 µmol·L^{-1}, respectively.

▶ *Distinct responses of phytoplankton to AR additions in surface and SCM seawaters*

AR additions generally stimulated phytoplankton growth and shifted the phytoplankton size structure towards larger cells in surface incubated seawaters (Fig. 5). This is justified because of the established supplementary relationship between nutrients (primarily N) supplied by AR and phytoplankton requirements. Such fertilization effect has also been widely reported in previous studies[3,53,54]. In contrast, AR additions had a limited effect on phytoplankton size structure at U3$_{SCM}$, although its trophic status was similar to those in surface seawaters (Fig. 5). This is ascribed to the photoacclimation of phytoplankton under the condition of low irradiance in SCM layer[55,56]. With the abrupt enhancement of light intensity (from SCM to surface), phytoplankton in the incubated seawater need to readjust to the new environment and thus showed a limited response to AR additions.

Interestingly, AR addition had a significant fertilization effect on phytoplankton growth at U1$_{SCM}$ (Fig. 4), which was characterized by the higher trophic status among these sites. Note that there was a shift in dominant phytoplankton from picosized cells to large cells during the incubations at U1$_{SCM}$, which was different from the sustaining

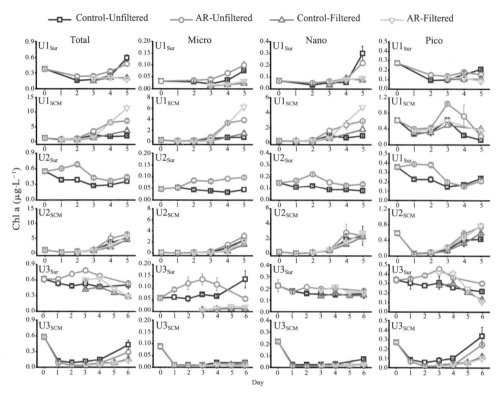

The error bar represents the standard deviation of three parallel samples.

Fig. 4　The concentration of Chl a in the control and AR treatments incubated with surface and SCM (subsurface chlorophyll a maximum) seawaters

dominance of picosized cells at $U3_{SCM}$ (Fig. 5). Moreover, large phytoplankton can better acclimate to the abrupt increase in light intensity compared with picosized cells, due to their stronger self-shading capacity by the pigment (package effect) to reduce light absorption[12]. At $U2_{SCM}$, the stimulation effect of AR addition was not as obvious as that at $U1_{SCM}$ (Fig. 4). This is because there was an obvious shift of dominant algae from Coscinodiscophyceae in the control to Dinophyceae in the AR treatments at $U2_{SCM}$ (Fig. 6). The obvious succession in phytoplankton community while slight change in Chl a under the condition of aerosol enrichment was also observed in eutrophic seawaters of the ECS[57]. The substantial input of N relative to P supplied by AR addition increased the N:P ratio from 24:1 in the original seawater to 28:1 in the AR treatments at $U2_{SCM}$, which was more favorable for the growth of dinoflagellates[3] (Table 2). At $U1_{SCM}$, in contrast, the N:P ratio ranged between 10:1 and 16:1 in the control and AR treatments, leading to the increase in relative abundance of diatoms (primarily Coscinodiscophyceae[3]). Collectively, in contrast to the consistent phytoplankton response to AR addition in surface seawater, the impact of AR addition in SCM seawater is complicated, which is closely related to nutrient concentration and structure in seawater.

▶ *Utilization of DOP enhanced by AR addition*

The substantial N supplied by AR had the potential to alleviate and even alter N pressure of phytoplankton in the incubated seawater (Fig. 2). As a result, the relatively P deficient environment created by AR additions made it possible for phytoplankton to induce acclimatization mechanisms to copy with P stress[15]. As we saw in Fig. 3, the

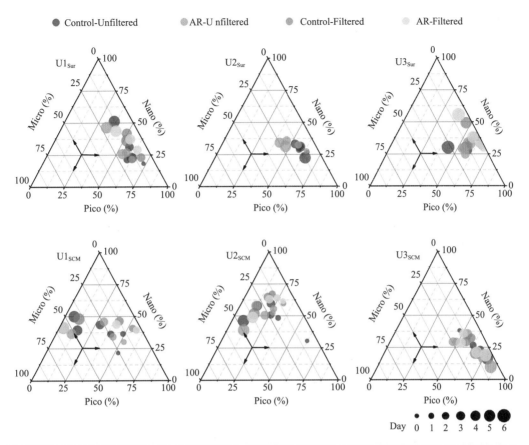

Fig. 5 Changes in size structure of phytoplankton in the control and AR treatments incubated with surface and SCM (subsurface chlorophyll a maximum) seawaters

APA value in the AR treatments was higher than that of the control at the end of the incubations at all sites, indicating that AR could enhance the utilization of DOP to increase P bioavailability in the incubated seawater. Such phenomenon was supported by the good correlation between PO_4^{3-} and APA (Fig. 7). In apart from the establishment of P deficient environment, AR additions also provided a considerable amount of soluble Fe and Zn, which acted as cofactors of phosphohydrolytic enzymes[21,58]. The promotion effect of AR addition on the utilization of DOP shows the acclimation of phytoplankton to the overwhelming N input relative to P in coastal waters[31,59,60], and is conductive to understanding P biogeochemical cycles in the perspective of atmospheric deposition.

In terms of the phytoplankton community composition, Dinophyceae generally dominated the community in UFS enriched with AR at all sites (Fig. 6). With relatively high tolerance to nutrient-deficient environments and the potential to utilize DOP by inducing the expression of the gene for the synthesis of AP[37], dinoflagellates showed an advantageous growth in the AR treatments. This is also the reason why nutrient limitation had a slight impact on the growth of Dinophyceae based on RDA (Fig. 7). In addition, due to the selective feeding of micrograzers, picosized phytoplankton suffer from a higher grazing pressure[53,61]. In contrast, large dinoflagellates have the ability to keep themselves away from the prey of the dominant zooplankton species *Paracalanus parvus* (Hexanauplia) through particle rejection behavior (reject particles as food, Fig. S4[62,63]).

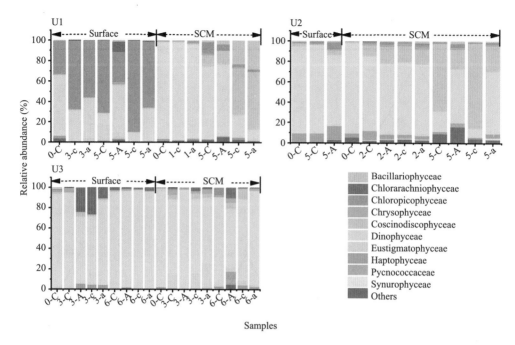

-C, -A, and -c, -a refer to the control, and AR treatments in the unfiltered and filtered seawater, e.g., 0-C means on day 0 in the unfiltered seawaters for the control treatment. "Others" refers to the groups outside of the top 10.

Fig. 6　Relative abundances of dominant eukaryotic phytoplankton classes (level 3 of the taxonomic hierarchy in SILVA 132) in the control and AR treatments during the incubation experiments

▶ *Role of microsized phytoplankton in affecting the growth of nanosized and picosized phytoplankton*

As described in Section "Changes in total and size-fractionated chlorophyll a", at $U1_{Sur}$, $U3_{Sur}$ and $U3_{SCM}$ characterized by lower trophic statuses, the nanosized and picosized Chl a concentrations in FS enriched with AR were lower than those in UFS. In contrast, we did not observe similar results at $U1_{SCM}$ and $U2_{SCM}$ characterized by higher trophic statuses (Fig. 4).

Nutrients, irradiance and temperature are considered the three major factors that affect phytoplankton growth[64,65]. There was no difference in light and temperature between UFS and FS, and thus, nutrients play a key role in causing the lower nanosized and picosized Chl a concentrations in FS. At $U1–3_{Sur}$ and $U3_{SCM}$, there were no obvious differences in N+N and PO_4^{3-} between UFS and FS

(Fig. 2). Meanwhile, we found that the APA values in UFS were 1.6–7.3 times higher than those in FS (Fig. 3), indicating that microsized cells played an important role in increasing P bioavailability in the incubated seawater[44]. This was supported by the positive nonlinear relationship between APA and the contribution of microsized cells to the total Chl a (Fig. 8A). As an extracellular enzyme, AP enters the environment through autolyzing or organisms excreting[66]. Besides, in contrast to FS, nanosized and picosized Chl a concentration in UFS increased linearly with relative change of APA (Fig. 8B). Therefore, under the impact of AR addition, microsized cells have the ability to favor the growth of nanosized and picosized cells by increasing P bioavailability in seawater. The result at $U1_{SCM}$ characterized by higher trophic statuses could also support this argument. On days 1–2 of the incubations, there was no difference in the

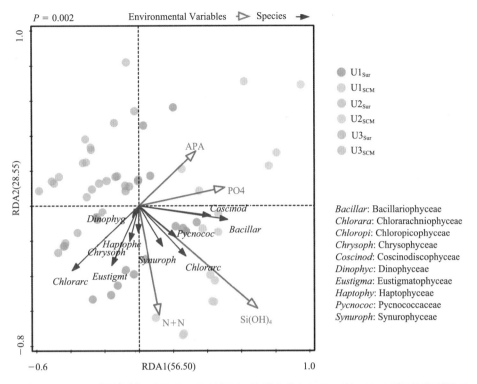

Fig. 7　Correlations between species (level 3) and environmental factors based on redundancy analysis (RDA)

nanosized and picosized Chl a concentrations in the AR treatments between UFS and FS when PO_4^{3-} was sufficient in the seawater, but lower nanosized and picosized Chl a concentrations in FS were measured when PO_4^{3-} was exhausted on day 3 (Figs. 2, 4). Besides, dinoflagellates and green algae with the capacity of utilizing DOP also showed advantageous growth in the P-deficient condition (Fig. 6).

There are other factors that might have caused the mismatch between UFS and FS in the concentration of nanoszied and picosized Chl a. For example, the biodiversity of the community decreased after removal of microsized cells, which increased the difficulty for the community to reestablish a new balance[67]. Nanosized and picosized phytoplankton may adopt a strategy to survive in unstable habitats, e.g., by producing spores that can be dormant temporarily and revive at an appropriate time[68]. However, these inferences cannot account for the change in nanoszied and picosized Chl a at $U1_{SCM}$ and $U2_{SCM}$ characterized by higher trophic statuses (Fig. 4). Therefore, our study provides a new clue from the perspective of nutrient utilization to illustrate how microsized phytoplankton affect the growth of nanoszied and picosized ones.

Conclusion

In this study, we conducted three onboard incubation experiments using surface and SCM seawaters under the condition of sea surface light intensity in the Yellow Sea and East China Sea. AR addition generally stimulated phytoplankton growth in surface incubations, and had a stimulatory or slight impact in SCM incubations, which primarily depends on the nutrient concentration and structure in seawater. We also found that AR addition could alleviate P limitation by promoting the utilization of DOP in both surface and SCM incubations. Especially in seawater with lower trophic status, microsized cells have the ability to

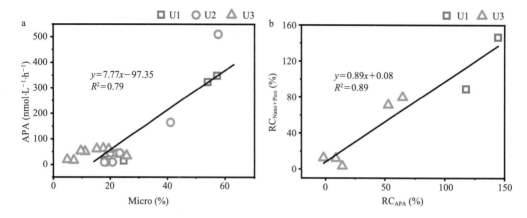

RCAPA and RC$_{Nano+Pico}$ was calculated as ((mean in the unfiltered seawater − mean in the filtered seawater) / mean in the filtered seawater) × 100. The contribution of microsized Chl a to the total Chl a in the control was less than 15% throughout the experiments, and thus there was no incubations with filtered seawater at U2$_{Sur}$.

Fig. 8   The relationship between APA and the contribution of microsized Chl a to the total Chl a in the AR treatments (a) and the relationship between relative change in APA (RC$_{APA}$) and nanosized+picosized Chl *a* (RC$_{Nano+Pico}$) in the unfiltered seawater relative to the filtered seawater (b)

promote the growth of nanosized and picosized cells by increasing P bioavailability in the incubated seawater. Considering the lower contribution of microsized cells in the oligotrophic areas of the open oceans[4,69], such a promotion effect of microsized cells induced by anthropogenic aerosol deposition may focus on coastal waters, and thus can be regarded as a result of anthropogenic influences to a large extent. With the enhanced influence of human activities in recent years, atmospheric deposition characterized by high N:P ratios has intensified the prevailing P limitation in offshore waters[70,71]. The acclimation mechanism of different sized phytoplankton to P limitation under the influence of atmospheric deposition deserves to be further investigated.

Supplementary Material

The Supplementary Material for this article can be found online at: https://www.frontiersin.org/articles/10.3389/fmicb.2022.915255/full#supplementary-material.

Acknowledgements

This work was financially supported by National Natural Science Foundation of China (NSFC) (41876125 and 41906119), and NSFC-Shandong Joint Fund (U1906215). Data and samples were collected onboard of R/V "Beidou" implementing the open research cruise NORC2019-1.

Author Contributions

Qin Wang: Conceptualization, investigation, methodology, data curation, formal analysis, visualization, software, and writing (original draft). Chao Zhang: Conceptualization, investigation, writing (review and editing), data curation, and funding acquisition. Haoyu Jin: Investigation. Ying Chen and Xiaohong Yao: Writing (review and editing). Huiweng Gao: Supervision, methodology, resources, writing (review and editing), and funding acquisition and also was responsible for ensuring that the descriptions are accurate and agreed by all authors. All authors contributed to the article and approved the submitted version.

Competing Interests

The authors declare that the research was conducted in the absence of any commercial or financial relationships that could be construed as a potential conflict of interest.

References

1. Hooper J, Mayewski P, Marx S, et al. Examining links between dust deposition and phytoplankton response using ice cores[J]. Aeolian Res., 2019, 36: 45-60.
2. Jickells T D, An Z S, Andersen K K, et al. Global iron connections between desert dust, ocean biogeochemistry, and climate[J]. Science, 2005, 5718(308): 67-71.
3. Zhang C, Yao X, Chen Y, et al. Variations in the phytoplankton community due to dust additions in eutrophication, LNLC and HNLC oceanic zones[J]. Sci. Total Environ., 2019, 669: 282-293.
4. Marañón E, Cermeño P, Latasa M, et al. Resource supply alone explains the variability of marine phytoplankton size structure[J]. Limnol. Oceanogr., 2015, 60(5): 1848-1854.
5. Okin G S, Baker A R, Tegen I, et al. Impacts of atmospheric nutrient deposition on marine productivity: Roles of nitrogen, phosphorus, and iron[J]. Global Biogeochem. Cy., 2011, 25: GB2022.
6. Krishnamurthy A, Moore J K, Zender C S, et al. Effects of atmospheric inorganic nitrogen deposition on ocean biogeochemistry[J]. J. Geophys. Res. Biogeosci., 2007, 112(G2): G02019.
7. Hutchins D A, Witter A E, Butler A, et al. Competition among marine phytoplankton for different chelated iron species[J]. Nature, 1999, 400(6747): 858-861.
8. Stolte W, Riegman R. A model approach for size-selective competition of marine phytoplankton for fluctuating nitrate and ammonium[J]. J. Phycol., 1996, 32(5): 732-740.
9. Finkel Z V. Does phytoplankton cell size matter? The evolution of modern marine food webs[M]//Evolution of Primary Producers in the Sea. Falkowski P G, Knoll A H. Burlington: Academic Press, 2007: 333-350.
10. Agusti S, Kalff J. The influence of growth conditions on the size dependence of maximal algal density and biomass[J]. Limnol. Oceanogr., 1989, 34(6): 1104-1108.
11. Wei Y, Sun J, Zhang X, et al. Picophytoplankton size and biomass around equatorial eastern Indian Ocean[J]. MicrobiologyOpen, 2019, 8(2): e00629.
12. Marañón E. Cell size as a key determinant of phytoplankton metabolism and community structure[J]. Ann Rev Mar Sci., 2015, 7: 241-64.
13. Pasciak W J, Gavis J. Transport limitation of nutrient uptake in phytoplankton[J]. Limnol. Oceanogr., 1974, 19(6): 881-888.
14. Chisholm S W. Phytoplankton size[M]//Primary Productivity and Biogeochemical Cycles in the Sea. Falkowski P G, Woodhead A D, Vivirito K. Boston, MA: Springer US, 1992: 213-237.
15. Moore C M, Mills M M, Arrigo K R, et al. Processes and patterns of oceanic nutrient limitation[J]. Nature Geoscience, 2013, 6: 701-710.
16. Zhang C, Gao H, Yao X, et al. Phytoplankton growth response to Asian dust addition in the northwest Pacific Ocean versus the Yellow Sea[J]. J. Geophys. Res. Biogeosci., 2018, 15(3): 749-765.
17. Boyd P W, Jickells T, Law C S, et al. Mesoscale iron enrichment experiments 1993–2005: Synthesis and future directions[J]. Science, 2007, 315(5812): 612-617.
18. Kim T, Lee K, Duce R, et al. Impact of atmospheric nitrogen deposition on phytoplankton productivity in the South China Sea[J]. Geophys. Res. Lett., 2014, 41(9): 3156-3162.
19. Wu Y, Zhang J, Liu S, et al. Aerosol concentrations and atmospheric dry deposition fluxes of nutrients over Daya Bay, South China Sea[J]. Mar. Pollut. Bull., 2018, 128: 106-114.
20. Browning T J, Achterberg E P, Yong J C, et al. Iron limitation of microbial phosphorus acquisition in the tropical North Atlantic[J]. Nat. Commun., 2017, 8(1): 15465.
21. Mahaffey C, Reynolds S, Davis C E, et al. Alkaline phosphatase activity in the subtropical ocean: Insights from nutrient, dust and trace metal addition experiments[J]. Front. Mar. Sci., 2014, 1.
22. Zheng L, Zhai W. Excess nitrogen in the Bohai and Yellow seas, China: distribution, trends, and source apportionment[J]. Sci. Total Environ., 2021, 794: 148702.
23. Chu Q, Liu Y, Shi J, et al. Promotion effect of Asian dust on phytoplankton growth and potential dissolved organic phosphorus utilization in the South China Sea[J]. J. Geophys. Res. Biogeosci., 2018, 123(3): 1101-1116.
24. Zhang K, Gao H. The characteristics of Asian-dust storms during 2000–2002: From the source to the sea[J]. Atmos. Environ., 2007, 41(39): 9136-9145.
25. Wang S, Arthur Chen C, Hong G, et al. Carbon dioxide and related parameters in the East China Sea[J]. Cont. Shelf Res., 2000, 20(4): 525-544.
26. Xiao H W, Xiao H Y, Shen C Y, et al. Chemical composition and sources of marine aerosol over the Western North Pacific Ocean in winter[J]. Atmosphere-Basel, 2018, 9(8).
27. Kang M, Yang F, Ren H, et al. Influence of continental

organic aerosols to the marine atmosphere over the East China Sea: Insights from lipids, PAHs and phthalates[J]. Sci. Total Environ., 2017, 607-608: 339-350.
28. Fu H, Zheng M, Yan C, et al. Sources and characteristics of fine particles over the Yellow Sea and Bohai Sea using online single particle aerosol mass spectrometer[J]. J. Environ. Sci.-China, 2015, 29: 62-70.
29. An Z, Huang R, Zhang R, et al. Severe haze in northern China: A synergy of anthropogenic emissions and atmospheric processes[J]. Proc. Natl. Acad. Sci. U.S.A., 2019, 116(18): 8657-8666.
30. Du J, Cheng T, Zhang M, et al. Aerosol size spectra and particle formation events at urban Shanghai in Eastern China[J]. Aerosol Air Qual. Res., 2012, 12(6): 1362-1372.
31. Xing J, Song J, Yuan H, et al. Fluxes, seasonal patterns and sources of various nutrient species (nitrogen, phosphorus and silicon) in atmospheric wet deposition and their ecological effects on Jiaozhou Bay, North China[J]. Sci. Total Environ., 2017, 576: 617-627.
32. Tan S, Shi G. Transpot of a severe dust storm in March 2007 and impacts on chlorophyll a concentration in the Yellow Sea[J]. SOLA, 2012, 8: 85-89.
33. Tan S, Wang H. The transport and deposition of dust and its impact on phytoplankton growth in the Yellow Sea[J]. Atmos. Environ., 2014, 99: 491-499.
34. Gong X, Jiang W, Wang L, et al. Analytical solution of the nitracline with the evolution of subsurface chlorophyll maximum in stratified water columns[J]. J. Geophys. Res. Biogeosci., 2017, 14(9): 2371-2386.
35. Hodges B A, Rudnick D L. Simple models of steady deep maxima in chlorophyll and biomass[J]. Deep Sea Res. Part I Oceanogr. Res. Pap., 2004, 51(8): 999-1015.
36. Guo C, Jing H, Kong L, et al. Effect of East Asian aerosol enrichment on microbial community composition in the South China Sea[J]. J. Plankton Res., 2013, 35(3): 485-503.
37. Lin X, Zhang H, Cui Y, et al. High sequence variability, diverse subcellular localizations, and ecological implications of alkaline phosphatase in Dinoflagellates and other eukaryotic phytoplankton[J]. Front. Microbiol., 2012, 3.
38. Heisler J, Glibert P M, Burkholder J M, et al. Eutrophication and harmful algal blooms: A scientific consensus[J]. Harmful Algae, 2008, 8(1): 3-13.
39. Granéli E, Carlsson P, Legrand C. The role of C, N and P in dissolved and particulate organic matter as a nutrient source for phytoplankton growth, including toxic species[J]. Aquat. Ecol., 1999, 33(1): 17-27.
40. Shi J, Gao H, Zhang J, et al. Examination of causative link between a spring bloom and dry/wet deposition of Asian dust in the Yellow Sea, China[J]. J. Geophys. Res. Atmos., 2012, 117(D17).
41. Sin Y, Hyun B, Jeong B, et al. Impacts of eutrophic freshwater inputs on water quality and phytoplankton size structure in a temperate estuary altered by a sea dike[J]. Mar. Environ. Res., 2013, 85: 54-63.
42. Zhang C, He J, Yao X, et al. Dynamics of phytoplankton and nutrient uptake following dust additions in the northwest Pacific[J]. Sci. Total Environ., 2020, 739: 139999.
43. Shi J, Gao H, Qi J, et al. Sources, compositions, and distributions of water-soluble organic nitrogen in aerosols over the China Sea[J]. J. Geophys. Res. Atmos., 2010, 115(D17).
44. Sebastián M, Arístegui J, Montero M F, et al. Alkaline phosphatase activity and its relationship to inorganic phosphorus in the transition zone of the North-western African upwelling system[J]. Prog. Oceanogr., 2004, 62(2-4): 131-150.
45. Liu Q, Zhao Q, McMinn A, et al. Planktonic microbial eukaryotes in polar surface waters: Recent advances in high-throughput sequencing[J]. J Mar Sci Technol., 2021, 3(1): 94-102.
46. Hernández-Ruiz M, Barber-Lluch E, Prieto A, et al. Response of pico-nano-eukaryotes to inorganic and organic nutrient additions[J]. Estuar. Coast. Shelf Sci., 2020, 235: 106565.
47. Martin M. Impacts of atmospheric deposition on phytoplankton community structure in the Yellow Sea[J]. EMBnet. journal, 2011, 17(1): 10-12.
48. Callahan B J, Mcmurdie P J, Rosen M J, et al. DADA2: High-resolution sample inference from Illumina amplicon data[J]. Nat. Methods, 2016, 13(7): 581-583.
49. Bokulich N A, Kaehler B D, Rideout J R, et al. Optimizing taxonomic classification of marker-gene amplicon sequences with QIIME 2's q2-feature-classifier plugin[J]. Microbiome, 2018, 6(1): 90.
50. Quast C, Pruesse E, Yilmaz P, et al. The SILVA ribosomal RNA gene database project: Improved data processing and web-based tools[J]. Nucleic Acids Res., 2013, 41(D1): D590-D596.
51. Andersen I M, Williamson T J, González M J, et al. Nitrate, ammonium, and phosphorus drive seasonal nutrient limitation of chlorophytes, cyanobacteria, and diatoms in a hyper-eutrophic reservoir[J]. Limnol. Oceanogr., 2020, 65(5): 962-978.
52. Wang D, Huang B, Liu X, et al. Seasonal variations of phytoplankton phosphorus stress in the Yellow Sea Cold Water Mass[J]. Acta Oceanol. Sin., 2014, 33(10):

53. Cottingham K L. Nutrients and zooplankton as multiple stressors of phytoplankton communities: evidence from size structure[J]. Limnol. Oceanogr., 1999, 44(3): 810-827
54. Liu Y, Zhang T R, Shi J H, et al. Responses of chlorophyll a to added nutrients, Asian dust, and rainwater in an oligotrophic zone of the Yellow Sea: Implications for promotion and inhibition effects in an incubation experiment[J]. J. Geophys. Res. Biogeosci., 2013, 118(4): 1763-1772.
55. Fu M, Sun P, Wang Z, et al. Structure, characteristics and possible formation mechanisms of the subsurface chlorophyll maximum in the Yellow Sea Cold Water Mass[J]. Cont. Shelf Res., 2018, 165: 93-105.
56. Fujiki T, Taguchi S. Variability in chlorophyll a specific absorption coefficient in marine phytoplankton as a function of cell size and irradiance[J]. J. Plankton Res., 2002, 24(9): 859-874.
57. Meng X, Chen Y, Wang B, et al. Responses of phytoplankton community to the input of different aerosols in the East China Sea[J]. Geophys. Res. Lett., 2016, 43: 7081-7088.
58. Mills M M, Ridame C, Davey M, et al. Iron and phosphorus co-limit nitrogen fixation in the eastern tropical North Atlantic[J]. Nature, 2004, 429(6989): 292-294.
59. Zhang J, Chen S Z, Yu Z G, et al. Factors influencing changes in rainwater composition from urban versus remote regions of the Yellow Sea[J]. J. Geophys. Res. Atmos., 1999, 104: 1631-1644.
60. Zamora L M, Landolfi A, Oschlies A, et al. Atmospheric deposition of nutrients and excess N formation in the North Atlantic[J]. J. Geophys. Res. Biogeosci., 2010, 7: 777-793.
61. Strom S L, Macri E L, Olson M B. Microzooplankton grazing in the coastal gulf of Alaska: Variations in top-down control of phytoplankton[J]. Limnol. Oceanogr., 2007, 52(4): 1480-1494.
62. Tiselius P, Saiz E, Kiørboe T. Sensory capabilities and food capture of two small copepods, *Paracalanus parvus* and *Pseudocalanus* sp.[J]. Limnol. Oceanogr., 2013, 58: 1657-1666.
63. Huntley M, Sykes P, Rohan S, et al. Chemically-mediated rejection of dinoflagellate prey by the copepods *Calanus pacificus* and *Paracalanus parvus*: Mechanism, occurrence and significance[J]. Mar. Ecol. Prog. Ser., 1986, 28: 105-120.
64. Litchman E. Resource competition and the ecological success of phytoplankton[M]//Evolution of Primary Producers in the Sea. Falkowski P G, Knoll A H. Burlington: Academic Press, 2007: 351-375.
65. Laws E A, Falkowski P G, Smith Jr. W O, et al. Temperature effects on export production in the open ocean[J]. Global Biogeochem. Cy., 2000, 14(4): 1231-1246.
66. Štrojsová A, Vrba J, Nedoma J, et al. Seasonal study of extracellular phosphatase expression in the phytoplankton of a eutrophic reservoir[J]. Eur. J. Phycol., 2003, 38(4): 295-306.
67. Dyke J, Mcdonald-Gibson J, Di Paolo E, et al. Increasing Complexity Can Increase Stability in a Self-Regulating Ecosystem[M]. Almeida E Costa F, Rocha L M, Costa E, et al. Berlin, Heidelberg: Springer Berlin Heidelberg, 2007: 133-142.
68. Nayaka S, Toppo K, Verma S. Adaptation in algae to environmental stress and ecological conditions[M]//Shukla V, Kumar S, Kumar N. Plant Adaptation Strategies in Changing Environment,. Singapore: Springer Singapore, 2017: 103-115.
69. López-Urrutia Á, Morán X A G. Temperature affects the size-structure of phytoplankton communities in the ocean[J]. Limnol. Oceanogr., 2015, 60(3): 733-738.
70. Harrison P J, Hu M H, Yang Y P, et al. Phosphate limitation in estuarine and coastal waters of China[J]. J. Exp. Mar. Biol. Ecol., 1990, 140(1): 79-87.
71. Xu J, Yin K, He L, et al. Phosphorus limitation in the northern South China Sea during late summer: Influence of the Pearl River[J]. Deep Sea Res. Part I Oceanogr., 2008, 55(10): 1330-1342.
72. Zhao R, Han B, Lu B, et al. Element composition and source apportionment of atmospheric aerosols over the China Sea[J]. Atmos. Pollut. Res., 2015, 6(2): 191-201.

A low-cost in-situ CO₂ sensor based on a membrane and NDIR for long-term measurement in seawater

Li Meng[1], Du Baolu[1], Guo Jinjia[1*], Zhang Zhihao[1], Lu Zeyu[2], Zheng Rong'er[1]

1 College of Information Science and Engineering, Ocean University of China, Qingdao 266100, China
2 R & D Center for Marine Instruments and Apparatuses, Pilot National Laboratory for Marine Science and Technology (Qingdao), Qingdao 266200, China
* Corresponding author: Guo Jinjia (opticsc@ouc.edu.cn)

Abstract

The multi-point simultaneous long-term measurement of CO_2 concentration in seawater can provide more valuable data for further understanding of the spatial and temporal distribution of CO_2. Thus, the requirement for a low-cost sensor with high precision, low power consumption, and a small size is becoming urgent. In this work, an in-situ sensor for CO_2 detection in seawater, based on a permeable membrane and NDIR (non-dispersive infrared) technology, is developed. The sensor has a small size (Φ 66 mm×124 mm), light weight (0.7 kg in air), low power consumption (<0.9 W), low cost (<$1000), and high pressure tolerance (<200 m). After laboratory performance tests, the sensor was found to have a measurement range of 0–2000×10⁻⁶ (volume fraction), and the gas linear correlation R^2 is 99.8%, with a precision of about 0.98% at a sampling rate of 1 s. A comparison measurement was carried out with a commercial sensor in a pool for 7 days, and the results showed a consistent trend. Further, the newly developed sensor was deployed in Qingdao nearshore water for 35 days. The results proved that the sensor could measure the dynamic changes of CO_2 concentration in seawater continuously, and had the potential to carry out long-term observations on an oceanic platform. It is hoped that the sensor could be applied to field ocean observations in near future.

Keywords: in-situ sensor; dissolved CO_2; long-term measurement; permeable membrane; NDIR; low-cost

Introduction

The ocean is a huge reservoir of carbon and have the capacity for absorbing and retaining CO_2[1]. The oceanic uptake of anthropogenic CO_2 causes pronounced changes to the marine carbonate system[2]. Since the 1980s, 20% to 30% of CO_2 from human activity has been absorbed by the ocean, which has caused ocean acidification[3]. High quality pCO_2 measurements with good temporal and spatial coverage are required to monitor the oceanic uptake, identify regions with pronounced carbonate system changes, and observe the effectiveness of CO_2 emission mitigation strategies[2]. Therefore, measuring dynamic changes of CO_2 in seawater is of great significance to understanding the ocean carbon cycle and ocean acidification.

---

① 本文于2022年2月发表在*Journal of Oceanology and Limnology*第40卷，https://doi.org/10.1007/s00343-021-1133-7。

In the past few decades, underwater in-situ CO_2 sensors have attracted more and more attention[2]. In 2009, the Coastal Technology Alliance (ACT) undertook detailed performance tests on commercial sensors in Hood Canal, Washington, and Kaneohe Bay, Hawaii, for a month, including Contros HydroC™/CO_2, PMEL MAPCO$_2$/Battelle Seaology pCO_2 monitoring system, Pro-Oceanus Systems Inc. PSI CO_2-Pro™, and Sunburst Sensors SAMI-CO_2[40]. Meanwhile, water samples were collected to measure pCO_2 in the laboratory by two traditional methods, and in-situ pCO_2 measurements were compared to these references, and estimates of analytical and environmental variability were reported[4,38,39]. The extensive time-series data provided by these sensors at both test sites revealed patterns in pCO_2, and captured a significantly greater dynamic range and temporal resolution than could be obtained from discrete reference samples. Aliasing of water sampling missed some of the extreme and rapid changes in pCO_2 often observed in these environment[4,5,38,39]. The results indicate the feasibility of these sensors for underwater applications, and the importance of continuous in-situ measurements. In addition, some new pCO_2 sensors have been produced and applied in recent years, including Pro-Oceanus company's Mini CO_2 sensor[6], Solu-Blu CO_2 probe[6], and Turner-Designs company's C-Sense probe[7], among others.

Commercial CO_2 sensors play an important role in in-situ measurements based on various underwater platforms. Take the Contros HydroC™/CO_2 sensor, for example. In 2011, Fietzek et al. improved the HydroC™ (CO_2/CH_4) sensors and successfully deployed them on a variety of fixed and mobile platforms, including water sampler rosette, surface drifter measuring platform, large research Autonomous Underwater Vehicle (AUV), small lander, profile float, ultra-heavy duty Remote Operated Vehicle (ROV), and more, demonstrating the feasibility of the use of this series of sensors on underwater platforms[8]. In 2013, Fiedler et al. fixed a HydroC™/CO_2 sensor equipped with an SBE 5M pump on an Argo-type profiling float, and carried out four consecutive deployments with regular pCO_2 sensor zeroings near the Cape Verde Ocean Observatory (CVOO) in the eastern tropical North Atlantic[9]. In 2015, Qiannan Hu et al. measured in-situ CO_2 concentrations dissolved in seawater near the hydrothermal vent (within ten meters from the seafloor) in the mid-Okinawa Trough using HydroC™ (CO_2) sensors based on the ROV, and the results showed that the maximum values of CO_2 as high as $12,000 \times 10^{-6}$ (volume fraction) occur near active hydrothermal vents in Iheya North area[10]. In 2020, Totland et al. carried out submarine CO_2 leakage detection using the HydroC™/CO_2 sensor deployed on an AUV, although the response of the sensor was too slow (about 2 min with the pump) to satisfy the fast-moving measurement requirements of the AUV through the plume (about 10–15 s), so no significant change of pCO_2 was directly detected[11]. Apart from the above mentioned, other commercial sensors have also been widely used in in-situ CO_2 measurements. For example, in 2018, Park et al. carried the Pro-Mini CO_2 sensor on a buoy to study the pCO_2 dynamics of a stratified reservoir in a temperate zone, and CO_2 pulse emissions during turnover events[12].

In addition to commercial sensors, there are also some home-made sensors for use in specific environments. For example, Blackstock et al.[13] developed a low-cost (US$250–300) Arduino monitoring platform (CO2-LAMP) for recording CO_2 variability in electronically harsh conditions: humid air, soil, and aquatic environments. A relatively inexpensive CO_2 gas analyzer was waterproofed using a semi-permeable, expanded polytetrafluoroethylene membrane without additional support and putted in a plastic case housing. The performance and parameters of the CO2-LAMP for detecting the dissolved CO_2 are shown in Table 1. The CO2-LAMP was deployed at Blowing Springs Cave, and operated alongside a relatively greater-cost CO_2 monitoring platform. Over the monitoring period, measured values between the two systems covaried linearly ($R^2 = 0.99$ for cave stream dissolved CO_2). Although the CO2-LAMP has a good performance in the field measurement, it can not withstand higher hydrostatic pressure due to its simple packaging, and can not accurately rapidly measure microvariations of the CO_2 concentration due to its low precision and long response time[13].

With the development and wide application of new underwater vehicles, such as AUVs, gliders, Argo Floats, and so forth, the acquisition of CO_2 data with spatial and temporal variability has become more convenient, and new requirements for in-situ CO_2 sensors have emerged in response. In order to be suitable for these cable-less underwater vehicles, the sensor must fulfill several requirements: (1) low production cost; (2) low power consumption and long-term operation ability; (3) small size; (4) robust against pressure[14]. Among these requirements, the production cost of the CO_2 sensor is an important consideration, especially for disposable floats or multi-point simultaneous measurement. The commercial sensors mentioned above have good performances for in-situ CO_2 measurements, as shown as Table 1, however the price of these commercial sensors is expensive (much more than US$10,000); consequently, it is difficult to be used as disposable sensors or for multi-point simultaneous measurement. According to the Defense Advanced Research Program Agency (DARPA) Ocean of Things (OoT) program[15], sensors with a small size, low power, and low cost will be the trend in near future. In order to fulfill these requirements for these new platforms and programs, realizing observations of large-scale, long-term measurements of dissolved CO_2 in seawater, a CO_2 sensor with low power consumption, a small size, acceptable measurement accuracy, and a price of less than US$1000 would be a good choice. In this paper, a miniature, low power consumption, low cost in-situ CO_2 sensor based on a membrane and Non-Dispersive Infrared (NDIR) technology was developed. Both laboratory experiments and field experiments were undertaken for the CO_2 sensor performance evaluation.

Table 1　Parameters of some in-situ CO_2 sensors

Manufacturer and model	Principle	Detection range	Precision	Response time	Pressure rating	Size/weight	Power consumption
Pro-Oceanus CO_2-Pro	Membrane, NDIR	$0–600\times10^{-6}$ (volume fraction)	±0.5%	τ_{63} = 2.5 min (with water pump)	50 m	Φ190 mm×330 mm, 6.5 kg in air	4 W
Pro-Oceanus CO_2-Pro™ CV	Membrane, NDIR	$0–600\times10^{-6}$ (volume fraction)	±0.5%	τ_{63} = 50 s (with water pump)	600 m	Φ100 mm×380 mm, 2.8 kg in air	3 W
Pro-Oceanus Mini CO_2	Membrane, NDIR	$0–2000\times10^{-6}$ (volume fraction)	±2%	τ_{63} = 3 min	600 m	Φ53 mm×280 mm, 0.53 kg in air	85 mW
Pro-Oceanus Solu-Blu CO_2	Membrane, NDIR	0–2000 μatm	±3%	τ_{63} = 4 min	50 m	Φ50 mm×200 mm, 0.28 kg in air	88 mW
Contros Hydro-CO_2	Membrane, NDIR	200–1000 μatm	±0.5%	τ_{63} = 1 min (with water pump)	2000 m	Φ89 mm×380 mm, 4.5 kg in air	3.6 W
BATELLE MAPCO$_2$	Bubble equilibrator, NDIR	$0–1000\times10^{-6}$ (volume fraction)	0.03%	20 min in water 17 min in air	Surface	about 80 kg	3 W

to be continued

Manufacturer and model	Principle	Detection range	Precision	Response time	Pressure rating	Size/weight	Power consumption
Turner-designs C-sense	Membrane, NDIR	$0–1000\times10^{-6}$ (volume fraction)	3%	τ_{63} = 4 min	600 m	Φ50 mm$\times$203 mm, 0.43 kg in air	0.48 W
Sunburst SAMI-CO_2	Membrane, spectrophotometry	150–700 µatm	±3 µatm	about 5 min	600 m	Φ152 mm$\times$550 mm, 7.6 kg in air	4.8 W
CO_2-LAMP	Membrane, NDIR	0–10%	300×10^{-6} (volume fraction)	27–38 min	1.4 m	–	–

Material and Method

▶ *Material*

Due to the particularity of in-situ detection of dissolved CO_2 in seawater, it is necessary to consider the sensor as a whole in order to improve its adaptability. The configuration of the newly developed CO_2 sensor is shown in Fig. 1. The sensor includes three parts: gas-liquid separation, gas detection, and electronics. The CO_2 detection part and electronic part are packaged in a pressure vessel. A permeable membrane for gas-liquid separation is installed in the front end cap of the vessel, and an 8-pin connecting port is installed in the rear end cap of the vessel.

Gas detection

In order to achieve accuracy of the long-term measurement data, a high-precision CO_2 detector and a temperature, humidity, and pressure sensor were selected. The CO_2 detector (NE Sensor Technologies, Ltd, 7NE/CO_2), based on NDIR technology with a 2000×10^{-6} (volume fraction) full scale detection range and 1×10^{-6} (volume fraction) resolution, has good selectivity and no oxygen dependence. It has an inner optical cavity with multiple reflection structures and dual-channel detectors. This cavity can achieve spatial dual optical path reference compensation, leading to a stable performance and small fluctuations for CO_2 detection. What's more, the CO_2 detector is compensated by temperature (0–50 °C). The temperature of sea water ranges from 0 to 30 °C approximately. In practical applications, the heat inside the in-situ CO_2 sensor is constantly exchanged with the heat in the seawater surrounding the sensor. Considering the heat dissipation of the devices, the temperature inside the sensor is approximately 5–35 °C, within the temperature compensation range, so the selected CO_2 detector is suitable for our application requirement, and does not need extra temperature correction theoretically. In addition, the detector measures the CO_2 absorption band at 4.3 µm, while water vapor has no absorption at 4.3 µm, so it is not affected by humidity theoretically. The high-precision temperature, humidity, and pressure sensor (BOSCH, BME680) was used to monitor the condition inside the in-situ sensor and correct the data from the CO_2 detector. Its temperature measurement range is 0–65 °C, with an accuracy of ± 1 °C and a resolution of 0.01 °C; the humidity measurement range is 20%–80% relative humidity (RH), with an accuracy of ± 3%RH and a resolution of 0.008%RH; the pressure measurement range is 300–1100 hPa, with an accuracy of 0.6 hPa and a resolution of 0.18 Pa. Furthermore, the compact structure and size of the CO_2 detector and temperature, humidity, and pressure sensor are suitable for underwater sensor encapsulation, to maximize the utilization of space inside the sensor.

Electronics

In order to obtain data with a high spatial and temporal resolution, a high sampling frequency can be set as 1 Hz (1 s). However, for uncabled platforms such as buoys, it is difficult to send data in

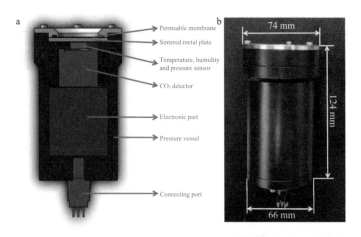

Fig. 1 Structure diagram (a) and photograph (b) of the newly developed CO_2 sensor

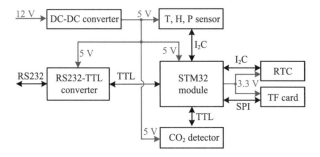

RTC: real time clock; TF: trans flash; SPI: serial peripheral interface; TTL: transistor-transistor logic.

Fig. 2 Connection diagram of each module inside the sensor

real time to a shore-based system, so a data storage module is essential. The connection of each module inside the sensor is shown in Fig. 2. The STM32 module, as the main controller of the sensor, records the time from the RTC (real time clock) module, environmental parameters (temperature, humidity and pressure), and CO_2 concentration into the TF (Trans Flash) card for storage through the SPI (Serial Peripheral Interface) bus. The communication module converts TTL (Transistor-Transistor Logic) to the RS232 to obtain more stable and reliable data. Each data will be recorded and saved as the format of "xxxx/xx/xx xx:xx:xx xx.xx degC xxxxxx.xx Pa xx.xx%RH xxxx ppm" with the capacity of 66 bytes. Thus, it can be calculated that if the sensor works continuously for 1 year with the sampling frequency of 1 s, the data will just take up 1.94 GB of storage space. The TF card selected here has a data reading speed of up to 100 Mb/s and a total capacity of 16 Gb, which fully meets the requirements of high-frequency continuous long-term observation. The power conversion module was used to avoid the situation where the sensor would not work normally due to an excessive cable pressure drop. As a result, the CO_2 sensor has two working modes: interactive mode and automatic mode. When working in the interactive mode, the obtained data is directly stored and displayed in the deck computer via a waterproof cable, and the data is also stored inside as a backup. When working in the automatic mode, an additional pressure vessel with 12 V batteries inside was used for the power supply, and the CO_2 sensor operates intermittently according to the initial setup. Considering the integration of these modules above, the electronic part with a 50 mm long by 50 mm diameter was developed.

Gas-liquid separation and pressure vessel

The response time of the sensor is an important parameter for underwater in-situ measurement. Although the change of CO_2 concentration is a slow process which will be no sudden change in a short time for fixed-point long-term measurement in seawater. In order to measure the CO_2 concentration in real time and accurately, the response time should be as short as possible without affecting other parameters and performance. The response time of the sensor depends on several factors, including the gas-liquid separation efficiency of the membrane, the time for gas to fill the chamber, and the response time of the CO_2 detector. In order to improve the efficiency of gas-liquid separation, the effective area of the permeable membrane should be enlarged as much as possible. To realize the measurement of dissolved CO_2 in water, a 70 μm thickness Teflon AF2400 membrane with good permeability to CO_2 was selected[16]. Its high mechanical strength and slight pressure effect make it very suitable for measuring dissolved CO_2 in seawater. Teflon AF membrane has good compressive resistance, the hydrostatic pressure on the outside of the membrane has little effect on the pressure on the inside[17], so the CO_2 detector does not need pressure correction. However, the larger the effective area of the permeable membrane, it is the easier to rupture because of the influence of external liquid pressure underwater, so the effective area of the permeable membrane should be suitable, and a sintered stainless steel plate was included to support the membrane. Considering the size of the CO_2 detector and a shorter response time, the effective diameter of the membrane is designed to be 34 mm, which is consistent with the diameter of the internal CO_2 detector.

In addition, the aperture and thickness of the sintered metal plate will not only affect the time for gas penetration, but also affect its compression resistance. The aperture of the sintered metal plate is usually 0.22–100 μm. The larger the aperture is, the more rough the surface of the sintered metal plate is, so as to the more easily the membrane is damaged. The smaller the aperture is, the longer the time for the gas to pass through the metal plate, the slower the overall sensor response. Therefore, a sintered stainless steel plate with a moderate diameter of 50 μm was selected.

According to the parameters such as design pressure, inner diameter of the pressure vessel and tensile strength of the material, the size of the corresponding pressure vessel and the thickness of the sintered metal plate can be designed. The wall thickness δ of the pressure vessel can be calculated by Eq. 1, and the thickness of the end cap δ_1 and the thickness of the sintered metal plate δ_1' can be calculated by Eq. 2[18].

$$\delta \geq \frac{P \cdot D}{2\sigma_b/n} \quad (1)$$

$$\delta \geq 0.433D\sqrt{\frac{P}{\sigma_b/n}} \quad (2)$$

where δ is the wall thickness of the pressure vessel (mm), δ_1 is the thickness of the end cap (mm), P is the design pressure (MPa), D is the inner diameter of the pressure vessel (mm), σ_b is the tensile strength (MPa), and n is the safety coefficient.

Table 2 Design results of the pressure cabin

Place	Material	σ_b (MPa)	P (MPa)	D (mm)	n	δ (mm)	δ_1 (mm)	δ_1' (mm)
Vessel	POM	70	2	50	5	3.6→8		
End cap	POM	70	2	50	5		8.2→20	

to be continued

Place	Material	σ_b (MPa)	P (MPa)	D (mm)	n	δ (mm)	δ_1 (mm)	δ_1' (mm)
Sintered metal plate	Stainless steel	250	2	34	5			2.9→3.0

Note: σ_b. tensile strength; P. Design pressure; D. The inner diameter of the pressure vessel; n. Safety coefficient; δ. The wall thickness of pressure vessel; δ_1. The thickness of end cap; δ_1'. The thickness of the sintered stainless steel plate.

It should be noted that δ and δ_1 depends on the size of the electronic module (D = 50 mm), while δ_1' depends on the diameter of the CO_2 detector (D = 34 mm). As shown as Table 2, if the material is polyoxymethylene (POM) whose tensile strength is 70 MPa, and the stress resistance of pressure vessel is 2 MPa (water depth is about 200 m), the wall thickness δ should be no less than 3.6 mm, and the thickness of the end cap δ_1 should be no less than 8.2 mm under 5 times the safety factor by formula calculation. To facilitate the fixing of the end cap and the pressure vessel, the thickness of the pressure vessel δ is thickened to 8 mm, thus the diameter of the pressure vessel is 66 mm. To fit the waterproof connector, the thickness of the end cap δ_1 is thickened to 20 mm, as same as the screw thread length of the waterproof connector. Since the tensile strength of the permeable membrane which material is Teflon and sintered metal plate are unknown, it is impossible to accurately calculate the specific correspondence between the effective diameter of the membrane and the thickness of the sintered metal plate through the formula. As a result, we use half of the tensile strength of 316 L stainless steel (500 MPa) to estimate the tensile strength of the sintered stainless steel plate (250 MPa). According to Eq. 2, the thickness of the sintered stainless steel plate δ_1' should be no less than 2.9 mm under 5 times the safety factor, so it was designed as 3 mm. Considering the length of the inner devices, the total length of the pressure cabin is 124 mm. Then, a corresponding pressure cabin was made, and the success of the pressure test proved that it can withstand underwater pressure of 2 MPa.

The size and weight of each part are shown in Table 3. The total weight is 0.7 kg in air and 0.25 kg in water. The power consumption of the sensor is below 0.9 W. Although the membrane material with high permeability and the CO_2 detector with high precision were chosen, the cost of the newly developed in-situ sensor was kept under US$1000, about a twentieth to thirtieth of the price of similar commercial sensors shown in Table 1 (except $MAPCO_2$).

Table 3 The size and weight of each part, and the assembled sensor

Component		Size (mm)	Weight
Gas-liquid separation module	Permeable membrane	Φ50 mm×0.07 mm	19 g
	Sintered metal plate	Φ50 mm×3 mm	
Gas detection module	CO_2 detector	Φ33.5 mm×31 mm	64 g
	Temperature, humidity, and pressure sensor	16 mm×14 mm×4.5 mm	
Electronic module	TF card	Φ50 mm×50 mm	53 g
	RTC module		
	STM32 module		
	Communication module		
	Power supply		

to be continued

Component	Size (mm)	Weight
Pressure vessel	Φ66 mm×124 mm	500 g
Waterproof connector	Φ15.5 mm×53.2 mm	70 g
Total	Φ66 mm×124 mm / Φ66 mm×158 mm with connector	0.7 kg in air, 0.25 kg in water

▶ *Concentration calculation method*

While measuring the concentration of dissolved CO_2 in seawater, it is necessary that convert the concentration from the gas-phase to the aqueous-phase. For the special case of this sensor, the gas-phase concentration (xCO_2, $\times 10^{-6}$, volume fraction) in the gas cell could be expressed in terms of partial pressure in the gas-phase (pCO_2, μatm) whilst under equilibrium state using the Eq. 3 or Eq. 3′[19-21].

$$pCO_2 = xCO_2 \times p_{dry} = xCO_2 \times (p_{b,in} - P_{H_2O,in}) \quad (3)$$
$$pCO_2 = wCO_2 \times p_{wet} = wCO_2 \times p_{b,in} \quad (3')$$

where xCO_2 is the CO_2 mole fraction in dry gas that equilibrated with water sample and the barometric pressure ($p_{b,in}$, μatm) in gas cell after correcting for the vapor pressure ($P_{H_2O,in}$, μatm) at 100% relative humidity[19]. wCO_2 is the CO_2 mole fraction in wet gas, can be obtained through the CO_2 detector incapsulated in the sensor. In addition, the value of the vapor partial pressure is calculated by Eq. 4 at in-situ temperature (T_w, K) and salinity (S)[22]. Finally, the concentration of CO_2 dissolved in the seawater ($CO_2(aq)$) can be acquired by Eq. 5[23-25].

$$\ln(P_{H_2O,in}) = 24.4543 - 67.4509(100/T_w) - 4.8489\ln(T_w/100) - 0.000544S \quad (4)$$

$$CO_2(aq) = K_0 \times pCO_2 \quad (5)$$

where the solubility coefficient (K_0, mol·kg^{-1}·atm^{-1}) is the function of in-situ temperature (T_w, K) and the in-situ salinity (S), and it can be obtained using the Eq. 6[21].

$$\ln(K_0) = -60.2409 + 93.4517(100/T_w) + 23.3585\ln(T_w/100) + S(0.023517 - 0.023656(T_w/100) + 0.0047036(T_w/100)^2) \quad (6)$$

In summary, to calculate the concentration of dissolved CO_2 in seawater, the solubility and partial pressure of the gas are required to be known. The gas solubility can be calculated by Eq. 6, seawater temperature T_w and salinity S, and the partial pressure of CO_2 can be calculated by Eq. 3′, measured value of the CO_2 detector wCO_2 and measured value of the pressure sensor P_{wet}. The concentration of dissolved CO_2 in seawater can be calculated by the Eq. 7.

$$CO_2(aq) = K_0 \times wCO_2 \times P_{wet} \quad (7)$$

Result and Discussion

To test the long-term measuring ability of the newly developed in-situ CO_2 sensor, in the first place the performances of the CO_2 detector based on NDIR technology were evaluated in the laboratory, including the experiments of its accuracy, linearity, response time and precision by different concentration of CO_2 standard gas, evaluation of its temperature compensation effect, and verification of the issue that if the changes in humidity will affect its measured values. Then the in-situ CO_2 sensor and similar commercial instruments were placed in the pool for comparison to verify the overall measurement accuracy and precision of the sensor. Finally, a long-term nearshore experiment was carried out, and the data of the in-situ CO_2 sensor were analyzed reasonably through the changes of seawater temperature and tide, so as to verify the actual long-term measurement ability of the newly developed in-situ CO_2 sensor. The following content will introduce the experiment process and analyze the results one by one.

▶ *Calibration experiments*

The newly developed CO_2 sensor was calibrated in the laboratory with a series of different concentrations of dry and certified standard CO_2 gases, including 0×10^{-6}, 202.8×10^{-6}, 398×10^{-6}, 503×10^{-6}, 601×10^{-6}, 808×10^{-6}, 1006×10^{-6}, and

$2019×10^{-6}$ (volume fraction). As the CO_2 sensor is passive diffusion type, a gas chamber with a sealing ring was installed on the front end cover of the sensor before the experiments. Each standard gas was flushed into the gas chamber with flow of 400 mL/min by a mass flow controller (Flows Instruments Co, Ltd, AIR-500sccm-b01) to keep the pressure in the gas chamber at about 1 atm. The time intervals between each gas concentration are 10 minutes or so, and each measured value of the sensor is recorded per second to evaluate the response time. Fig. 3 is the results of the calibration experiments. The dynamic measurement results for different gas concentrations are shown in Fig. 3a. From Fig. 3a we can see that the dynamic responses for gas concentration changing are fast. For example, we can see the sensor took the same 15 s from air to reach 63% of the step change of $0×10^{-6}$ and from $503×10^{-6}$ to reach 63% of the step change of $601×10^{-6}$ (volume fraction) (τ_{63} = 15 s by exponential function fitting). The calibration curve between the measured values and the actual values of different concentrations is shown in Fig. 3b. Where, the actual values (abscissa) are the standard gas concentrations, the measured values (ordinate) are the measured mean values of each concentration, and the corresponding standard deviations are expressed as the light blue error bar. From Fig. 3b we can see that the measured values of the newly developed CO_2 sensor and the actual values has a good linear correlation, with R^2 = 0.99 over the range of $0–2000×10^{-6}$ (volume fraction).

▶ *Precision experiments*

To evaluate the measurement precision of the CO_2 sensor, one-hour continuous measurements were performed with a $528×10^{-6}$ (volume fraction) CO_2 standard gas. The standard gas was flushed into the gas chamber with flow of 400 mL/min by a mass flow controller for 5 minutes, then two valves on the gas chamber were closed to keep the concentration of CO_2 gas in chamber constantly. The precision experimental results are shown in Fig. 4. For clarity, the scatter plots have been converted into a frequency distribution histogram, which is fitted using a Gaussian function. From Fig. 4a we can see that the concentration values are mainly distributed in the range of $(528.28 ± 5.16)×10^{-6}$ (volume fraction). From Fig. 4b we can see that the frequency distribution of concentration value shows a roughly normal distribution. Taking the ratio of the half width at half maximum (HWHM) to the average

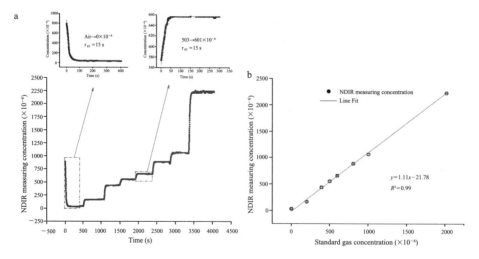

a. Dynamic responses for different gas concentrations, τ_{63} is the time taken for the signal to reach 63% of the next concentration span; b. calibration curve between real values and measured values.

Fig. 3 Calibration results of the newly developed CO_2 sensor with standard gases of different concentrations

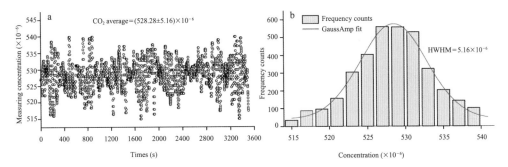

a. Measured concentration values with 1 s intervals; b. histogram and Gaussian distribution.

Fig. 4 Continuous monitoring of 528.28×10^{-6} (volume fraction) CO_2 with a duration of 60 min

concentration value as the precision, we obtain a precision of 0.98% for the CO_2 sensor at a sampling rate of 1 s.

▶ *Influence evaluations of temperature and humidity*

As what mentioned before, although the adopted CO_2 detector theoretically does not need temperature and humidity correction, some experiments were still carried out for evaluation and verification. The data from the CO_2 detector with temperature change were measured firstly, to evaluate the temperature compensation effect. The CO_2 detector was placed in the climate chamber (Vötschtechnik, VC³ 7034). The humidity in the chamber was set at a constant value of 70%RH, and the temperature was set to decrease gradually from 40 °C to 10 °C, to observe if the data from the CO_2 detector change. The temperature and the data from CO_2 detector in the chamber are shown in the Fig. 5. The blanks in the temperature and CO_2 data were caused by an accidental power failure. It can be seen from Fig. 5 that with the increase of temperature, the data from the CO_2 detector almost have no change. Therefore, it is proved that the adopted CO_2 detector has excellent temperature compensation effect, and does not need extra temperature correction practically.

Since long-term measurements in seawater will inevitably lead to an increase in humidity inside the in-situ CO_2 sensor, we conducted simulation tests in the laboratory. The in-situ CO_2 sensor was placed in a sealed tank filled with water to test the humidity change inside the sensor. After determining the range of humidity variation, the influence of humidity on the data from the CO_2 detector was evaluated. The CO_2 detector was placed in the climate chamber mentioned above. The temperature in the chamber was set at a constant value, and the humidity was set according to the range of humidity variation in the last test, to observe the changes in the data from the CO_2 detector. The test of the humidity change inside the sensor lasted for about 5 days with the sampling frequency of 1 s, and the results are shown in Fig. 6a. From the Fig. 6a we can see that the humidity inside the CO_2 sensor shows an exponential growth trend, and it can be predicted that the humidity will stabilize at 68.17%RH through exponential fitting. Next, the temperature in the chamber was set at 10 °C constantly, and the humidity was set at 10%RH, 30%RH, 50%RH, 70%RH, respectively, to observe the changes in the data from the CO_2 detector. The humidity and the data from CO_2 detector in the chamber are shown in the Fig. 6b. It can be seen that with the increase of humidity, the data from the CO_2 detector fluctuated within the range of $(517.55 \pm 4.02)\times10^{-6}$ (volume fraction), without significant change. Therefore, it can be considered that humidity has no effect on the CO_2 detector.

▶ *Stability measurement in the pool*

After evaluating the basic performances of our CO_2 sensor with standard gases, a 7-day stability measurement was carried out in a pool.

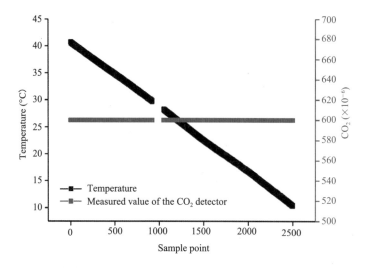

Fig. 5　The evaluation of the temperature compensation effect of the CO_2 detector

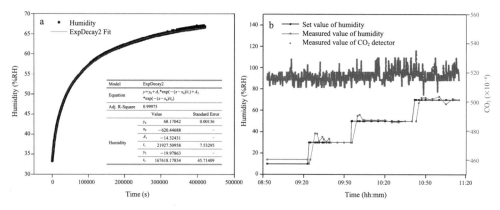

Fig. 6　The humidity change inside the sensor for a long-term test (a) and the influence of humidity on the data from the CO_2 detector (b)

A commercial CO_2 sensor (Pro Oceanus, Mini CO_2[37]) was used simultaneously for comparison. The data of the newly developed sensor were recorded per second, and the data of commercial sensor were recorded per two seconds. Because the newly developed sensor and the commercial sensor have different sampling frequencies, to facilitate the comparison, we average the raw data from two sensors to one value per minute. The 7-day comparison results of the commercial sensor and the newly developed CO_2 sensor are shown in Fig. 7. From Fig. 7, we can see the two sensors' results have good consistency, with R^2 of 0.87. With the same NDIR principle, our CO_2 sensor shows better precision compared with the commercial sensor. The results indicate our CO_2 sensor has good stability for dissolved CO_2 measurements in water.

▶ *Field experimental results at the Qingdao nearshore*

Field experiments were carried out at a depth of about 1 m in Qingdao nearshore waters from 17 May, 2019 to 21 June, 2019. The continuously 35-day CO_2 concentration measurement results were obtained. Meanwhile, in order to make a reasonable explanation for the CO_2 measurement results, the

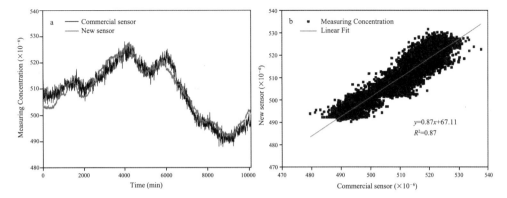

(a) The measuring concentration change of the two CO_2 sensors over time, (b) the relationship between the two CO_2 sensors.

Fig. 7   The 7-day comparison results between the newly developed CO_2 sensor and the commercial CO_2 sensor in the pool

seawater temperature was detected by a commercial multi-parameter water quality sonde (YSI, EXO2). The 35-day CO_2 concentration measurement results are shown in Fig. 8, and the tidal heights and seawater temperature data are also given. The tidal heights data observed at the Qingdao Station were downloaded from the China Maritime Services Network. Several blanks in the CO_2 and temperature data were caused by accidental power failures, and equipment maintenance, especially biofouling checking and cleaning regularly to ensure the accuracy of the CO_2 measurement results.

From the 35-day data, we can see an interesting phenomenon: the CO_2 concentration showed a "double peak" distribution within a day, like a half-day tide. There is an obvious negative correlation between the CO_2 concentration and the tidal heights. Fig. 9 shows zoomed data in the week from June

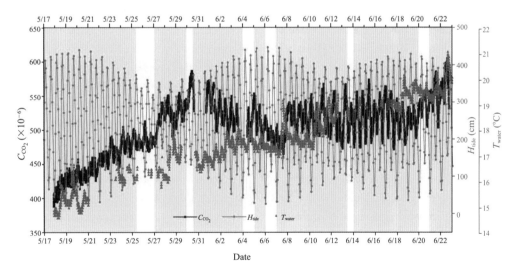

The orange background represents sunny and cloudy weather, the gray background represents overcast weather, and the blue background represents rainy weather.

Fig. 8   The 35-day measurement results of the newly developed CO_2 sensor, tidal heights data from the website, and seawater temperature data from the commercial sensor

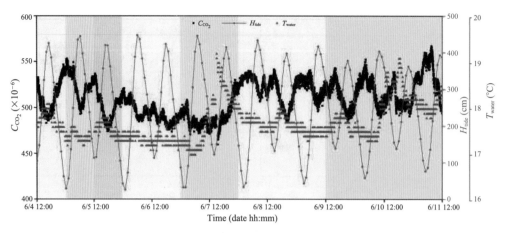

Fig. 9　Changes of the CO_2 concentration, tidal heights, and seawater temperature within a week

4 to 11. Due to the field experiment location being in a dock in Qingdao, which is close to the city, the measured concentration of dissolved CO_2 in coastal seawater is affected by hydrological[26,27], biological[28], surface runoff, and terrestrial input factors[29]. Therefore, it is very difficult to comprehensively explain the CO_2 concentration data obtained from fixed-point observations in the Qingdao nearshore. The obvious correlation between the CO_2 concentration and the tides, and the seawater temperature, needs to be explored further.

We speculate that the correlation between the CO_2 concentration and the tidal height may be related to submarine groundwater discharge (SGD) because in coastal zones, SGD is an important pathway for terrestrial materials to be delivered into the sea[30-32]. Dissolved inorganic carbon concentrations in groundwater are often much higher than those in surface waters, leading groundwater seepage plays a significant role in carbon budgets in aquatic ecosystems[33-35]. SGD fluxes usually show an inversely correlated pattern with the tides[36]; we therefore speculate the semidiurnal pattern of CO_2 we observed was possibly caused by the SGD process in the studied coastal zone. As to the relative correlation between the CO_2 concentration and the seawater temperature, we speculate that this phenomenon is related to the solubility of CO_2 in seawater. With the increase (or decrease) of the seawater temperature, the solubility of CO_2 decreases (or increases), leading to a decrease (or increase) of the CO_2 concentration in seawater. It also can be affected by the weather, because the sensors were located close to the sea surface. For example, on 6th June, there was a heavy rain/shower accompanied by a southeast wind of magnitude 6–7. The rain brought CO_2 in the air into the sea water. As the CO_2 concentration in the air is usually lower than that in the sea water, and the strong wind accelerated the mixing of air and the sea surface, the intraday CO_2 concentration on the sea surface showed an overall downward trend. In addition, rainfall will enrich the groundwater and promote the discharge of groundwater into the sea. However, this process takes a period of time, so the CO_2 concentration on the sea surface showed an upward trend during the period after the rain stopped (June 7–8).

The 35-day field experiment proved the performance of the newly developed CO_2 sensor. It can be seen that our sensor measured the dynamic changes of the CO_2 concentration in seawater continuously, and had the potential to carry out long-term observations on an oceanic platform.

Conclusion

In order to realize the miniaturization, low power consumption, and low cost of in-situ CO_2 sensors in the ocean, we developed a CO_2 sensor based on a permeable membrane and NDIR technology in this paper. The sensor has small dimensions (Φ66 mm×124 mm), low power consumption (<0.7 W), a light weight (0.7 kg in air and 0.25 kg in water), low cost (<US$1000), and high pressure tolerance (<200 m). It is suitable for a variety of offshore platforms and mobile platforms in the sea. After laboratory performance tests, the sensor showed a measurement range of $0-2000\times10^{-6}$ (volume fraction), and the gas linear correlation R^2 was 99.8%, with a precision of about 0.98%. To evaluate the performance of the newly developed sensor, a comparison measurement was carried out with a commercial sensor in a pool for seven days. The experimental results showed consistent trends, and our CO_2 sensor showed better precision compared with the commercial sensor. The newly developed sensor was also deployed in seawater at a depth of about 1 m in the Qingdao nearshore for 35 days. Some interesting phenomena were found from the results of the field experiment, and some reasonable explanations for these were given. The experiment proved that the newly developed sensor could measure the dynamic changes of CO_2 concentration in seawater continuously, and had the potential to carry out long-term observations on an oceanic platform. It is hoped that the sensor could be applied to field ocean observations in near future.

Data Availability Statement

The datasets generated during and/or analyzed during the current study are available from the corresponding author on reasonable request.

Acknowledgement

The authors would like to thank Wangquan Ye and Ning Li for their helpful discussion of the experiments, and thank the IOISAS (Institute of Oceanographic Institution, Shandong Academy of Sciences) for providing the sea trial platform.

References

1. Yin J P, Wang Y S, Xu J R, et al. Adavances of studies on marine carbon cylce[J]. Acta Ecologica Sinica, 2006, 26(2): 565-575. (in Chinese with English abstract)
2. Clarke J S, Achterberg E P, Connelly D P, et al. Developments in marine $p CO_2$ measurement technology; towards sustained in situ observations[J]. Trends in Analytical Chemistry, 2017, 88: 53-61.
3. Bindoff N L, Cheung W W L, Kairo J G, et al. Changing Ocean, Marine Ecosystems, and Dependent Communities[R]//IPCC Special Report on the Ocean and Cryosphere in a Changing Climate, 2019: 477-587.
4. Schar D, Atkinson T, Johengen T, et al. Performance Demonstration Statement for Pro-Oceanus Systems Inc. PSI CO_2-ProTM[Z/OL]. Alliance for Coastal Technology. ACT DS10-03, 2009. http://dx.doi.org/10.25607/OBP-343.
5. Tamburri M N, Johengen T H, Atkinson M J, et al. Alliance for coastal technologies: advancing moored $p CO_2$ instruments in coastal waters[J]. Marine Technology Society Journal, 2011, 45: 43-51.
6. Pro Oceanus. Solu-Blu™ Dissolved CO_2 Probe[Z/OL]. 2021. https://pro-oceanus.com/products/solu-blu-series/solu-blu-co2. Accessed on 2021-02-08.
7. Turner Designs. C-sense in situ $p CO_2$ Sensor[Z/OL]. 2021. https://www.turnerdesigns.com/c-sense-in-situ-pco2-sensor. Accessed on 2021-02-08.
8. Fietzek P, Kramer S, Esser D. Deployments of the HydroC™ (CO_2/CH_4) on stationary and mobile platforms—Merging trends in the field of platform and sensor development[R]. OCEANS'11 MTS/IEEE KONA, Waikoloa, HI, USA, 2011: 1-9.
9. Fiedler B, Fietzek P, Vieira N, et al. In Situ CO_2 and O_2 Measurements on a Profiling Float[J]. Journal of Atmospheric & Oceanic Technology, 2013, 30(1): 112-126.
10. Hu Q N, Zhang X, Wang B, et al. In situ detection of CO_2/CH_4 dissolved in vent-associated seawater at the CLAM and Iheya North hydrothermal vents area, Okinawa Trough[R]. OCEANS 2015-Genova. Genova, Italy, 2015: 1-6.
11. Totland C, Eek E, Blomberg A E A, et al. The correlation between $p O_2$ and $p CO_2$ as a chemical marker for detection of offshore CO_2 leakage[J]. International Journal of Greenhouse Gas Control, 2020, 99: 103085.
12. Park H, Chung S. $p CO_2$ dynamics of stratified reservoir in temperate zone and CO_2 pulse emissions during turnover events[J]. Water, 2018, 10(10): 1347.
13. Blackstock J M, Covington M D, Perne M, et al. Monitoring atmospheric, soil, and dissolved CO_2 using a low-cost, arduino monitoring platform (CO2-LAMP): theory, fabrication, and operation[J]. Frontiers in Earth Science, 2019, 7: 313.
14. Fritzsche E, Staudinger C, Fischer J P, et al. A

14. validation and comparison study of new, compact, versatile optodes for oxygen, pH and carbon dioxide in marine environments[J]. Marine Chemistry, 2018, 207: 63-76.
15. Waterston J, Rhea J, Peterson S, et al. Ocean of Things: Affordable Maritime Sensors with Scalable Analysis[R]. OCEANS 2019-Marseille, Marseille, France, 2019: 1-6.
16. Biogeneral. Teflon™ AF 2400[Z/OL]. 2021. https://www.biogeneral.com/teflon-af/.
17. Chua E J, Savidge W, Short R T, et al. A review of the emerging field of underwater mass spectrometry[J]. Frontiers in Marine Science, 2016, 3: 209.
18. Cheng D X. Handbook of Mechanical Design[M]. 5th ed. Beijing: Chemical Industry Press, 2010: 310-311. (in Chinese)
19. Wu Y X, Dai M H, Guo X H, et al. High-frequency time-series autonomous observations of sea surface $p$$CO_2$ and pH[J]. Limnology and Oceanography, 2020, 66: 588-606.
20. Takahashi T, Sutherland S C, Wanninkhof R, et al. Climatological mean and decadal change in surface ocean $p$$CO_2$, and net sea–air CO_2 flux over the global oceans[J]. Deep Sea Research Part II: Topical Studies in Oceanography, 2009, 56: 554-577.
21. Weiss R F. Carbon dioxide in water and seawater: the solubility of a non-ideal gas[J]. Marine Chemistry, 1974, 2(3): 203-215.
22. Weiss R F, Price B A. Nitrous oxide solubility in water and seawater[J]. Marine Chemistry, 1980, 8(5): 347-359.
23. Johnson J E. Evaluation of a seawater equilibrator for shipboard analysis of dissolved oceanic trace gases[J]. Analytica Chimica Acta, 1999, 395: 119-132.
24. Pro Oceanus. Technical Note 1.1: Dissolved CO_2 and Units of Measurement[Z/OL]. 2019. https://pro-oceanus.com/images/pdf/PSITechnicalNote1.1-DissolvedCO2_andUnitsofMeasurement2019.pdf. Accessed on 2021-02-08.
25. Zhang Z H, Li M, Guo J J, et al. A portable tunable diode laser absorption spectroscopy system for dissolved CO_2 detection using a high-efficiency headspace equilibrator[J]. Sensors, 2021, 21(5): 1723.
26. Wanninkhof R, Pickers P A, Omar A M, et al. A surface ocean CO_2 reference network, SOCONET and associated marine boundary layer CO_2 measurements[J]. Frontiers in Marine Science, 2019, 6: 400.
27. Takahashi T, Olafsson J, Goddard J G, et al. Seasonal variation of CO_2 and nutrients in the high-latitude surface oceans: A comparative study[J]. Global Biogeochemical Cycles, 1993, 7(4): 843-878.
28. Millero F J. Thermodynamics of the carbon dioxide system in the oceans[J]. Geochimica et Cosmochimica Acta, 1995, 59(4): 661-677.
29. Zhai W D, Dai M H, Cai W J, et al. High partial pressure of CO_2 and its maintaining mechanism in a subtropical estuary: The Pearl River estuary, China[J]. Marine Chemistry, 2005, 93(1): 21-32.
30. Moore W S. Large groundwater inputs to coastal waters revealed by 226Ra enrichments[J]. Nature, 1996, 380: 612-614.
31. Burnett W C, Aggarwal P K, Aureli A, et al. Quantifying submarine groundwater discharge in the coastal zone via multiple methods[J]. Science of the Total Environment, 2006, 367: 498-543.
32. Zhang Y, Santos I R, Li H L, et al. Submarine groundwater discharge drives coastal water quality and nutrient budgets at small and large scales[J]. Geochimica et Cosmochimica Acta, 2020, 290: 201-215.
33. Santos I R, Maher D T, Eyre B D. Coupling automated radon and carbon dioxide measurements in coastal waters[J]. Environmental Science and Technology, 2012, 46(14): 7685-7691.
34. Charette M A. Hydrologic forcing of submarine groundwater discharge: Insight from a seasonal study of radium isotopes in a groundwater-dominated salt marsh estuary[J]. Limnology and Oceanography, 2007, 52(1): 230-239.
35. Santos I R, Maher D T, Larkin R, et al. Carbon outwelling and outgassing vs. burial in an estuarine tidal creek surrounded by mangrove and saltmarsh wetlands[J]. Limnology and Oceanography, 2019, 64(3): 996-1013.
36. Burnett W C, Dulaiova H. Estimating the dynamics of groundwater input into the coastal zone via continuous radon-222 measurements[J]. Journal of Environmental Radioactivity, 2003, 69: 21-35.
37. Pro Oceanus. Mini CO_2 Submersible $p$$CO_2$ Sensor[Z/OL]. 2021. https://pro-oceanus.com/products/mini-series/mini-co2. Accessed on 2021-02-08.
38. Schar D, Atkinson T, Johengen T, et al. Performance Demonstration Statement for Contros HydroC™/CO_2[Z/OL]. Alliance for Coastal Technology. ACT DS10-01, 2009. http://dx.doi.org/10.25607/OBP-341.
39. Schar D, Atkinson T, Johengen T, et al. Performance demonstration statement PMEL MAPCO$_2$/Battelle Seaology $p$$CO_2$ monitoring system[Z/OL]. Alliance for Coastal Technology. ACT DS10-02, 2009. http://dx.doi.org/10.25607/OBP-342.
40. Schar D, Atkinson T, Johengen T, et al. Performance demonstration statement for sunburst sensors SAMI-CO_2. Alliance for Coastal Technology[Z/OL]. ACT DS10-03, 2009. http://dx.doi.org/10.25607/OBP-344.

第四篇
海洋碳交易市场标准与气候变化评估

Part IV
Standardization of Ocean Carbon Trading and Climate Change Assessment

Examining the social pressures on voluntary CSR reporting: The roles of interlocking directors[①]

Xueji Liang[1], Lu Dai[2], Sujuan Xie[3*]

1 School of Business, Sun Yat-Sen University, Guangzhou, China
2 Business School, Renmin University of China, Beijing, China
3 Management College/China Business Working Capital Management Research Center, Ocean University of China, Qingdao, China
* Corresponding author: Sujuan Xie(xiesujuan@ouc.edu.cn)

Abstract

Purpose: Corporate social responsibility (CSR) reporting is a widely accepted procedure for firms to disclose their performance in multiple domains, including environmental protection, labour welfare, protection of human rights, community services, contribution to society, and pursuit of product safety. This study investigates whether and how board interlocks affect firms' decisions with respect to CSR reporting. We argue that board interlocks act as an important source of social pressure, and firms are influenced by their peer firms to adopt CSR reporting.

Design/methodology/approach: This paper sampled listed companies on China's Shanghai and Shenzhen Stock Exchanges from 2009 – 2015. The data were collected from Runling database and China Stock Market and Accounting Research database. A multiperiod logit model was used to conduct the main regression analysis, and propensity score matching (PSM) method was used in the robustness checks.

Findings: A study based on a sample of Chinese publicly listed firms from 2009–2015 confirms the argument and shows that sharing a common director on the board with a previous CSR reporter facilitates the firm's engagement in CSR reporting. Furthermore, we show that the influence of board interlocks on CSR reporting depends on three characteristics: status of the interlocking director, size of the linked CSR reporter, and performance implications of previous CSR activities.

Practical implications: The findings of this study have important implications for practitioners. First, the messaging role of interlocking directors suggests that director selection should consider the effectiveness of information transfer. Knowing and analysing specific interlock and its links with the firm's strategy is very important. Meanwhile, firms should be vigilant that the balance between the access to information and loss of autonomy, because searching for information related to firms' strategic decisions might challenge current strategy. Second, the results of our study suggest that in order to effectively urge companies to engage in CSR reporting, government and policy makers should consider beyond institutional pressure, but also be sensitive to the social pressure exerted upon the companies.

Social implications: The positive role of board interlocks on corporate voluntary CSR reporting can

---

① 本文于2022年1月发表在 *Sustainability Accounting, Management and Policy Journal* 第13卷，https://doi.org/10.1108/SAMPJ-05-2021-0166。

not only make valuable contributions to the Chinese society but also, as an important participants of global economy and trade, the Chinese interlocking directors' contribution to CSR reporting have global benefits.

Originality/value: This study extends the institutional perspective on CSR reporting by uncovering the effect of social pressure. It advances the literature on the antecedents of CSR reporting by linking board interlocks to CSR reporting. Finally, the study enriches the broader interlock literature by delineating three specific characteristics of interlocks that influence CSR reporting.

Keywords: voluntary CSR reporting; board interlocks; corporate social responsibility; interlocking directors

Introduction

Corporate social responsibility (CSR) reporting is a widely accepted procedure for firms to disclose their performance in multiple domains, including environmental protection, labour welfare, protection of human rights, community services, contribution to society, and pursuit of product safety. While researchers largely agree that regulatory pressures positively affect firms' decisions to voluntarily issue CSR reports[1-5], the variation in CSR reporting of firms under the same regulatory pressure remains unexplained. The different responses to the same regulatory pressure suggest that firms may interpret and make sense of requirements differently. As interpretations and making sense of regulatory signals depend greatly on social peers[6-8], the social pressure exerted by peers affects the focal firm's decisions on whether to issue a CSR report.

In this study, we investigate the social pressure generated by firm peers and its influence on firms' voluntary CSR reporting. Our theoretical framework follows that the *social aspect* of institutional pressures largely results from the firm's normative need to align its activities with those of its peer firms[9,10]. Social pressures arise from individuals' incentives to look and/or behave like social peers, and this generates norms in their decision-making, such as CSR reporting. Herein, we focus on board interlock, defined as a governance structure when a person is on the board of directors of two or more corporations[11] and a key mechanism in which firms refer to social peers. We argue that the focal firm is influenced by its peer firms linked through board interlocks when deciding on CSR reporting.

Board interlocks act as mechanisms through which joint control is exercised among firms[12] and as conduits to convey information between firms[13]. Such information can influence various important firm decisions[14-19]. The received information serves as a type of social pressure that drives firms to copy the practices of their interlocking partners. A taken-for-granted rationalisation of certain strategic choices emerges from social interactions, a process through which participating firms exchange information and interpret, translate, and make sense of regulations. In case of voluntary CSR reporting, where the government does not enforce the policy, the social pressure from interlocking directors is responsible for the variations in firms' decisions regarding whether or not to issue a CSR report because the directors on the board affect the degree to which CSR reporting resonates with and is prioritised by management[20]. Thus, we draw attention to the impact of board interlocks, a key force that assists the firm in internalising social pressures. By investigating whether and how interlocking directors cause variations in voluntary CSR reporting, our study pinpoints the cause for different CSR reporting decisions.

To examine the specific firm-level mechanisms through which social pressures affect firms' CSR reporting, we consider three characteristics of interlocking directors. We contend that the status of the interlocking directors (i.e., executive vs. non-executive director), the size of the linked CSR reporter, and the performance implications of previous CSR activities moderate the relationship between board interlocks and CSR reporting.

We test our theoretical framework by examining

the CSR reporting of Chinese publicly listed firms between 2009 and 2015. The Chinese setting is appropriate for testing our arguments for three reasons. First, interlocking directors are relatively common in Chinese firms[14]. In fact, they even appeared in government ministries when a representative was assigned to the boards of more than one company[21]. Regarding the regulatory settings for board directors in Chinese-listed firms, they are elected to three-year terms but can serve consecutive terms. A listed firm is required to have a minimum of 5 and a maximum of 19 directors. Since June 30, 2003, a firm is required to have at least one-third of its board consisting of independent directors. An independent director cannot (1) be related to the manager, (2) be one of the top 10 shareholders or hold more than 1% of the company shares, or (3) have a business relationship with the firm[22]. Second, CSR reporting, a recent development in China, provides an ideal setting to observe firms' initiation of CSR reports[23]. At the end of 2007, the Chinese government issued the first guidelines for public firms to publish their CSR activities in their annual reports. In 2008, Chinese public firms began issuing stand-alone CSR reports[3]. Lastly, the government guidelines for CSR reporting are not mandatory, leaving room for firms to interpret institutional signals. When formal institutions and their enforcement are weak[24], firms seek social cues to guide their behaviours. As board interlocks provide a structure to access and understand social cues[16] and are relatively common among Chinese firms[14], examining the impact of Chinese board interlocks on their CSR reporting is a suitable empirical setting to investigate the extent to which social pressures lead to variations in CSR reporting.

Our empirical findings support our main hypothesis that observing an interlocking firm that has already issued a CSR report drives the focal firm to engage in CSR reporting. Furthermore, we find that the focal firm's initiation of CSR reporting is dependent on (1) status of the interlocking director, (2) size of the linked CSR reporter, and (3) benefits from previous CSR activities.

We aim to make three theoretical contributions with this study. First, we extend the institutional perspective on CSR reporting by drawing attention to social pressure. The social aspect is a key component of institutional pressure[9,10]. Recent studies on CSR reporting have focused on the social/normative impacts[25], but research devoted to the mechanism of how social pressure affects CSR reporting is scant. We argue that the mechanism of social pressure explaining the variation in CSR reporting can be tested in the context of board interlocks. Our finding that board interlocks indeed drive CSR reporting fills this research gap. Second, we advance the literature on the antecedents of CSR reporting by linking board interlocks to CSR reporting. Prior research indicates that the quality of corporate governance[26], especially board characteristics, affects CSR disclosure[27,28]. Our study offers new insights into the mechanism of board interlocks by investigating how sharing board memberships with firms that previously issue a CSR report can affect the focal firm's decision on CSR disclosure. The results show that a connection with a previous CSR reporter leads to new CSR reporting. Third, we enrich the broader interlock literature by delineating three specific characteristics of previous CSR reporters in the interlocks that drive CSR reporting. In addition to showing that the existence of board interlocks positively relates to CSR reporting, we investigate which types of interlocks are more likely to generate social pressures. Although existing research has conceptually proposed different types of influences among social peers[29], it has not empirically examined them to understand their mechanisms. We complement this literature by showing that when the previous CSR reporter firm has an executive director in the interlock, is a large firm, and has benefited from its prior CSR activities, its influence on other firms' CSR reporting is more likely to occur than when these characteristics are absent.

Theories and Hypotheses

▶ *Social pressure and CSR reporting*

Notably, CSR reporting tends to be a response to regulatory pressures[1,5,30]. Firms follow regulations and guidelines to gain legitimacy from key resource providers[31]. The institutional perspective of a firm states that institutional pressures consist of both regulatory and social aspects[32], and the previously argued institutional effect on CSR reporting applies to both aspects. However, the extant literature predominantly focuses on regulatory pressure while ignoring the social pressure on firms with regard to CSR reporting. It is important to note that social pressure differs from regulatory pressure and, therefore, the effect of social pressure on CSR reporting needs to be examined separately from that of regulatory pressure.

The social aspect of institutional pressure differs from the regulatory aspect in terms of its influence on firm behaviour. First, the evaluating audiences are different. While the demands of the two pressures are both influential, regulatory pressure entails coercive influence from regulators (e.g., governments, licencing, accreditation organisations), which can influence the existence, survival, growth, and performance of organisations with standardised criteria[33], while social pressure largely results from the normative need for the focal firm to align its activities with those of the reference firms[9,10]. The choice is grounded in a social context that considers a firm's relationship with others. Firms in a community may develop institutional beliefs and structures that normalise 'new' practices[34], and these beliefs can be transmitted through common activities between firms. Therefore, the firm feels the pressure to conform to the norms and values of the general society, typically reflected in the behaviours of the peer firms, such as interlock partners.

Second, regulatory and social pressures function differently in the institutionalisation process[7,32,35]. The pressure from regulators is a top-down force, in which standardised requirements are given to firms, and responses to these requirements are expected by the regulators. Judgments and possible penalties are both made use of by regulators to assess the extent to which institutional demand is met. In contrast, social pressures arise mostly from the firm itself, from the cognitive need to recognise the identity of a social group. As information is costly to access and analyse, firms gather information from the parties they have relationships with or consider valuable. The logic of appropriateness limits the firm's behaviour. Firms may consult their reference firms on important strategic decisions, such as market entry[36], foreign entry[37], and innovation[38].

▶ *Board interlock-induced social pressure and CSR reporting*

Research documents that board interlocks induce social pressure because they play a role in exchanging information among firms. The information transferred through interlocks may lead firms to engage in the same activities as their interlocked peers. For example, firms are likely to imitate the interlocking firm in decisions related to acquisitions[16], multidivisional structure[18], corporate governance practices[19], foreign market expansion[14], charitable recipients of corporate donations[15], and firm performance[17]. Moreover, the interlock is a useful channel through which firms collect information, interpret external signals, and experience change[36]. The importance of such channels is greater when institutional environments are uncertain and complex[32].

Consistent with the conclusion based on Western evidence, board interlocks in Chinese listed firms are common and usually used to manage a firm's relationship with the external environment. However, these interlocks differ from their Western counterparts in terms of the extent to which they can affect the firm. Western interlocks reduce uncertainty by providing information on crucial external parties and being a form of co-optation[17]. Chinese interlocks mainly serve as information providers and social connections that link the firm to the broader society[21]. However, in an emerging market like China, where information asymmetry is relatively high, interlocks may serve as a type of social pressure that firms pay more attention to and, hence, are more important in affecting firm activities. The Chinese context allows us to examine the theoretical implications of board interlock-induced social pressure.

The institutional environment of Chinese firms underscores the effect of interlocks on CSR reporting. In China, the regulation of CSR disclosure, which came into existence in late 2007, is not mandatory, and offers firms few historical materials to refer to. Thus, to guide their CSR strategies, Chinese firms are incentivised to seek relevant information from other firms with which they have connections. These connections offer firms channels to gather key information that helps resolve the uncertainty surrounding the benefits of CSR reporting or the penalties associated with not reporting CSR activities. Many interfirm connections are created and developed in board interlocks[16], and the individual decisions of a firm influence peer firms mostly through social pressures[39]. We argue that this is especially true for Chinese firms when they make decisions on voluntary CSR reporting.

Social pressure through board interlocks exists to the extent that the strategies or behaviours of the imitated firm are taken as normatively appropriate[8]. While the government takes the initiative of CSR regulation and enforcement[29,30], groups of firms serve as carriers of normative decisions. Notably, CSR reporting is likely to become a widely accepted strategy through social interaction among firms that share directors on boards. Specifically, when the focal firm knows or observes the decision of a previous CSR reporter through their common directors, they may reflexively engage in CSR reporting. Directors not only offer information on other firms' strategies but also possess private knowledge about how such decisions are made and what factors need to be considered. They may actively offer relevant information and knowledge when the other firms they serve decide on the same strategy, thereby influencing the focal firm to make a similar choice. This social pressure through board interlocks is even more likely to occur when the firm is deciding on CSR reporting for the first time, as prior information on the outcome of the decision to be made is unavailable.

Regarding the unavailability of reliable prior knowledge on CSR reporting, as mentioned previously, the CSR guideline for public firms in China is a recent policy; therefore, firms have little historical information to rely on when deciding on CSR disclosure. Historical performance serves as a key reference for determining firm behaviour[36]. In China, CSR regulation began in 1998, and only a small number of firms issued CSR reports in the first two years after the initial publication of the regulation. This absence of hard evidence on the consequences of CSR reporting in China suggests that the next best kind of information is from peer firms with which the focal firm has connections. Knowing that peer firms engage in CSR reporting will turn the decision maker's attention to the practice when information on the motivation for and expectations of CSR reporting are exchanged by the directors. Regarding the complexity of CSR performance implications, while CSR-related activities require resource commitments and a cost is associated with fulfilling them, these activities are considered to be driving factors for positive firm outcomes, including superior performance[40], high firm reputation[6], lower cost of capital[41], moral capital[42], and market reaction[43]. However, the net effect of CSR-related activities on firms remains rather unclear, especially regarding decision-making by an individual firm. Calculating the expected costs and benefits associated with CSR reporting poses challenges for the focal firm. Taken together, with the unavailability of prior information and the complex outcome implications of CSR reporting, Chinese firms imitate their peer firms to make decisions on CSR reporting. Therefore, we propose as follows:

Hypothesis 1a*: Sharing an interlocking director with a previous CSR reporter increases the* *likelihood of the focal firm to initiate CSR reporting.*

Hypothesis 1b: *Higher the number of interlocking directors shared with a previous CSR reporter on the board, higher the likelihood of the focal firm to initiate CSR reporting.*

▶ *Characteristics of previous CSR reporter and CSR reporting*

We have already argued the merits of interlocking directors in motivating CSR reporting. However, interlocking directors can sometimes constrain CSR reporting: particularly, when directors serve on too many boards and are being "busy directors"[44]. Busy directors sitting on multiple boards have, on an average, less time and attention for the boards, which is considered as the main reason for weaker governance[45] and reduced financial performance of these firms[46,47].

As such, we examine specific characteristics associated with board interlocks that counter the shortcomings of interlocking directors in motivating CSR reporting. More specifically, although the existence of board interlocks generally increases the tendency of the focal firm to issue a CSR report, different characteristics associated with board interlocks exert distinctive effects. Therefore, to deepen our understanding of the social pressure on CSR reporting, we investigate three characteristics of the previous CSR reporter, with which the focal firm shares a common director.

The first characteristic is the status of the interlocking director in the previous CSR reporter. We argue that when a director who serves on multiple boards is an executive director in the previous CSR reporter, he or she is more influential in driving the decision than when the director is a non-executive director.

Executive directors typically have more knowledge about firm operations than non-executive directors, who have no formal affiliation with the firm[47]. Firms tend to copy the practices of "model firms" without fully understanding the outcomes of doing so. The modelling influence of the interlocking firm varies at different levels. At the individual level, the type of director in terms of his or her affiliation with the firm determines the degree of social pressure. Specifically, the type of director who sits on multiple boards determines the strength of social pressure. The most commonly used classification of director type on the board is the executive versus non-executive director[48]. Consequently, we argue that when the interlocking director is an executive director in the previous CSR reporter, his/her influence on CSR reporting of the focal firm is stronger. Executive directors possess more private information about the internal decision-making process of CSR reporting primarily because they have official appointments in the previous CSR reporter. When influencing other firms' decisions, private knowledge permits executive directors to have greater influence than non-executive directors. Private knowledge includes detailed descriptions of how the CSR reporting decision is made, what factors are needed to be considered, and possibly the qualitative outcomes of CSR reporting in the previous CSR reporter. Conversely, non-executive directors are less likely to have the same kind of valuable information as executive directors because of their lack of formal affiliation with a previous CSR reporter and limited access to key information that guides the decision on CSR reporting in the focal firm. Therefore, we argue that when the director sitting on multiple boards is an executive director, he or she exerts a greater influence on the focal firm's CSR reporting. Thus:

Hypothesis 2a: *When the director (who serves on multiple boards) is an executive director, the likelihood of the focal firm to initiate CSR reporting is higher than when the director is a non-executive director.*

The second characteristic is the size of the linked CSR reporter. When the practice in use by some subset of firms stands out in terms of impact, the practice tends to be imitated by other firms. That is, if the firm that has previously issued a CSR report is a large firm, the social pressure is stronger than its smaller counterparts are. Firms experience social pressure to model themselves after peer firms in the field when they face uncertainty about goals related to CSR activities. Thus, large firms that have previously issued CSR reports are more likely to be imitated.

There are two reasons for the demonstrating effect of large firms. First, large firms are typically high-status firms identified as legitimate for their CSR reporting. Directors are increasingly involved in formulating strategic decisions[39]. As direct participation in decision-making is particularly likely to encourage social learning, firms are more likely to learn from other firms perceived as legitimate. Directors imitate successful models or avoid specific behaviours characterised as unsuccessful[49]. Therefore, in the absence of adequate information on a means–ends relationship, when directors seek models to imitate behaviours, large firms stand out. The CSR reporting of large firms is perceived as a strategy that peer firms can copy, regardless of the actual benefits of CSR reporting. Second, the CSR reporting of large firms elicits more influence by implying greater resources associated with the strategy. Firms disclose their CSR activities mainly to gain stakeholders' endorsement and acquire the resources necessary to grow and survive[50]. In general, large firms are more resourceful than small firms; thus, the issuing of a CSR report by firms with fewer resource constraints to gain stakeholder support heightens the importance of communicating with stakeholders by issuing a CSR report. Therefore:

Hypothesis 2b: *Larger the size of the linked CSR reporter, higher the likelihood of the focal firm to initiate CSR reporting.*

The third characteristic is the performance of CSR activities in the previous CSR reporter. When the focal firm deems that CSR reporting benefits the firm that has chosen to engage in CSR reporting, it is more likely to report its own CSR-related activities. The decision to adopt a practice or otherwise depends on its expected returns. When facing multiple options, firms tend to weigh the predicted outcomes of different strategies and choose the one with the best outcome. Regarding competing strategies, firms may abandon one strategy in favour of another that has emerged in a given reference group. In particular, when the firm observes that CSR reporting has led to positive performance, it is more likely to engage in CSR reporting.

It is difficult to gauge the precise effect of CSR reporting on firm performance[43]. However, as a means of communicating CSR-related activities to stakeholders, CSR reporting is directly associated with firms' CSR strategies. That is, when the firm spends resources on CSR, it will issue a CSR report to advertise its engagements, including activities in community building, environmental protection, employee welfare, and product improvement and safety. Thus, from the perspective of the peer firm, the link between CSR activities and firm performance of the previous CSR reporter can serve as a proxy for the link between CSR reporting and firm performance in general. In China, information on firm CSR activities is generally available to board members. Therefore, interlocking directors can acquire the performance of CSR-related activities in the linked firm and use it as a reference when it comes to the focal firm's decision to report CSR. Thus:

Hypothesis 2c: *Greater the performance associated with CSR activities by the previous CSR reporter, higher the likelihood of the focal firm to initiate CSR reporting.*

In Fig. 1, we illustrate the logical relationship among the hypotheses.

Methods

▶ *Sample and data sources*

Data were collected from two primary sources. First, we gathered information on firms' CSR reporting from Runling (also known as RKS in English), an independent CSR report rating agency in China. RKS rates the annual CSR reports issued

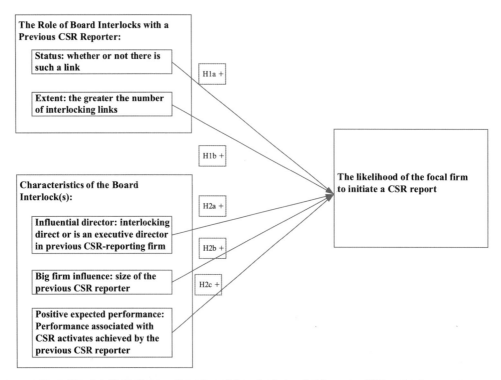

Fig. 1　Hypotheses about board interlocks and voluntary CSR reporting

by Chinese public firms in all industries based on four components: macrocosm, content, technique, and industry. The macrocosm component assesses firms' disclosure of CSR strategy, governance structure, and stakeholder relationships, including whether they have a stated CSR strategy and engage in actual activities to fulfil that strategy. The content component assesses the detailed disclosure on financial performance, labour and human rights, environmental protection, fair operation, customer satisfaction, and community engagement. The technique component assesses the quality of CSR reporting, such as the coverage of content, comparability of information, innovation in reporting, credibility, and transparency. We used the overall rating scores for this study, which is an integrated score of the four components[3]. Second, we gathered information on the directors and other firm-level data from the China Stock Market and Accounting Research (CSMAR) database. From corporate annual reports, the CSMAR provides full coverage of firms' boards of directors, including a biographical sketch of the directors, their current board appointments, and their experiences in industry and government[51].

The sample for this research covered all Chinese-listed firms from 2009 to 2015. We chose 2009 as our starting year because it was the first year when information on whether a firm issued a CSR report was available. The initial sample for 2009–2015 has 20,609 firm-year observations. First, we eliminated 3766 firm-year observations that are mandated to disclose CSR information because we are interested in the choice of voluntary CSR reporting. Second, considering the statistical model used in this research (introduced in detail subsequently), we excluded firm-year observations that follow the first CSR reporting year (7168 observations). We dropped firm observations from 2010, 2013, and 2015 because no firms issued a new CSR report in those years (4582 firm-year observations). We dropped additional 306 observations with missing information on key variables. The final sample contained 4787 firm-year observations, covering 1546 firms.

▶ *Statistical model*

This study investigates how CSR reporting practices are determined by a firm's board links to previous CSR reporters. Thus, we needed to test whether a firm that issued its first CSR report in year (t) did so because of its board link to a previous CSR reporter in year ($t + 1$) and onward. Because a firm cannot be identified as an initial CSR reporter in multiple years, the traditional lead–lag model was not feasible for our data analysis. Alternatively, we adopted a multi-period logit regression model.

The multi-period logit regression model is equivalent to a hazard model. According to Shumway[52] (p. 111), "each year in which the firm survives is included in the logit program's 'sample' as a firm that did not fail.... Time-varying covariates are incorporated simply by using each firm's annual data for its firm-year logit observations". However, the numbers of firm years in the logit program are not independent of each other; that is, a firm's characteristics in year (t) are correlated with its characteristics in year ($t + 1$). This violates the hazard model's assumption of independence between firm-year observations. A firm that fails in year (t) cannot fail again in year ($t + 1$). Therefore, the sample size of the logit program must be adjusted. Following Shumway[52], we exclude firms identified as new CSR reporters in year (t) from the sample for year ($t + 1$) and onward. For the coefficients in the model, we observed robust standard errors clustered by firms. To test the hypothesis H1a and H1b, we run the multi-period logit regression with the initiating of CSR reporting variable against board interlock variable; to test the hypothesis H2a, H2b and H2c that about the characteristics of previous CSR reporting and voluntary CSR reporting, we run the main multi-period log regression in subsamples of firms that are grouped based on specific characteristics of previous CSR reporters. By comparing the effects of the board interlock on voluntary CSR reporting between different subsamples, we can observe whether a specific characteristic of the previous CSR reporter indeed moderates the relationship between the board interlock and voluntary CSR reporting. We provide detailed measurements for variables in the following session.

▶ *Measures*

Dependent variable

The dependent variable—New CSR Reporting—is equal to 1 if a firm issues its first CSR report in year (t), and (0) otherwise. We adopted a multi-period logit regression to investigate the spread of CSR reporting through board interlocks. This regression is equivalent to a hazard model[52] and requires that observations identified as 1 in year (t) are dropped from the sample in year ($t + 1$) and onward (we present more details on the multi-period logit model in the section on model specification). For example, in 2009, we identified 81 firms as new CSR reporters and 826 firms as non-new CSR reporters. These 81 firms were removed from the 2010 sample to avoid biasing the investigation into the spread of CSR reporting by board interlocks in that year. Therefore, the dependent variable is equal to 0 for firm-year observations before the year when a firm is identified as a new CSR reporter or for firm-year observations that have never issued any CSR reports.

Independent variables

The first independent variable is Board Links to CSR Reporter, which measures whether a firm shares a board seat with a firm that has previously issued a CSR report. It is equal to 1 when the firm is linked to a CSR reporter that has already issued CSR reports before year (t), and 0 otherwise. The second independent variable is a continuous measure of Board Links to CSR Reporter, which is measured by the natural logarithm of 1 plus the summed links to previous CSR reporters.

Fig. 2 illustrates the coding for the dependent variable New CSR Reporting and the independent variable Board Links to CSR Reporter. Here, firms A

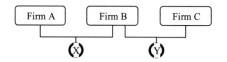

Year	Firm A Links to CSR reporter/New CSR	Firm B Links to CSR reporter/New CSR	Firm C Links to CSR reporter/New CSR
2009	No/Yes	No/No	No/No
2010	Dropt	Yes/Yes	Yes/No
2011	Dropt	Dropt	No/No

Example of New CSR Reporting and Links to CSR Reporter coding. Firms are represented by rectangular boxes and board members by lettered circles.

Fig. 2 Interlocked boards

and B are linked through director X, and firms B and C are linked through director Y. We are interested in the extent to which firm B's CSR reporting decision is influenced by its link to firm A through the shared director X, provided that firm A has already issued CSR reports. As shown in Fig. 2, firm A is a new CSR reporter in 2009. Because 2009 is the first year when Chinese firms began to voluntarily issue CSR reports, we code the dependent variable New CSR Reporting as 1 for firm A in 2009. Firms B and C did not report CSR in 2009 (New CSR Reporting = 0), and thus, their links to a previous CSR reporter are valued at 0. In 2010, firm B began reporting CSR through board interlock to firm A, and is therefore defined as a new CSR reporter (Board Links to CSR Reporter = 1). In 2010, firm C had no board links to a previously identified CSR reporter because firm B did not report CSR in 2009.

Furthermore, to test the hypothesis H2a-H2c, we measure the characteristics about previously identified CSR reporters in the following ways. First, H2a suggests that the focal firm is more likely to adopt CSR reporting when the interlocking director is an executive director in the previous CSR reporter. We created the Executive interlocking director dummy variable to indicate whether the interlocking directors were executive directors in the previous CSR reporter, and next we group our sample firms into two subsamples based on the executive interlocking director dummy.

H2b suggests that larger the size of the previous CSR reporter, the higher the likelihood of the focal firm initiating a CSR report. We created the Size of previous CSR reporter dummy to indicate whether the previous CSR reporter is a large firm; it is equal to 1 when the firm size of a previous CSR reporter is greater than the median firm size of the same industry in a given year, and 0 otherwise. We further divided the sample into two groups based on the firm size of previous CSR reporter: one group with large firms, and other with small firms.

H2c proposes that the benefits derived from previous CSR reporting or CSR-related activities are important signals for potential CSR reporters. First, we estimated a linear prediction for the benefits of CSR reporting or prior donation activity obtained by previously identified CSR reporters. Specifically, we ran two separate regressions with previous CSR reporters' return on assets (ROA) in year (t) against their CSR reporting dummy in year ($t-1$), and firms' donation in year ($t-1$). Then, we estimated the coefficient of CSR reporting dummy/firms' donation for each firm; a positive value on the estimated coefficient denotes benefits from a firm's previous CSR reporting or previous CSR-related activities, while a negative value or zero indicates no benefits. Finally, we grouped the total sample into two sets of subsamples based on the signs of the estimated coefficient of each firm: the first set of subsamples contain one group of firms that have links to previous CSR reporter that benefits from its CSR reporting (CSR – ROA prediction = 1), and the other group of firms that have links to previous CSR reporter that does not benefit from its CSR reporting (CSR – ROA prediction = 0). The second set of subsamples contain one group of firms that have links to previous CSR reporter that benefits from its donation activities (donation – ROA prediction = 1), and the other group of firms that have links to previous CSR reporter that does not benefit from its donation activities (donation – ROA prediction = 0).

Control variables

To rule out confounding factors, we included multiple control variables in the analyses. To examine this effect and to rule out the impact of having a board interlock in general, we introduce Interlock and Total Number of Interlocks. *Interlock* is a dummy variable which is equal to 1 when the firm has board links to any other firm, including CSR reporters, and 0 otherwise. The Total Number of Interlocks is the natural logarithm of 1 plus the summed board links to any other firm, including CSR reporters (ln(1+summed board links)). We measure firm Size using the natural logarithm of a firm's market value of equity. Return on assets (*ROA*) is the ratio of net profit to total assets. Leverage is the ratio of long-term debt to total assets. We measure Liquidity as the number of shares traded divided by the total outstanding shares and Boardsize as the number of directors on the board. Independence is the percentage of independent directors on the board. State-owned enterprises (SOEs) equal 1 when a firm's largest shareholder is the state, and 0 otherwise. Directors from sensitive industry equals 1 when a firm has links, through board interlocks, to firms in environmentally sensitive industries, which include paper manufacturing, oil, natural gas, metal, and chemical industries[13]. Related-party equals 1 when a firm's links to other firms are through related-party transitions. Industry CSR Score is the standardised median of the CSR reporting score for a certain industry in a given year. We add this industry median to control for the influences caused by shocks to industry in each year. Within-industry interlock is equal to 1 when a firm's links, through interlock directors, to other firms are within the same industry, and 0, otherwise. Within-region interlock is equal to 1 when corporate links, through interlock directors, are within the same province in China. The Appendix Ⅰ provides the definitions of the variables. To exclude the effects of outliers, we winsorized all the control variables at the 99% level.

Results

▶ *Summary statistics on interlocks and CSR reporting*

To provide initial evidence of the relationship between board interlocks and the new voluntary CSR reporting, Panel A of Table 1 lists the descriptive statistics of firms identified as having initiated CSR reports and their board characteristics. Columns 2 and 3 report the total number of firms identified as CSR reporters in each year (*N*) and the number of firms identified as new CSR reporters. Column 2 clearly shows that voluntary CSR reporting was spread unevenly across the period. In 2009, 89 firms were identified as having voluntarily issued CSR reports that year, of which, 81 were new CSR reporters. The number of new CSR reporters reduced to 42 in 2011 but quickly rose to 86 and 71 in 2012 and 2014, respectively. Panel B of Table 1 presents the data on the number of firms in each year that do not report CSR.

Table 1 also reports the statistics on board size and board interlocks for the new CSR reporters and non-new CSR reporters. Firms identified as new CSR reporters have a slightly larger board size on average. Consistent with the idea that director interlocks play a role in the spread of CSR reporting practices, new CSR reporters have a greater number of links to firms previously identified as CSR reporters. For example, as Panel A of Table 1 shows, 17.9% of firms identified as new CSR reporters are linked through board interlocks to another firm that previously reported CSR, compared to 13.4% of non-new CSR reporters.

Table 1 Distribution of new CSR reporting firms, board interlock characteristics by year

Year	N	Number of new CSR reporter	Board size	Interlocks (%)	Average interlocks (%)	Links to CSR (%)	Average links to CSR (%)
Panel A CSR reporting firms							
2009	89	81	8.852	50.6	1.185	0	0
2010	96	Year observation removed					
2011	122	42	9.048	81	1.690	14.3	0.214
2012	197	86	9.118	87.1	2.094	22.4	0.353
2013	199	Year observation removed					
2014	230	71	9.268	90.1	2.768	35.2	0.493
2015	244	Year observation removed					
Total	1177	279	9.068	76.3	1.935	17.9	0.264
Panel B non-CSR reporting firms							
2009	1089	826	9.008	60.3	1.185	0	0
2010	1385	1058	Year observation removed				
2011	1604	1212	8.831	70	1.506	8.17	0.0998
2012	1654	1223	8.768	83.3	1.931	14.6	0.181
2013	1690	1233	Year observation removed				
2014	1720	1247	8.498	84.6	2.194	26.1	0.376
2015	1800	1265	Year observation removed				
Total	10,924	8064	8.754	75.9	1.749	13.4	0.178

Table 2a provides the descriptive statistics of the new CSR reporting by industry. We followed 19 industry classifications by the Chinese Security Regulatory Commission in 2012. Of the firms, 61.6% are from the manufacturing industry. To obtain a clearer picture of the CSR reporting distribution within the same industry, we further categorised the manufacturing industry into four sub-industries: food manufacturing, chemical material, electronic/metal/non-metal/automobile manufacturing, and other equipment manufacturing. We removed two industries that did not have any observations in CSR reporting, leaving 20 industries in our sample. Notably, the health, construction, conglomerate, professional service, and food manufacturing industries have the highest new CSR reporting frequencies (12.5%, 7.38%, 7.14%, 6.9%, and 6.82%, respectively). This is possibly due to their higher social and environmental impacts. What is important in the distribution is that we observed voluntary CSR reporting in 18 of the 20 industries in our sample. The widespread nature of the practice across industries was appropriate to test our hypotheses that board links play a role in influencing voluntary CSR reporting practices.

Table 2a Distribution of new voluntary CSR reporting firms, board interlock characteristics, by industry

Industry	N	Number of new CSR reporter	New CSR (%)	Interlocks (%)	Average interlocks (%)	Links to CSR (%)	Average links to CSR (%)
Agriculture	88	5	5.68	82.95	1.690	9.09	0.103
Mining	124	7	5.65	73.39	1.866	12.10	0.151
Food M	337	23	6.82	77.74	1.714	13.35	0.181
Chemical M	822	50	6.08	73.72	1.699	12.29	0.154
Electronic M	1702	98	5.76	75.97	1.697	14.22	0.195
Equipment M	86	4	4.65	66.28	1.561	10.47	0.159
Power supply	193	11	5.70	76.68	1.995	17.10	0.221
Construction	149	11	7.38	75.17	1.828	15.44	0.200
Wholesale	279	17	6.09	76.70	1.845	9.68	0.140
Transportation	145	8	5.52	80.00	1.895	14.48	0.196
I.T.	305	17	5.57	76.72	1.890	18.69	0.268
Finance	9	0	0	55.56	1.250	0.00	0
Real estate	265	15	5.66	80.75	1.813	11.32	0.149
Business Service	56	3	5.36	71.43	1.887	16.07	0.302
Professional service	29	2	6.90	79.31	2.069	24.14	0.241
Public facility	58	2	3.45	75.86	1.702	10.34	0.175
Health	8	1	12.50	87.50	3	12.50	0.125
Publication	90	2	2.22	68.89	1.791	14.44	0.151
Conglomerate	42	3	7.14	78.57	1.974	11.90	0.154
Total	4787	279	5.83	75.91	1.760	13.62	0.183

Note: M stands for manufacturing.

Table 2b provides descriptive statistics about control variables. Averagely about 14% of the total sample firms have general links to other listed firms through board directors. The average (log) number of the interlock directors is 3. An average (log) firm size in our sample is 15.17. The mean value of ROA is 0.04, and the average leverage ratio is 0.42. Averagely shares traded is 5% of its total outstanding shares. The average board size is about 9 people with about 37% of the board members are independent directors. 40% of firms in our sample are SOEs. 29% of firms have directors from sensitive industries, and the standardised median of the CSR reporting score for is 0.01. About 1% of the firms have same-industry board interlocks, and 3% of firms have same-region board interlocks.

Table 2b Descriptive statistics for control variables

Variable	N	Mean	S.D.	Min	Median	Max
Interlock	4787	0.14	0.34	0	0	1
Total number of interlocks	4787	3	3.23	0	2	26
Size	4787	15.17	0.81	13.79	15.06	17.52
ROA	4787	0.04	0.05	−0.17	0.04	0.20
Leverage	4787	0.42	0.23	0.04	0.41	0.97
Liquidity	4787	0.05	0.15	−0.21	0	0.64
Boardsize	4787	8.77	1.7	5	9	15
Independence	4787	0.37	0.05	0.31	0.33	0.56
SOEs	4787	0.40	0.49	0	0	1
Directors from sensitive industry	4787	0.29	0.45	0	0	1
Industry CSR score	4787	0.01	0.94	−2.45	−0.30	2.97
Within-industry interlock	4787	0.01	0.11	0	0	1
Within-region interlock	4787	0.03	0.17	0	0	1

▶ *Regression results*

Board links to CSR reporter and new CSR reporting

H1a suggests that sharing a board director with a previous CSR reporter increases the likelihood of the focal firm issuing a CSR report. Table 3 reports the results for the H1 test. Model 1 includes the control variable, and Model 2 adds the Board Links to CSR Reporter and general interlock control variable. The Board Links to CSR Reporter variable has a positive impact on new CSR reporting (0.426, $P < 0.01$). Therefore, H1a is supported. H1b proposes that the number of interlocking directors serving at a previous CSR reporter has a positive effect on the likelihood of the focal firm's CSR reporting. To test this hypothesis, in Model 3 of Table 3, we replaced Board Links to CSR Reporter with the total number of such links (the natural logarithm of 1 plus the summed links to previous CSR reporters), as well as Interlocks with Total Number of Interlocks (the natural logarithm of 1 plus the summed board interlocks). As the results in Model 3 of Table 3 indicate, the coefficient of the new board link variable is positive and significant (0.384, $P < 0.01$), lending support to H1b.

Table 3 New CSR reporting and board interlocks

Variable	(1) New CSR reporting	(2) New CSR reporting	(3) New CSR reporting
Board links to CSR reporter		0.426** [0.185]	
Interlock		−0.029 [0.154]	
ln(1+summed links to CSR)			0.384** [0.187]
ln(1+summed board links)			0.137 [0.090]
Firm variables			
Size	0.445*** [0.066]	0.517*** [0.081]	0.513*** [0.081]

to be continued

Variable	(1) New CSR reporting	(2) New CSR reporting	(3) New CSR reporting
ROA	1.878 [1.334]	0.124 [1.257]	0.094 [1.263]
Leverage	−0.817** [0.379]	−0.198 [0.338]	−0.200 [0.339]
Liquidity	−0.618*** [0.238]	−0.607*** [0.234]	−0.603*** [0.233]
Governance variables			
Boardsize	0.019 [0.041]	0.026 [0.041]	0.017 [0.042]
Independence	−0.808 [1.331]	−0.969 [1.336]	−1.047 [1.343]
SOEs	0.246 [0.155]	0.312** [0.153]	0.300** [0.152]
Directors from sensitie industry	−0.126 [0.151]	−0.213 [0.152]	−0.272+ [0.153]
Other control variables			
Related-party	0.361+ [0.199]	0.340+ [0.198]	0.334+ [0.197]
Industry CSR score	0.308 [0.215]	−0.316+ [0.170]	−0.320+ [0.170]
Within-industry interlocks	−0.216 [0.262]	0.085 [0.509]	0.058 [0.506]
Within-region interlocks	0.520* [0.295]	0.469 [0.293]	0.410 [0.291]
Industry fixed	Yes	Yes	Yes
Year fixed	Yes	Yes	Yes
Constant	−2.421*** [0.461]	−10.174 [1.319]	−10.104*** [1.326]
N	4787	4787	4787
Pseudo R2	0.056	0.059	0.060

Note: + $P<0.1$; ** $P<0.05$; *** $P<0.01$. Robust standard errors clustered by firm are reported in parentheses.

Characteristics of the previous CSR reporter and new CSR reporting

Director Type and New CSR Reporting. H2a suggests that when an interlocking director shared between a firm previously identified as a CSR reporter and a firm without such a practice is an executive director in the firm that has already issued CSR reports, another firm is more likely to adopt CSR reporting. To test this hypothesis, we first created a dummy variable, executive interlocking director, to indicate whether a focal firm has directors who serve as executive directors in the previous CSR reporter. Second, we divided the sample into two subsamples by the executive interlocking director, one including firms with one or more interlocking directors who are executive directors in the previous CSR reporter, and the other consisting of firms whose interlocking directors are non-executive directors in the interlock. We then ran the main tests for the two subsamples. Models 1 and 2 in Table 4 show the regression results. The coefficient of Board Links to CSR Reporter is positive and significant (0.669, $P < 0.01$) for the subsample of executive interlocking directors. By contrast, the coefficient of Board Links to CSR Reporter in the subsample of non-executive interlocking directors is not significant. Having board links to CSR reporter exerts a positive impact on CSR reporting only when the director who sits on multiple boards is an executive director in the previous CSR reporter. Therefore, H2a is supported.

Size of Previous CSR Reporter and New CSR Reporting. H2b suggests that larger the size of the previous CSR reporter, more likely it is for the focal firm to initiate CSR reporting. We then applied the multi-period logit regression to two subsamples of the firms, one indicating the group of firms with large firm size and the other indicating the group of firms with small firm size. Models 3 and 4 of Table 4 show the results. Model 3 shows that in the subsample of large previous CSR reporters, Board Links to CSR Reporter has a positive and significant effect on New CSR Reporting (1.41, $P < 0.01$). By contrast, in the group of small previous CSR reporters (Model 4), the Board Links to CSR Reporter coefficient is insignificant. Having links to CSR reporters exerts a positive impact on new CSR reporting when the firm previously identified as a CSR reporter is a large firm. Therefore, H2b is fully supported.

Performance Implication of Previous CSR Activities and New CSR Reporting. H2c proposes that the benefits derived from previous CSR reporting or CSR-related activities are important signals for potential CSR reporters to decide about initiating CSR reporting. We applied the main multi-period logit regression in two sets of subsamples. The first set of subsamples contain a group of previous CSR reporter that benefit from their CSR reporting (CSR reporting – ROA prediction = 1) and a group that does not benefit from their CSR reporting (CSR reporting – ROA prediction = 0). The second set of subsamples contain a group of previous CSR reporter that benefit from their donation activity (Donation – ROA prediction = 1), and a group that does not (Donation – ROA prediction = 0).

Models 5–8 in Table 4 report the results of these tests. In the subsamples of CSR reporting/donation–positive performance in Models 5 and 7, *Board Links to CSR Reporter* has a positive effect on firms' *New CSR Reporting* (CSR reporting: 0.773, $P < 0.01$; donation: 0.727, $P < 0.01$). Comparatively, in the subsamples of CSR reporting/donation–negative/no performance in Model 6 in Table 4, the coefficient of the links to previous CSR reporters is insignificant. Having links to previous CSR reporters exerts a positive impact on CSR reporting only when prior CSR-related activities are associated with good performance. The results in Models 7 and 8 in Table 4 show a similar pattern. The coefficient of links to CSR reporters is positive and significant in Model 7, while the coefficient of the same variable is insignificant in Model 8. Therefore, H2c is well supported.

Table 4 Three characteristics of board interlocks and new CSR reporting

New CSR reporting	(1) Executive interlocking director=1	(2) Executive interlocking director=0	(3) Large previous CSR reporter=1	(4) Large previous CSR reporter=0	(5) CSR-ROA prediction=1	(6) CSR-ROA prediction=0	(7) Donation-ROA prediction=1	(8) Donation-ROA prediction=0
Board links to CSR	0.669** [0.280]	0.308 [0.264]	10412*** [0.411]	−0.243 [0.833]	0.773*** [0.273]	0.550 [0.475]	0.727*** [0.279]	0.563 [0.443]
Reporter interlock	0.138 [1.008]	0.741 [1.022]	0.283 [0.223]	−0.016 [0.262]	Omitted	0.339 [0.225]	Omitted	0.303 [0.223]
Firm variables size	0.478***	0.482***	0.074	0.508+	0.398**	0.282+	0.401**	0.291**
ROA	0.248 [2.135]	1.668 [2.338]	4.941 [3.616]	4.529+ [2.464]	1.619 [2.768]	4.468*** [1.496]	1.136 [2.838]	4.569*** [1.429]
Leverage	−0.443 [0.633]	0.187 [0.533]	0.303 [0.953]	−1.501** [0.668]	−0.800 [0.682]	−0.107 [0.492]	−1.006 [0.691]	0.010 [0.484]
Liquidity	−0.496 [0.366]	−0.749** [0.363]	−0.669 [0.517]	−0.400 [0.563]	−1.233*** [0.434]	−0.623 [0.412]	−1.148*** [0.439]	−0.734+ [0.412]
Boardsize	0.100 [0.078]	0.006 [0.062]	0.009 [0.108]	−0.084 [0.133]	0.022 [0.043]	0.030 [0.042]	0.100 [0.078]	0.006 [0.062]

to be continued

New CSR reporting	(1) Executive interlocking director=1	(2) Executive interlocking director=0	(3) Large previous CSR reporter=1	(4) Large previous CSR reporter=0	(5) CSR-ROA prediction=1	(6) CSR-ROA prediction=0	(7) Donation-ROA prediction=1	(8) Donation-ROA prediction=0
Independence	−1.939 [2.542]	−1.454 [2.194]	0.822 [3.488]	4.213 [3.237]	−1.193 [1.364]	−1.121 [1.376]	−1.939 [2.542]	−1.454 [2.194]
SOEs	0.450+ [0.250]	0.063 [0.239]	0.331 [0.337]	−0.520 [0.404]	0.310** [0.154]	0.361** [0.156]	0.450+ [0.250]	0.063 [0.239]
Directors from sensitive	−0.163 [0.264]	−0.136 [0.211]	−0.556 [0.344]	0.070 [0.402]	−0.238 [0.155]	−0.201 [0.156]	−0.163 [0.264]	−0.136 [0.211]
Related-party	−0.138 [0.437]	0.552** [0.276]	0.164 [0.483]	0.514 [0.680]	0.396** [0.197]	0.353+ [0.200]	−0.138 [0.437]	0.552** [0.276]
Industry CSR score	−0.012 [0.239]	7.488*** [0.683]	−0.810*** [0.257]	0.137 [0.198]	−0.319+ [0.170]	−0.309+ [0.171]	−0.012 [0.239]	7.488*** [0.683]
Within-industry interlock	−0.342 [1.035]	0.227 [0.629]	−0.207 [1.138]	1.148 [1.408]	0.453 [0.312]	Omitted	−0.342 [1.035]	0.227 [0.629]
Within-region interlock	0.550 [0.404]	0.335 [0.410]	−0.884 [0.980]	−0.033 [1.305]	Omitted	0.006 [0.635]	0.550 [0.404]	0.335 [0.410]
Industry fixed	YES	YES	YES	YES	YES	YES	YES	YES
Year fixed	YES	YES	YES	YES	YES	YES	YES	YES
Constant	−9.852*** [2.424]	−2.721 [2.335]	−6.773 [5.135]	−13.138** [5.925]	−9.925*** [1.328]	−10.039*** [1.353]	−9.852*** [2.424]	−2.721 [2.335]
N	1724	2005	1706	1857	4733	4649	1724	2005
Pseudo-R2	0.064	0.081	0.123	0.166	0.059	0.057	0.064	0.081

Note: + $P<0.1$; ** $P<0.05$; *** $P<0.01$. Robust standard errors clustered by firm are reported in parentheses. Omitted: No available observations for the noted variable in the sub-sample regression.

▶ *Additional analyses*

The existence of board interlocks can also indicate the greater resources of the firm[53]. Thus, investigating the influence on the firm to engage in CSR activities should rule out the resource influence of interlocks on CSR reporting. This supplementary analysis serves this purpose.

We argue that if the resources gained through the board interlocks affect firm CSR decisions, they will enhance both CSR reporting and quality of the CSR reports. Firms' CSR quality can be constrained by limited firm resources and is positively related to firms' capabilities[54]. Meanwhile, as interlock directors are typically more resourceful and knowledgeable and have higher reputations than their peers[49], boards with interlock directors should play a better resource provider role. In the context of CSR reporting, firms with interlock directors on the board should issue higher-quality CSR reports than those with no interlock directors on the board.

We tested this hypothesis by regressing the CSR reporting score of the focal firm in year (t) against the CSR report score of its linked firms that were previously identified as CSR reporters in year ($t-1$), with an identical set of control variables. Notably, CSR report scores indicate the quality of CSR activities. While CSR disclosure is a milestone of firm social performance, the CSR quality is more closely related to the specific CSR activities in which the firm engages in. The idea is that if interlocks serve as a resource provider for CSR reporting, we should observe a positive effect of interlocks on the CSR report score. In the absence of such an effect, the positive effect of interlocks on CSR disclosure is likely to be a social effect, which is the mechanism we propose herein. The results of this analysis, shown in Table 5, indicate that the CSR scores of the previous CSR reporters do not significantly affect the focal firms' CSR report score, lending no support to the resource provider's argument. Therefore, this supplementary analysis shows that the influence of the interlock leads to CSR disclosure.

Table 5 Interlock directorship and CSR reporting quality

Variable	(1) Standardised CSR score (year *t*) OLS	(2) Standardised CSR score (year *t*) fixed effect
Standardised CSR score_Linked firms (*t*–1)	–0.202*** [0.076]	–0.167** [0.083]
Firm variables		
Size	0.456*** [0.104]	0.153 [0.209]
ROA	1.751 [1.564]	2.104 [1.804]
Leverage	–1.517*** [0.562]	–0.032 [0.760]
Liquidity	–0.252 [0.266]	–0.391 [0.254]
Governance variables		
Boardsize	0.029 [0.060]	0.081 [0.072]
Independence	–0.254 [1.509]	2.038 [1.822]
SOEs	0.237 [0.200]	–0.845 [0.538]
Directors from sensitive industry	–0.236+ [0.141]	–0.173 [0.132]
Other control variables		
Related-party	0.156 [0.212]	0.177 [0.270]
Within-industry interlocks	0.140 [0.178]	0.318+ [0.184]
Within-region interlocks	–0.079 [0.244]	–0.035 [0.232]
Year fixed	Yes	Yes
Industry fixed	Yes	No
Constant	–10.056*** [2.320]	–4.853 [4.645]
N	238	238
Adjusted R^2	0.4395	0.1251

Note: + $P<0.1$; ** $P<0.05$; *** $P<0.01$. Robust standard errors clustered by firm are reported in parentheses.

▶ *Robustness checks*

Propensity score matching method

To reduce potential sample selection bias associated with certain firm-level characteristics, we employed the propensity score matching (PSM) method to select a group of firms (i.e., firms without any board links to firms already voluntarily reporting CSRs) that match with the group of treatment firms (i.e., firms with board links to firms already voluntarily reporting CSRs). To implement this method, we employed the original full sample of firms before adjusting for the multi-period logit regression and removed observations with missing values on key variables, resulting in 13,374 firm-year observations for the period 2009–2015.

First, we ran a logistic regression to estimate the probability of board links to already CSR-reporting firms using the characteristics of linked firms as independent variables. We then used this model to create two propensity-matched samples of firms, in which the treatment firms have board links to firms that already voluntarily report CSR, while the control firms do not. Model (1) of Table 6 shows the results of the first-stage logistic regression used to estimate the firm's probability of having board interlocks. The results suggest that this model is fairly well specified, as evidenced by the area under the receiver operating characteristic curve of 76.2%[55]. To ensure that the matched observations are similar, we imposed a requirement that the propensity scores for the treatment and control firms are within a fixed distance of each other (i.e., calliper matching, with calliper width equal to 0.02 times the standard deviation of the logit of the propensity scores).

Model (2) of Table 6 reports the logit regression results using voluntary CSR reporting as the dependent variable with the propensity-matched sample. The coefficient of the board links to the already CSR-reporting firms (Board Links to CSR Reporter) is positive and statistically significant (1% level). This result suggests that with a reduced sample-selection bias, we can still observe a fairly strong positive effect of board links to already CSR-reporting firms on the focal firm's probability of reporting its CSR activities.

Table 6 Robustness test: board links to CSR reporter and new voluntary CSR reporting, PSM sample

Model	(1) Board links to CSR reporter		(2) Voluntary CSR reporting	
Variable	ln(1+summed board links)	0.144*** [0.008]	Board links to CSR reporter	0.617*** [0.129]
	Size (LF)	0.150*** [0.033]	Interlock	0.016 [0.018]
	ROA (LF)	−1.269** [0.571]	Size	0.347*** [0.082]
	Leverage (LF)	−0.385** [0.192]	ROA	1.838 [1.383]
	Liquidity (LF)	−1.175*** [0.195]	Leverage	−0.059 [0.439]
	Boardsize (LF)	−0.009 [0.022]	Liquidity	−0.429 [0.468]
	Independence (LF)	0.250 [0.692]	Boardsize	−0.025 [0.059]
	SOEs (LF)	−0.017 [0.076]	Independence	0.198 [1.570]
	Directors from sensitive industry (LF)	1.015*** [0.063]	SOEs	−0.051 [0.193]
	Industry CSR score (LF)	0.147+ [0.080]	Directors from sensitive industry	−0.408*** [0.137]
	Within-industry interlock (LF)	0.767*** [0.206]	Industry CSR score	0.040 [0.175]
	Within-region interlock (LF)	0.601*** [0.149]	Within-industry interlock	−0.297 [0.510]
	Region (LF)	0.147+ [0.080]	Within-region interlock	0.072 [0.304]
	Industry FE	Yes	Region	0.040 [0.175]
	Year FE	Yes	Industry FE	Yes
	Constant	−5.06*** [0.744]	Year FE	Yes
			Constant	−10.01*** [1.987]

to be continued

Model	(1) Board links to CSR reporter	(2) Voluntary CSR reporting
N	13,374	4116
Pseudo R^2	0.23	0.06
Area under ROC curve	0.762	

Note: + $P<0.1$; ** $P<0.05$; *** $P<0.01$. Robust standard errors clustered by firm are reported in parentheses.

Other links between the focal firm and the interlocking firm

Apart from the board links, there might be other connections between the focal and interlocking firms. For example, directors from both firms know each other because they were colleagues at the same organisation before. This connection is likely to facilitate the procurement of information on voluntary CSR reporting to transfer among firms, which could be a confounding factor for the observed effect of the main independent variable. To control for this potential effect, we created a new dummy variable, ColleagueConnection. It is equal to 1 if directors or other top executives of the focal firm are former colleagues with directors or other top executives of the interlocking firm, and 0 otherwise.

We collected data on colleague information from the CSMAR database. As shown in Model (1) of Table 7, the ColleagueConnection coefficient is insignificant. In addition, the coefficient of the independent variable Board Links to CSR Reporter remains positive and statistically significant, suggesting that even after controlling for other potential channels of information transfer between the focal and interlocking firms, board interlocks still play an important role in helping the focal firms initiate voluntary CSR reporting. In Model (2) of Table 7, we replace the dummy variable measuring colleague connections with the natural logarithm of the total number of colleague connections that the board of directors in a firm has, and the results are consistent with our main findings.

Table 7　Robustness test: controlling for other potential connections between the focal firm and the locking firm

Variable	(1) New CSR reporting	(2) New CSR reporting
Board links to CSR reporter	0.385** [0.185]	0.393** [0.185]
ColleagueConnetion	0.140 [0.150]	
ln(total ColleagueConnection)		0.081 [0.089]
Interlock	0.031 [0.019]	0.032+ [0.019]
Size	0.488*** [0.068]	0.489*** [0.069]
ROA	2.025 [1.363]	2.011 [1.365]
Leverage	−1.090*** [0.373]	−1.098*** [0.372]
Liquidity	−0.414 [0.465]	−0.414 [0.465]
Boardsize	0.005 [0.042]	0.005 [0.042]
Independence	−0.960 [0.374]	−0.928 [1.371]
SOEs	0.168 [0.151]	0.159 [0.152]

to be continued

Variable	(1) New CSR reporting	(2) New CSR reporting
Directors from sensitive industry	−0.277+ [0.153]	−0.277+ [0.153]
Industry CSR score	−0.278 [0.183]	−0.277 [0.183]
Within-industry interlock	0.095 [0.490]	0.070 [0.494]
Within-region interlock	0.315 [0.293]	0.328 [0.291]
Region	−0.278 [0.183]	0.032+ [0.019]
Constant	−12.41*** [1.47]	−12.44*** [1.48]
N	4787	4787
Pseudo R^2	0.059	0.059

Note: + $P<0.1$; ** $P<0.05$; *** $P<0.01$. Robust standard errors clustered by firm are reported in parentheses.

Conclusion

In this study, we propose a theoretical framework that considers CSR reporting an outcome of board interlock-induced social pressure. Interlocks that share a common directorate are governance structures in which social modelling take place. In particular, they provide critical social references when firms make decisions under uncertainty. Our analyses, in the context of Chinese publicly listed firms' CSR reporting, lends support to this framework. The findings show that when the interlocking firm has previously issued a CSR report, the focal firm is more likely to issue its first CSR report. Similarly, linking to more interlocking directors who serve at a previous CSR reporter increases the likelihood of the focal firm initiating a CSR report. In addition, the three characteristics of the board interlock increase social pressure. First, if the director who sits on multiple boards is an executive director rather than a non-executive director in the previous CSR reporter, the focal firm is more likely to adopt CSR reporting. Second, when the previous CSR reporter is a large firm, its CSR reporting decision is more likely to be imitated by the focal firm. Third, when the previous CSR reporter had good returns from prior CSR-related activities, the focal firm is more likely to make the same CSR reporting decision.

▶ *Theoretical implications*

The framework and empirical findings of this study have important implications for the management literature. First, we extend the institutional perspective on CSR reporting by showing that the social pressure faced by firms influences their decision to report CSR. The institutional perspective is a core theoretical argument for various firm decisions, and the social aspect of institutional pressure is equally important to its regulatory counterpart[9,10]. The motivation of this study is the dearth of research efforts to examine the social aspect of institutional pressure. To achieve our research goal, we argue that the relationship between social pressure and variation in CSR reporting can be tested in the context of board interlocks. Our finding that social pressure through board interlocks indeed drives CSR reporting fills this research gap.

Second, we advance the literature on the antecedents of CSR reporting by linking board interlocks to CSR reporting. Corporate governance, including its overall quality[26] and key characteristics, can affect the decision to disclose CSR[27,28]. Board interlock is a corporate governance structure by which firms connect with external parties, and it conveys its power by influencing key strategic decisions. The combined role of monitoring and advising is heightened when directors sit on multiple boards. Our empirical findings show that this role is also important in CSR reporting. That is, a connection with a prior CSR reporter drives CSR reporting decisions. Future research should control for the effects of board interlocks when studying voluntary CSR disclosure.

Finally, we enrich the broader interlock literature by delineating three specific characteristics of the previous CSR reporters that enhance social pressure. While interlocks have merits in motivating CSR reporting, they may also have shortcomings. The 'busy director' does not have sufficient time and attention for the boards[45], which may be a hinder for firm performance. As such, we strive to discover the board characteristics that encounter such negative effects. More specifically, we go beyond the baseline model by investigating specific types of interlocks. To the best of our knowledge, different types of social pressures have not been empirically studied. Therefore, we propose a more balanced view of board interlocks by showing that when the previous CSR reporter has an executive director in the interlocks, is a large firm, or has good returns from its prior CSR activities, the focal firm is more likely to adopt CSR reporting than when these characteristics are absent.

▶ *Practical implications*

The findings of this study have important implications for practitioners. First, the messaging role of interlocking directors suggests that director selection should consider the effectiveness of information transfer. Knowing and analysing specific interlock and its links with the firm's strategy is very important. Firms should recognize that their directors have a messaging role and pay attention to the networks of their directors, value that each director can contribute. Meanwhile, firms should be vigilant that the balance between the access to information and loss of autonomy, because searching for information related to firms' strategic decisions might challenge current strategy. Besides director expertise, the director's personalities, communication skills, and styles should also be taken into account. Directors who are willing and able to convey information produce more effective communication and engagement.

Second, the results of our study suggest that in order to effectively urge companies to engage in CSR reporting, government and policy makers should consider beyond institutional pressure, but also be sensitive to the social pressure exerted upon the companies. Individualized messages often attract more attention from the decision makers than regulations. Policy makers should utilize the social influence through personalized communication channels. Furthermore, our study can help investors and regulators understand better potential mechanism(s) behind the voluntary CSR reporting.

▶ *Limitations*

The interpretation of the current findings should be considered in light of these limitations. First, while board interlocks are an important social aspect of institutional pressure, other types of social pressure exist. For example, industry competitors influence firm strategy. Although integrating other types of social pressure into our theoretical framework is beyond the scope of our study, we find that social pressure arising from the board is important. Chinese business networks are characterised by *guanxi*, in which the trust associated with the connection in the network influences members' business decisions. Board interlock is a typical type of business network, and directors on the board are the people's decision makers trust the most. Second, our focus is on the CSR reporting decisions. However, CSR reporting can also be symbolic[3,4], with little substantive quality to improve CSR-related activities. We do not investigate the details of the quality of CSR reports, but such an investigation would be a suitable lens through which to examine social pressure. To the extent that we find no evidence that CSR reporting is negatively related to actual CSR activities, we maintain that drivers of actual CSR activities also apply to CSR reporting. Third, we argue that both regulatory and social pressures influence the decision to report on CSR. However, we were unable to determine the weight of each pressure. That is, we do not offer any findings on the relative importance of the regulatory and social pressure firms face, although by emphasising the previously overlooked social pressure, we advance our understanding of the variation in CSR reporting. Future research should follow this direction. Finally, the influence of certain behaviours through interlocks is stronger in the initial stage of the institutionalisation process. During this stage, the practice did not spread widely through the interlock. This applies well to our setting, CSR reporting in China. Unfortunately, the generalizability of our findings may be limited to practices that are already institutionalised or have diffused to a certain degree[8]. Future research should continue to explore whether and how social pressure affects institutionalised CSR practices.

▶ *Summary*

In this paper, we argue that CSR reporting is an outcome of board interlock-induced social pressure. We test this argument in the context of Chinese listed firms. Our empirical results show that the focal firm is more likely to issue its first CSR report when the interlocking firm has previously issued a CSR report. Furthermore, when the interlocking director is an executive director and the previous CSR reporter is a large firm that has benefited from its historical CSR activities, the social pressure from the interlocking director on the focal firm is stronger. By linking board interlocks to CSR reporting, we have advanced the understanding of why variations in firms' decisions to issue CSR reporting persist in the same regulatory environment.

References

1. Aguilera R V, Rupp D E, Williams C A, et al. Putting the S back in corporate social responsibility: A multilevel theory of social change in organizations[J]. Academy of Management Review, 2007, 32(3): 836-863.
2. Ioannou I, Serafeim G. What drives corporate social performance? The role of nation-level institutions[J]. Journal of International Business Studies, 2012, 43(9): 834-864.
3. Luo X R, Wang D, Zhang J. Whose call to answer: Institutional complexity and firms' CSR reporting[J]. Academy of Management Journal, 2017, 60(1): 321-344.
4. Marquis C, Qian C. Corporate social responsibility reporting in China: Symbol or substance?[J]. Organization Science, 2014, 25(1): 127-148.
5. Wang W, Zhao C, Jiang X, et al. Corporate environmental responsibility in China: a strategic political perspective[J]. Sustainability Accounting, Management and Policy Journal, 2020, 12(1): 220-239.
6. Doh J P, Howton S D, Howton S W, et al. Does the market respond to an endorsement of social responsibility? The role of institutions, information, and legitimacy[J]. Journal of Management, 2010, 36(6): 1461-1485.
7. Hofman P S, Moon J, Wu B. Corporate social responsibility under authoritarian capitalism: Dynamics and prospects of state-led and society-driven CSR[J]. Business & Society, 2017, 56(5): 651-671.
8. Scott W R. Institutions and organizations: Ideas, interests and identities[M]. Thousand Oaks, CA: Sage, 1995.
9. Shabana K M, Buchholtz A K, Carroll A B. The institutionalization of corporate social responsibility reporting[J]. Business & Society, 2017, 56(8): 1107-1135.
10. Suchman M C. Managing legitimacy: Strategic and institutional approaches[J]. Academy of Management Review, 1995, 20(3): 571-610.
11. Pennings J M. Interlocking directorates: Origins and consequences of connections among organizations' board of directors[J]. Canadian Journal of Sociology-cahiers Canadiens De Sociologie, 1982, 7: 428.
12. Khanna T, Thomas C. Synchronicity and firm interlocks in an emerging market[J]. Journal of Financial Economics, 2009, 92(2): 182-204.
13. Ortiz-de-Mandojana N, Aragón-Correa J A, Delgado-Ceballos J, et al. The effect of director interlocks on firms' adoption of proactive environmental strategies[J]. Corporate Governance: An International Review, 2012, 20(2): 164-78.
14. Connelly B L, Johnson J L, Tihanyi L, et al. More than adopters: Competing influences in the interlocking directorate[J]. Organization Science, 2011, 22(3): 688-703.
15. Galaskiewicz J, Wasserman S. Mimetic processes within an interorganizational field: An empirical test[J]. Administrative Science Quarterly, 1989, 34(3): 454-479.

16. Haunschild P R, Beckman C M. When do interlocks matter? Alternate sources of information and interlock influence[J]. Administrative Science Quarterly, 1998, 43(4): 815-844.
17. Martin G P, Gozubuyuk R, Becerra M. Interlocks and firm performance: The role of uncertainty in the directorate interlock-performance relationship[J]. Strategic Management Journal, 2015, 36(2): 235-253.
18. Palmer D A, Jennings P D, Zhou X. Politics and institutional change: Late adoption of the multidivisional form by large U.S. corporations[J]. Administrative Sciences Quarterly, 1993, 38(1): 100-131.
19. Shipilov A V, Greve H R, Rowley T J. When do interlocks matter? Institutional logics and the diffusion of multiple corporate governance practices[J]. Academy of Management Journal, 2010, 53(4): 846-864.
20. Bundy J, Shropshire C, Buchholtz A K. Strategic cognition and issue salience: Toward an explanation of firm responsiveness to stakeholder concerns[J]. Academy of Management Review, 2013, 38(3): 352–376.
21. Keister L A. Engineering growth: Business group structure and firm performance in China's transition economy[J]. American Journal of Sociology, 1998, 104(2): 404-440.
22. Jiang W, Wan H, Zhao S. Reputation concerns of independent directors: Evidence from individual director voting[J]. The Review of Financial Studies, 2016, 29(3): 655-696.
23. Zhang Z, Chen H. Media coverage and impression management in corporate social responsibility reports: Evidence from China[J]. Sustainability Accounting, Management and Policy Journal, 2020, 11(5): 863-886.
24. La Porta R, Lopezdesilanes F, Shleifer A, et al. Investor protection and corporate governance[J]. Journal of Financial Economics, 2000, 58(12): 3-27.
25. Griffin P A, Sun E Y. Voluntary corporate social responsibility disclosure and religion[J]. Sustainability Accounting, Management and Policy Journal, 2018, 9(1): 63-94.
26. Chan M C, Watson J, Woodliff D. Corporate governance quality and CSR disclosures[J]. Journal of Business Ethics, 2014, 125(1): 59–73.
27. Jizi M I, Salama A, Dixon R, et al. Corporate governance and corporate social responsibility disclosure: Evidence from the US banking sector[J]. Journal of Business Ethics, 2014, 125(4): 601-615.
28. Prado-Lorenzo J, Garcia-Sanchez I. The role of the board of directors in disseminating relevant information on greenhouse gases[J]. Journal of Business Ethics, 2010, 97(3): 391-424.
29. Waddock S. Building a new institutional infrastructure for corporate responsibility[J]. Academy of Management Perspectives, 2008, 22(3): 87-108.
30. Marquis C, Toffel M W, Zhou Y. Scrutiny, norms, and selective disclosure: A global study of greenwashing[J]. Organization Science, 2016, 27(2): 483-504.
31. Arena C, Liong R, Vourvachis P. Carrot or stick: CSR disclosures by Southeast Asian companies[J]. Sustainability Accounting, Management and Policy Journal, 2018, 9(4): 422-454.
32. Bitektine A, Haack P. The "macro" and the "micro" of legitimacy: Toward a multilevel theory of the legitimacy process[J]. Academy of Management Review, 2015, 40(1): 49-75.
33. Rao H. Institutional activism in the early American automobile industry[J]. Journal of Business Venturing, 2004, 19(3): 359-384.
34. Earle J S, Spicer A, Peter K S. The normalization of deviant organizational practices: Wage arrears in Russia, 1991-98[J]. Academy of Management Journal, 2010, 53(2): 218-237.
35. Aguilera R V, Cuervo-Cazurra A. Codes of good governance[J]. Corporate Governance: An International Review, 2009, 17(3): 376–387.
36. Greve H R. Patterns of competition: The diffusion of a market position in radio broadcasting[J]. Administrative Science Quarterly, 1996, 41(1): 29–60.
37. Henisz W J, Delios A. Uncertainty, imitation, and plant location: Japanese multinational corporations, 1990-1996[J]. Administrative Science Quarterly, 2001, 46(3): 443-475.
38. Semadeni M, Anderson B S. The follower's dilemma: Innovation and imitation in the professional services industry[J]. Academy of Management Journal, 2010, 53(5): 1175-1193.
39. Westphal J D, Zajac E J. Defections from the inner circle: Social exchange, reciprocity, and the diffusion of board independence in U.S. corporations[J]. Administrative Science Quarterly, 1997, 42(1): 161-183.
40. McWilliams A, Siegel D, Wright P M. Corporate social responsibility: Strategic implications[J]. Journal of Management Studies, 2006, 43(1): 1-18.
41. Cheng B, Ioannou I, Serafeim G. Corporate social responsibility and access to finance[J]. Strategic Management Journal, 2014, 35(1): 1-23.
42. Godfrey P C. The relationship between corporate philanthropy and shareholder wealth: A risk management perspective[J]. Academy of Management Review, 2005, 30(4): 777-798.

43. Wang K T, Li D. Market reactions to the first-time disclosure of corporate social responsibility reports[J]. Journal of Business Ethics, 2016, 138(4): 661-682.
44. Core J E, Holthausen R W, Larcker D F. Corporate governance, chief executive officer compensation, and firm performance[J]. Journal of Financial Economics, 1999, 51(3): 371-406.
45. Fich E M, Shivdasani A. Financial fraud, director reputation, and shareholder wealth[J]. Journal of Financial Economics, 2007, 86(2): 306-336.
46. Fligstein N, Brantley P. Bank control, owner control, or organizational dynamics: Who controls the large modern corporation?[J]. American Journal of Sociology, 1992, 98(2): 280-307.
47. Li J, Ang J S. Quantity versus quality of directors' time: the effectiveness of directors and number of outside directorships[J]. Managerial Finance, 2000, 26(10): 1-21.
48. Westphal J D, Seidel M L, Stewart K J. Second-order imitation: Uncovering latent effects of board network ties[J]. Administrative Science Quarterly, 2001, 46(4): 717-747.
49. Haunschild P R, Miner A S. Modes of interorganizational imitation: The effects of outcome salience and uncertainty[J]. Administrative Science Quarterly, 1997, 42(3): 472-500.
50. Freeman R E. Strategic Management: A Stakeholder Approach[M]. Boston: Pitman Publishing, 1984.
51. Fan J P H, Wong T J, Zhang T. Politically connected CEOs, corporate governance, and Post-IPO performance of China's newly partially privatized firms[J]. Journal of Financial Economics, 2007, 84(2): 330–357.
52. Shumway T. Forecasting bankruptcy more accurately: A simple hazard model[J]. Journal of Business, 2001, 74(1): 101-124.
53. Hillman A J, Dalziel T. Boards of directors and firm performance: Integrating agency and resource dependence perspectives[J]. Academy of Management Review, 2003, 28(3): 383-396.
54. McWilliams A, Siegel D S. Creating and capturing value: Strategic corporate social responsibility, resource-based theory, and sustainable competitive advantage[J]. Journal of Management, 2011, 37(5): 1480-1495.
55. Hosmer D W, Lemeshow S, Sturdivant R X. Applied Logistic Regression[M]. New York: Wiley, 2000.

Appendix Definition for main variables

Dependent variable	Definition
New CSR reporting	1 = initial voluntary CSR disclosure in year t 0 = otherwise
Independent variable	
Board Links to CSR reporter	Dummy measure; 1 = a firm is linked to a previous CSR reporter 0 = otherwise
Board Links to CSR reporter	Continuous measure; the natural logarithm of 1 plus the summed links to previous CSR reporters
Controls	
Interlock	1 = a firm has board links to any other firm(s) in year t 0 otherwise
Total Number of Interlocks	The natural logarithm of 1 plus the summed board links to any other firm, including CSR reporters (ln(1+summed board links))
Size	Natural logarithm of a firm's market value of equity
ROA	Ratio of net profit to total assets
Leverage	Ratio of long-term debt to total assets
Liquidity	Number of shares traded divided by the total outstanding shares
Boardsize	Number of directors sitting on board

to be continued

Dependent variable	Definition
Independence	Percentage of independent directors in a board
SOEs	1 = a firm's largest shareholder is the state 0 = otherwise
Directors from sensitive industry	1 = if interlocked directors are from sensitive industry, including paper manufacturing, oil, natural gas, metal, and chemical industries 0 = otherwise.
Related-party	1 = when a firm's links to other firms are through related-party transitions 0 = otherwise
Industry CSR score	Standardized median of CSR reporting score for a certain industry
Within-industry interlock	1 = when a firm's links through interlock directors to other firms are within the same industry 0 = otherwise
Within-region interlock	1 when corporate links through interlock directors are within the same province 0 = otherwise

Carbon emission intensity and biased technical change in China's different regions: A novel multidimensional decomposition approach[①]

Lili Ding[1,2], Kaixuan Zhang[1], Ying Yang[1*]

1 School of Economics, Ocean University of China, Qingdao, China
2 Marine Development Studies Institute of OUC, Key Research Institute of Humanities and Social Sciences at Universities, Ministry of Education, Qingdao, China
* Corresponding author: Ying Yang (yangyingouc@126.com)

Abstract

The decomposition analysis has been employed to discover the driving factors of carbon emission intensity, but the current studies assume that production functions are under the condition of the neutral technical change. Grounded on biased technical change production theory, this paper proposes a novel multidimensional decomposition approach which combines production-theory decomposition analysis (PDA) and index decomposition analysis (IDA). This novel approach can illustrate how energy structure effect, element substitution effect, efficiency change effect, input biased technical change, output biased technical change and magnitude of technical change affect carbon emission intensity of China's 30 provinces. The results indicate that during the 11th FYP and 13th FYP, output biased technical change and the magnitude of technical change are the critical factors in China's carbon emission intensity, while other four drivers increase carbon emissions. But, during the 12th FYP, the role of six drivers has been reversed contrasting 11th FYP and 13th FYP. In addition, we also explore the impact of each driver from the perspective of regional heterogeneity.

Keywords: carbon emission intensity; multidimensional decomposition approach; biased technical change; energy structure; element substitution; efficiency change

Introduction

Climate change and global warming have attracted extensive attention to the mitigation of carbon emissions[1,2]. As one of the main carbon emissions in the world, China is actively involving the obligation of emission reduction[3]. The Chinese government promises to make the best effort to reach a peak in carbon emission by 2030 and achieve carbon neutrality by 2060. Carbon emission intensity target, i.e. carbon emission per unit gross domestic product will be cut by 60% to 65% from the 2005 level by 2030[4]. However, the dramatic development of China's economy tends to use more resources and energies, which essentially produces amounts of carbon emissions as a by-product[5,6]. China is facing heavy pressure in pursuit of lower

---

① 本文于2022年1月发表在 *Environmental Science and Pollution Research* 第29卷，https://doi.org/10.1007/s11356-021-18098-7。

carbon emission intensity, while maintaining good economic growth[7].

China has made outstanding efforts to achieve carbon emission intensity reduction. Since 2011, China has launched a carbon trading scheme, which allows energy enterprises to utilize market forces to optimize production decisions[8,9]. Meanwhile, the Chinese governments have raised a series of policies and regulations of clean energy, such as new energy vehicle subsidies and the Coal-to-Gas program[10,11]. These measures have been demonstrated to be efficient for carbon emission intensity reduction. It shows that China's carbon emission intensity decreases by 48% in 2019 compared to 2005 (The data is from the Ministry of Ecology and Environment of the People's Republic of China in 2019). But in geography, China is divided into 23 provinces, 5 autonomous regions, 4 municipalities and 2 special administrative regions (The data is from the National Bureau of Statistics of China). Since they have different resources endowment, cultures and economic structures, the carbon reduction strategies at the state level may not be suitable for these heterogeneous areas[12,13]. Hence, it is urgent to identify and understand the driving factors of carbon emission intensity not only at the national level but also at the regional level to formulate differentiated emission mitigation measures.

Many studies have focused on using the macro statistics methods to discover the driving elements of carbon emission intensity[14]. Spatial panel lag model, dynamic spatial econometric model, quantile regression model are currently the most popular macrometric methods for investigating the effects on carbon emission[15,67]. The results show that the elements such as technical change, energy structure and foreign domestic investment have the strongest influence on carbon emission intensity[16-19]. However, the macro statistics methods can only explain the comprehensive impact of the driving factors in the change of carbon emission intensity. They cannot qualify the specific contribution of each element on improving carbon emission intensity.

To break through the limitation of macro statistical methods, the decomposition analysis, which can decompose the change of carbon emission intensity into several parts and identify their contributions[20,21], has been extensively applied. The decomposition analysis consists of three major techniques, namely structural decomposition analysis (SDA), index decomposition analysis (IDA) and production-theory decomposition analysis (PDA). Based on the input-output (I–O) tables, SDA can interpret the change of carbon emission intensity embodied in trade and consumption-related carbon emissions[22,23]. They revealed that factors such as demand structure, production structure and energy intensity have different effects on the reduction of carbon emission intensity[24,25]. Spatial aggregation within a country[26,27] or multiple countries was explored; Su et al.[28] even linked China's 30 regions and 43 world countries to capture feedback effects in carbon emission. However, SDA has strict data requirements, which is invalid to analyze time-series data and small or medium-sized scale regions[29].

In contrast to SDA, IDA with loose data requirements is more feasible to use time-series data and have various application forms. There is a growing body of literature that investigates the factors affecting carbon emission intensity using the IDA at the regional level[30] as well as the sectorial level[31-33]. The economic output, industry structure, energy structure, energy intensity and population scale have been identified as the main drivers of carbon emission intensity change[34,35]. All these studies highlight the importance of energy structure and energy intensity on change of carbon emission. However, the current studies cannot decompose the energy intensity to deeply give a specific explanation bridge between energy intensity and carbon emission intensity[36,37]. Different from SDA and IDA, PDA utilizes the technology to measure the efficiency of energy consumption and carbon emissions based on the production theory[38]. Hence, the element substitution effect, efficiency changes effect and technical change effect have been widely proven to be the primary drivers to influence the energy intensity[39-41]. Technological gap[42], sectoral heterogeneity and regional heterogeneity[43] have been incorporated into the analytical framework. But, PDA neglects the influence of energy structure[66], which limits its application.

Hence, the combined methods between IDA and PDA are presented by Kim and Kim[68]. The advantage is to simultaneously identify the effect of structural change and production-related factors. In current years, there are few pieces of literature to study the combined methods based on IDA and PDA. For example, Yang and Li[44] found that carbon intensity changes can be decomposed into the structural effect, technology effect and economic effect, which made clear the contribution rate of technological progress to economic growth. Considered interregional technology differences, Zha et al.[45] combined the IDA and PDA approach with Meta-frontier environment production technology, which not only explained the effect of technology efficiency and technology progress but also explored the effects of technology gap on carbon emission change. Zhang et al.[46] proposed an extended IDA and PDA approach to qualify the effects of low-carbon technologies, which focused on the impart of efficiency change, i.e. potential energy efficiency effect and industrial energy carbon efficiency effect. Combining the Tapio model with LMDI and PDA, Yang et al.[47] revealed the decoupling process between global economy and its carbon emissions, and demonstrated the principal roles of technology progress in energy saving.

Although PDA, together with its combined methods, extends the traditional decomposition analysis by introducing production theory and demonstrates the role of technology progress in reducing emissions, they have a basic assumption that the production functions are under the condition of the neutral technical change. In reality, the use of production input factors, such as labor and capital, may lead to different directions of technical innovation[48]. Acemoglu[49] presented the theoretical framework of biased technical change, referring to changing the marginal rate of substitution among production input factors to save scarce resources. Energy as a production input factor plays a critical role on the path to industrialization for any countries. Hence, it is important to examine the biased technical change in carbon emission intensity. When the biased technical change incorporates energy and environmental factors, it can provide useful insights given the intensifying resource competition[50]. To fill this gap in the research, this paper extends the basic assumption about the traditional PDA method that production functions are not neutral technical change, but biased technical change. A novel multidimensional decomposition approach is presented to qualify accurately the contributions of various drivers of carbon emission intensity in a comprehensive framework.

The main contributions of this paper are presented in the following aspects. First, this paper extends the traditional PDA method by introducing the biased technology progress theory. It can clarify the influence of biased technology progress, explaining how the input biased technical and output biased technical change affect the carbon emission intensity. Second, a novel multidimensional decomposition method is constructed based on the IDA and extended PDA. The more comprehensive analysis framework is provided from the energy structure effect, element substitution effect, efficiency change effect, input biased technical change effect, output biased technical change effect and magnitude of technical change effect than the simple combined method. Third, this paper classifies Chinese 30 provinces over 2005–2019 into four clusters by the carbon emission intensity level using this novel combined approach to describe the characters of regional heterogeneity.

Methodology

This section introduces the methodology, consisting of the IDA decomposition method and the PDA decomposition method, as well as the extension to biased technical change. The decomposition method has been extensively applied to discover the driving factors of carbon emission. This paper extends the basic assumption about the traditional PDA method that production functions are not neutral technical change, but biased technical change. A novel multidimensional decomposition approach is proposed in Fig. 1, which can better explain how the energy structure, element substitution, magnitude of technical change, input biased technical change and output biased technical change on the change of carbon emission intensity. In the first stage, we use IDA method to decompose the influencing factors of carbon emission, i.e. energy structure effect and energy intensity effect. In the second stage, based on biased technical change theory, we construct an extend PDA method, which can further decompose the energy intensity effect into element substitution effect, efficiency change effect and biased technical change effect. In the third stage, the new indexes are presented to explain influences of the magnitude of technical change, input biased technical change and output biased technical change on carbon emission intensity. This paper provides a comprehensive framework to explain the driving factors about carbon emission to support the governments to implement the environmental policies.

▶ *Stage Ⅱ: the IDA decomposition*

Suppose that the whole economy system consists of I regions. Each region is regarded as a decision-making unit (DMU) in the production process. Each DMU uses labor force, energy and capital stock as inputs to produce, respectively denoted by L, E and K. The production has desirable output and undesirable output respectively denoted by Y and C, such as gross domestic product (GDP) and carbon emissions[51,52]. In order to explore the driving factors of regional carbon emissions, we apply the following Kaya identity[53] to describe the carbon emission intensity:

$$\text{CI} = \frac{C}{Y} = \frac{C}{E} \times \frac{E}{Y} = \text{CC} \times \text{EI} \qquad (1)$$

where CI is carbon emission intensity, i.e. carbon emission per unit desirable output[46]. CC = C/E represents the energy carbon emission coefficient, which can reflect the structure of energy consumption[54]. EI = E/Y denotes the energy intensity. We employ the Log-mean Divisia index (LMDI) approach to decompose the carbon emission intensity. Hence, using the IDA method, the increment of carbon emission intensity from period t to period T denoted by ΔCI can be constructed as

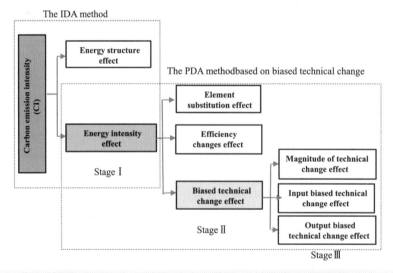

Fig. 1 A novel multidimensional decomposition approach

follows:

$$\Delta CI = CI_T - CI_t = \Delta CI^{cc} + \Delta CI^{ei} \quad (2)$$

According to the IDA method, the increment of carbon emission intensity over periods is decomposed into two components. The first decomposition term ΔCI^{cc} denotes the energy structure effect, illustrating how energy structure affects carbon emission intensity. The second decomposition term ΔCI^{ei} denotes the energy intensity effect, i.e. the influence of energy intensity on carbon emission intensity change. Thus, we evaluate each decomposition term in Eq. 2 by applying Eqs. 3–5.

$$\Delta CI^{cc} = L(CI_T, CI_t) \ln\left(\frac{CC_T}{CC_t}\right) \quad (3)$$

$$\Delta CI^{ei} = L(CI_T, CI_t) \ln\left(\frac{EI_T}{EI_t}\right) \quad (4)$$

$$L(CI_T, CI_t) = \begin{cases} \dfrac{CI_T - CI_t}{\ln CI_T - \ln CI_t}, & CI_T \neq CI_t \\ CI_T, & CI_T = CI_t \end{cases} \quad (5)$$

▶ *Stage II: the PDA decomposition*

The environmental production technology is denoted by EPT. At period t, EPT_t can be defined as

$$EPT_t(K_t, L_t, E_t) = \{(K_t, L_t, E_t, Y_t, C_t): \\ (K_t, L_t, E_t) \text{ can produce } (Y_t, C_t)\} \quad (6)$$

According to the production theory, the EPT is a closed set and satisfies the following assumptions: (1) strongly disposability of desirable outputs, i.e. if $(Y, C) \in EPT(K, L, E)$ and $C = 0$, then $Y = 0$. (2) Null jointness of desirable and undesirable outputs, i.e. if $(Y, C) \in EPT(K, L, E)$ and $Y' \leq Y$, then $(Y', C) \in EPT(K, L, E)$. (3) Weak disposability of undesirable outputs, i.e. if $(Y, C) \in EPT(K, L, E)$ and $0 \leq \theta \leq 1$, then $(\theta Y, \theta C) \in EPT(K, L, E)$.

For each region, the output distance function at period t denoted by D_t can be formulated as

$$D_t(K_t, L_t, E_t, Y_t, C_t) = \inf\{\theta: (K_t, L_t, E_t, Y/\theta C_t) \\ \in EPT^t(K^t, L^t, E^t)\} \quad (7)$$

The above distance function $D_t(K_t, L_t, E_t, Y_t, C_t)$ measures the distance from environmental production frontier, and $D_t(K_t, L_t, E_t, Y_t, C_t) \leq 1$. When $(K_t, L_t, E_t, Y_t, C_t)$ is on the frontier of EPT, $D_t(K_t, L_t, E_t, Y_t, C_t)$ is equal to 1. It needs to notice that the above Shephard output distance function is homogeneous of degree +1 in desirable outputs, while homogeneous of degree −1 in inputs as well as undesirable outputs[55].

To further analyze the influence of technical change, element substitution effect as well as efficiency change on energy intensity, production-theory decomposition analysis (PDA) is applied to decompose carbon emission intensity. Using EPT at period T as a reference, the change of energy intensity from period t to period T can be described as

$$\frac{EI_T}{EI_t} = \frac{E_T/Y_T}{E_T/Y_T} = \frac{\dfrac{E_T \times D_T(K_T, L_T, E_T, Y_T, C_T)}{Y_T}}{\dfrac{E_t \times D_T(K_t, L_t, E_t, Y_t, C_t)}{Y_t}} \times \\ \frac{D_t(K_t, L_t, E_t, Y_t, C_t)}{D_T(K_T, L_T, E_T, Y_T, C_T)} \times \frac{D_T^i(K_t, L_t, E_t, Y_t, C_t)}{D_t(K_t, L_t, E_t, Y_t, C_t)} \quad (8)$$

We notice that the above Shephard output distance function satisfies homogeneity[39]. Therefore, the following equation can be derived:

$$E_T \times D_T(K_T, L_T, E_T, Y_T, C_T) = D_T(k_T, l_T, 1, Y_T, c_T) \quad (9)$$

$$\frac{D_T(K_T, L_T, E_T, Y_T, C_T)}{Y_T} = D_T(K_T, L_T, E_T, 1, C_T) \quad (10)$$

where $k_T = K_T/E_T$, $l_T = L_T/E_T$ and $c_T = C_T/E_T$, respectively, named capital-energy ratio, labor-energy ratio and the carbonenergy ratio at the period T[39], so Eq. 8 can be expressed as

$$\begin{aligned}\frac{EI_T}{EI_t} &= \frac{E_T \times D_T(K_T, L_T, E_T, 1, C_T)}{E_t \times D_T^i(K_t, L_t, E_t, 1, C_t)} \times \frac{D_t(K_t, L_t, E_t, Y_t, C_t)}{D_T(K_T, L_T, E_T, Y_T, C_T)} \\ &\quad \times \frac{D_T(K_t, L_t, E_t, Y_t, C_t)}{D_t(K_t, L_t, E_t, Y_t, C_t)} \\ &= \frac{D_T(k_T, l_T, 1, 1, c_T)}{D_T(k_T, l_T, 1, 1, c_t^i)} \times \frac{D_t(K_t, L_t, E_t, Y_t, C_t)}{D_T(K_T, L_T, E_T, Y_T, C_T)} \\ &\quad \times \frac{D_T(K_t, L_t, E_t, Y_t, C_t)}{D_t(K_t, L_t, E_t, Y_t, C_t)} \\ &= PECH_T \times \frac{1}{EFF} \times \frac{1}{TECH_t}\end{aligned} \quad (11)$$

Similarly, taking EPT at period t as the production reference, energy intensity change can be expressed as the following Eq. 12.

$$\frac{\text{EI}_T}{\text{EI}_t} = \frac{\frac{E_T \times D_t(K_T, L_T, E_T, Y_T, C_T)}{Y_T}}{\frac{E_t \times D_t(K_t, L_t, E_t, Y_t, C_t)}{Y_t}} \times \frac{D_t(K_t, L_t, E_t, Y_t, C_t)}{D_T(K_T, L_T, E_T, Y_T, C_T)}$$

$$\times \frac{D_T(K_T, L_T, E_T, Y_T, C_T)}{D_t(K_T, L_T, E_T, Y_T, C_T)}$$

$$= \frac{E_T \times D_t(K_T, L_T, E_T, 1, C_T)}{E_t \times D_t(K_t, L_t, E_t, 1, C_t)} \times \frac{D_t(K_t, L_t, E_t, Y_t, C_t)}{D_T(K_T, L_T, E_T, Y_T, C_T)}$$

$$\times \frac{D_T(K_T, L_T, E_T, Y_T, C_T)}{D_t(K_T, L_T, E_T, Y_T, C_T)}$$

$$= \frac{D_t(k_T, l_T, 1, 1, c_T)}{D_t(k_t, l_t, 1, 1, c_t)} \times \frac{D_t(K_t, L_t, E_t, Y_t, C_t)}{D_T^t(K_T, L_T, E_T, Y_T, C_T)}$$

$$\times \frac{D_T(K_T, L_T, E_T, Y_T, C_T)}{D_t(K_T, L_T, E_T, Y_T, C_T)}$$

$$= \text{PECH}_t \times \frac{1}{\text{EFF}} \times \frac{1}{\text{TECH}_T} \tag{12}$$

To avoid the bias owing to the reference's selection of EPT[55], the geometric mean of Eqs. 11 and 12 is calculated. Therefore, energy intensity change can be shown as follows.

$$\frac{\text{EI}_T}{\text{EI}_t} = [\text{PECH}_T \times \text{PECH}_t]^{\frac{1}{2}} \times \frac{1}{\text{EFF}} \times \left[\frac{1}{\text{TECH}_T} \times \frac{1}{\text{TECH}_t}\right]^{\frac{1}{2}}$$

$$= \left[\frac{D_T(k_T, l_T, 1, 1, c_T)}{D_T(k_t, l_t, 1, 1, c_t)} \times \frac{D_t(k_T, l_T, 1, 1, c_T)}{D_t(k_t, l_t, 1, 1, c_t)}\right]^{\frac{1}{2}}$$

$$\times \frac{D_t(K_t, L_t, E_t, Y_t, C_t)}{D_T(K_T, L_T, E_T, Y_T, C_T)}$$

$$= \left[\frac{D_T(K_t, L_t, E_t, Y_t, C_t)}{D_t(K_t, L_t, E_t, Y_t, C_t)} \times \frac{D_T(K_T, L_T, E_T, Y_T, C_T)}{D_t(K_T, L_T, E_T, Y_T, C_T)}\right]^{\frac{1}{2}}$$

$$= \text{PECH} \times \frac{1}{\text{EFF}} \times \frac{1}{\text{TECH}} \tag{13}$$

In Eq. 13, the first component PECH is the maximum potential energy productivity change, which relies on the capital-energy ratio, labor-energy ratio and carbon-energy ratio. The maximum potential energy productivity change reflects the increase or decrease in production capacity caused by element substitution in the process of production adjustment. Obviously, the second and third terms in Eq. 13, i.e. 1/EFF and 1/TECH are the reciprocal of two Malmquist productivity indexes, which are efficiency change and technical change. This equation shows that an improvement in carbon emission efficiency change or technical change from period t to period T will reduce the energy intensity as well as carbon emission intensity.

In order to quantify the impact of element substitution, efficiency changes and technical change on carbon emissions, we substitute Eq. 13 into the energy intensity effect of $\Delta\text{CI}^\text{ei}$ and obtain the following equation.

$$\Delta\text{CI}^\text{ei} = \Delta\text{CI}^\text{pech} + \Delta\text{CI}^\text{eff} + \Delta\text{CI}^\text{tech} \tag{14}$$

In virtue of the PDA method, the energy intensity effect is refined into element substitution effect, efficiency change effect and technical change effect, denoted by $\Delta\text{CI}^\text{pech}$, $\Delta\text{CI}^\text{eff}$ and $\Delta\text{CI}^\text{tech}$ respectively. The element substitution effect reflects the changes in carbon emission intensity due to substitution between energy and other elements. The efficiency change effect quantified the influence originated from efficiency changes. The technical change effect evaluates the influence of technological progress on carbon emission intensity from period t to period T. Thus, we evaluate the effect of each factor in Eq. 14 by using Eqs. 15–17.

$$\Delta\text{CI}^\text{pech} = L(\text{CI}_T, \text{CI}_t) \ln \text{PECH} \tag{15}$$

$$\Delta\text{CI}^\text{eff} = L(\text{CI}_T, \text{CI}_t) \ln \frac{1}{\text{EFF}} \tag{16}$$

$$\Delta\text{CI}^\text{tech} = L(\text{CI}_T, \text{CI}_t) \ln \frac{1}{\text{TECH}} \tag{17}$$

▶ *Stage Ⅲ: the biased technical change decomposition*

Acemoglu[56] pointed out that technical change can alter the dependence of economic systems on certain elements, thus showing the non-homothetic shift of EPT, i.e. the biased characteristics of technical change. The technical change TECH is generally considered to be the synthesis of multiple factors: input biased technical change, output biased technical change and magnitude of technical change[57]. Hence, considering this biased feature, we redefine TECH as biased technical change. Similarly, the technical change effect $\Delta\text{CI}^\text{tech}$ is actually the comprehensive action of different biased components in technical change. Therefore, we further decompose technical change to analyze the role of different biased components in carbon emission reduction.

$$\text{TECH} = \left\{ \frac{D_t(K_t, L_t, E_t, Y_t, C_t)}{D_T(K_t, L_t, E_t, Y_t, C_t)} \times \frac{D_t(K_T, L_T, E_T, Y_T, C_T)}{D_T(K_T, L_T, E_T, Y_T, C_T)} \right\}^{\frac{1}{2}}$$

$$= \frac{D_t(K_t, L_t, E_t, Y_t, C_t)}{D_T(K_t, L_t, E_t, Y_t, C_t)} \times \left(\frac{D_T(K_t, L_t, E_t, Y_t, C_t)}{D_t(K_t, L_t, E_t, Y_t, C_t)} \right)$$

$$\times \frac{D_t(K_T, L_T, E_T, Y_t, C_t)}{D_T(K_T, L_T, E_T, Y_t, C_t)} \right)^{\frac{1}{2}} \times \left(\frac{D_t(K_T, L_T, E_T, Y_T, C_T)}{D_T(K_T, L_T, E_T, Y_T, C_T)} \right.$$

$$\times \frac{D_T(K_T, L_T, E_T, Y_t, C_t)}{D_t(K_T, L_T, E_T, Y_t, C_t)} \right)^{\frac{1}{2}}$$

$$= \text{MATC} \times \text{IBTC} \times \text{OBTC} \tag{18}$$

In Eq. 18, MATC denotes the magnitude of technical change, which measures the homothetic shift of EPT[57,58]. IBTC denotes the input biased technical change, which is the non-homothetic shift in the EPT created from the input elements. Symmetrically, OBTC is the output biased technical change, manifesting the positive effect of technical change on diverse outputs non-homothetic shift. Therefore, we can refine the impact of different biased components of technical change on carbon emission intensity. Combining Eqs. 13 and 18, the technical change effect based on biased technical change theory can be decomposed as

$$\Delta \text{CI}^{\text{tech}} = \Delta \text{CI}^{\text{matc}} + \Delta \text{CI}^{\text{ibtc}} + \Delta \text{CI}^{\text{obtc}} \tag{19}$$

In Eq. 19, $\Delta \text{CI}^{\text{matc}}$ describes the effect of the magnitude of technical change. When $\Delta \text{CI}^{\text{matc}}$ is greater than zero, it means that parallel movement of the production frontier caused by the technical change from period t to T has a negative impact on reducing carbon emissions. The terms $\Delta \text{CI}^{\text{ibtc}}$ and $\Delta \text{CI}^{\text{obtc}}$ depict the input biased technical change effect and output biased technical change effect, respectively. Similarly, a positive number indicates that changes in this factor will lead to an increase in the intensity of carbon emissions, which has a negative impact on reducing carbon emissions. Contrarily, negative numbers reflect that this factor will reduce the intensity of carbon emissions, thereby reducing carbon emissions.

$$\Delta \text{CI}^{\text{matc}} = L(\text{CI}_T, \text{CI}_t) \ln \frac{1}{\text{MATC}} \tag{20}$$

$$\Delta \text{CI}^{\text{ibtc}} = L(\text{CI}_T, \text{CI}_t) \ln \frac{1}{\text{IBTC}} \tag{21}$$

$$\Delta \text{CI}^{\text{obtc}} = L(\text{CI}_T, \text{CI}_t) \ln \frac{1}{\text{OBTC}} \tag{22}$$

Combining Eqs. 2, 14 and 19 gives the final decomposition as follows:

$$\Delta \text{CI} = \Delta \text{CI}^{\text{cc}} + \Delta \text{CI}^{\text{pech}} + \Delta \text{CI}^{\text{eff}} + \Delta \text{CI}^{\text{matc}} + \Delta \text{CI}^{\text{ibtc}} + \Delta \text{CI}^{\text{obtc}} \tag{23}$$

Equation 23 describes a comprehensive framework about driving factors influencing carbon emission intensity based on the above novel multidimensional decomposition method. The change of carbon emission intensity from period to period is decomposed into the six components: energy structure effect, element substitution effect, efficiency changes effect, the effect of magnitude of technical change, input biased technical change effect and output biased technical change effect.

Data Description

This paper presents the empirical analysis using 30 regions of China over 2005–2019 which includes the 11th FYP (2006–2010), the 12th FYP (2011–2015) and the 13th FYP (2016–2019) (FYP means five-year plan in China, which is a kind of state developing strategy. In fact, the 13th Five-Year Plan is 2016–2020, but due to data limitations, our study period only covers 2019). The FYP is China's overall national strategy for national economic or industrial development and has an important impact on China's carbon emission reduction. Each region is regarded as a decision-making unit (DMU). There are 30 DMUs to be analyzed in mainland China apart from Tibet because of the limitation of data availability and quality. The main indicators used in the multidimensional decomposition method relate to carbon emission, energy consumption, capital stock, labor employment and GDP. The data of carbon emission is mainly from China Emission Accounts and Datasets (CEADs), while the carbon emission in 2019 is calculated by IPCC method which multiplies the consumption of fuel type by the corresponding carbon emission coefficient[59]. Data on labor employment and GDP are directly collected from the State Statistical Bureau website. We use GDP deflators to eliminate the current time value (with the base year of 2005). Energy consumption data is from the *China Energy Statistical Yearbook*.

The capital stock is accounted for applying the following perpetual inventory method:

$$K_{it} = K_{it-1}(1 - \delta_{it}) + I_{it} \quad (24)$$

where K_{it} is the capital stock of the region i in the year t, I_{it} is the capital investment in the same period and δ_{it} is the capital depreciation rate which is equal to 10.96%[60]. The data of K_{it} and I_{it} are adjusted in terms of 2005 price levels to maintain data comparability with GDP.

Table 1 provides descriptive statistics for each indicator. It clearly shows a significant increase in all variables. The average carbon emission expectancy reaches 294.59 million tons per year. It also notices the huge heterogeneity in carbon emission, with a maximum of 1,106.03 million tons (Hebei in 2019) and a minimum of 16.5 million tons (Hainan in 2005) which is only 1.5% of the former. The average of 30 regions over 2005–2019 in the energy consumption, labor, capital stock and GDP are 131.29 million tons of standard coal, 24.28 million persons, 5,516.39 billion Yuan and 140.70 billion Yuan, respectively.

Table 1 Descriptive statistics analysis of the indicators

Notation	Labor force (L)	Capital stock (K)	Energy consumption (E)	Gross domestic product (GDP) (Y)	Carbon emission (C)
Unit	10^4 persons	10^9 Yuan	10^4 tons	10^8 Yuan	10^6 tons
Mean	2,428.03	5,516.39	13,129.46	1,407.02	294.59
Max	7,133.10	26,168.96	41,390.00	7,834.60	1,106.03
Min	161.70	329.81	822.20	49.94	16.50
Std. dev.	1,686.96	4,592.40	8,251.52	1,309.08	204.92

Results and Discussion

▶ *Carbon emission intensity in different regions*

Based on the average carbon emission intensity and the increment of carbon emission intensity from 2006 to 2019, we divided 30 regions into 4 groups, as shown in Fig. 2. The critical value is the median of each variable. It clearly shows in Fig. 2 that the majority of the high-carbon emission intensity regions (i.e. Cluster Ⅰ and Cluster Ⅳ) are located in western and central China, while most regions in Eastern China have lower-than-average emissions, especially in the coastal area of Southeast China. The highest carbon emission intensity of 9.74 ton per 10^4 Yuan is in Ningxia. The second-placed overall in the carbon emission intensity is Inner Mongolia with 6.09 ton per 10^4 Yuan. These regions in Cluster Ⅰ are all high-carbon emission intensity regions, which are mostly located in western China. Carbon reduction in Cluster Ⅰ, meanwhile, is also slightly inadequate. These areas will be key areas which will determine whether China can achieve the "Double Carbon Target". On the contrary, Tianjin has the lowest value of carbon emission intensity with 0.88 ton per 10^4 Yuan, which is only one-tenth that of Ningxia. This phenomenon shows that China's carbon emissions have a significant gap between regions. Moreover, the average carbon emission intensity of Zhejiang, Fujian, Shanghai and Guangdong are relatively low, which belong to Cluster Ⅱ, all below 2 ton per 10^4 Yuan. The Cluster Ⅲ includes four provinces (i.e. Henan, Hubei, Hunan, Sichuan) and two municipalities (Chongqing and Beijing). Compared with the Cluster Ⅱ, the Cluster Ⅲ has a significant effect of emission reduction while maintaining a lower carbon emission intensity. Cluster Ⅳ is also a highcarbon region, but its carbon emission reduction is relatively large in recent years, indicating China's emission reduction campaign has already been formed in nine regions, including Shanxi, Hebei and Shandong.

Fig. 3 depicts the general trend of China's carbon

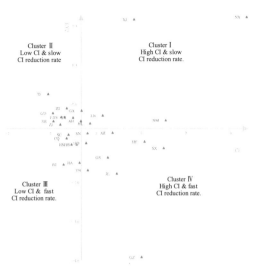

Fig. 2　Spatial distribution of carbon emission intensity in 30 regions (ton per 10^4 Yuan)

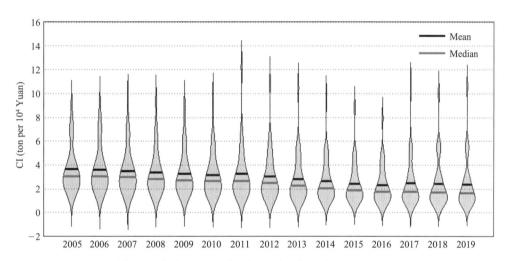

Fig. 3　Carbon emission intensity from 2005 to 2019

emissions intensity in 30 regions from 2005 to 2019. The shape of the violin chart reflects the distribution characteristics of carbon emission intensity in different provinces of China. The graph reveals that there has been a marked drop in the value of carbon emission intensity for the majority of regions in China. During the study period, the average carbon emission intensity dropped from 3.62 ton per 10^4 Yuan to 2.38 ton per 10^4 Yuan in 30 regions, at an annual average decrease of 0.09 ton per 10^4 Yuan. In addition, we also find that regional differentiation is also increasing. As time goes on, the upper arc of the violin curve increases gradually, indicating that although the national carbon emission intensity is mostly declining, there is still a small-scale rebound and agglomeration in some areas. The regions with the highest change in carbon emission intensity are Guizhou, Jilin and Yunnan with an annual average decrease of 0.32, 0.17 and 0.16 ton per 10^4 Yuan, respectively. Other provinces with the high change in carbon emission intensity include Beijing and Henan, in which the average annual reduction in carbon emissions intensity exceeds 0.15 ton per 10^4 Yuan. Relatively reduction rate in some regions is slow. In detail, from 2005 to 2019, carbon emission intensity in Inner Mongolia decreased from 6.83 to 5.82 ton per 10^4 Yuan, with a steady average annual reduction of 0.03 ton per 10^4 Yuan. We also noticed that the carbon emission intensity in Xinjiang and Ningxia has increased during the study period, from 4.01 ton per 10^4 Yuan in 2005 to 5.59 ton per 10^4 Yuan in 2019 and from 8.92 ton per 10^4 Yuan in 2005 to 10.52 ton per 10^4 Yuan in 2019 respectively, which reflects greater regional heterogeneity. This phenomenon illustrates the urgency of conducting this research. Though the Chinese government has announced the goal of reducing carbon emissions, it has shown greater regional heterogeneity in specific practices in recent years.

▶ *Factor decomposition of carbon emission intensity*

According to Eqs. 3–15, the proposed novel multidimensional decomposition approach is used to annually decompose the carbon emission intensity changes of the entire society from 2005 to 2019 based on the energy structure effect, element substitution effect, efficiency changes effect, magnitude of technical change effect, input biased technical change effect and output biased technical change effect. In order to ensure the accuracy of the data results, we deleted the DMU for which there is no feasible solution. Table 2 summarizes the decomposition results of carbon emission intensity via novel multidimensional decomposition approach.

Table 2 Overall multidimensional decomposition results (ton per 10^4 Yuan)

	ΔCI	ΔCI^{cc}	ΔCI^{ei}	ΔCI^{pec}	ΔCI^{eff}	ΔCI^{matc}	ΔCI^{ibtc}	ΔCI^{obtc}
2005–2006	0.01	0.07	–0.06	0.37	0.02	–0.43	0.46	–0.47
2006–2007	–0.12	0.00	–0.12	0.30	–0.03	–0.42	0.43	–0.40
2007–2008	–0.13	0.04	–0.17	0.16	–0.01	–0.37	0.33	–0.29
2008–2009	–0.13	0.03	–0.16	–0.07	0.01	–0.12	0.11	–0.09
2009–2010	–0.09	0.05	–0.14	0.00	0.05	–0.19	0.20	–0.20
2010–2011	0.08	0.16	–0.09	0.31	–0.03	–0.33	0.33	–0.36
2011–2012	–0.18	–0.05	–0.13	–0.61	–0.01	0.11	–0.15	0.53
2012–2013	–0.20	–0.02	–0.17	–0.09	–0.12	–0.07	0.05	0.05
2013–2014	–0.18	–0.09	–0.09	–0.32	–0.08	0.21	–0.18	0.28
2014–2015	–0.21	–0.05	–0.16	–0.18	–0.02	0.06	–0.05	0.04
2015–2016	–0.15	–0.07	–0.08	–0.13	–0.05	0.15	–0.11	0.07
2016–2017	0.02	0.02	–0.01	0.08	0.02	–0.11	0.11	–0.09
2017–2018	–0.09	0.03	–0.11	0.22	0.08	–0.52	0.46	–
2018–2019	–0.02	0.02	–0.04	–0.07	0.02	–0.01	0.02	0.00

Energy structure affected the change in carbon emission intensity differently by period. This conclusion is similar to the results in Wang and Feng[42]. The improvement of the dominant energy structure is conducive to the extenuation of carbon emissions intensity with a subtle extended effect. During the 11th FYP, China's industrialization and urbanization progressed rapidly with huge energy consumption. Hence, in the majority of the 11th FYP, the change of energy structure showed a negative effect on carbon emission. The energy structure effect (ΔCI^{cc}) shows a positive impact on carbon emission from 2012, even though its contribution was relatively limited, which is consistent with Gu et al.[36], one possible explanation is that China's 12th FYP proposed to optimize energy structure and develop new energy industries. Unfortunately, the carbon emission reduction effect of the energy structure could not be maintained for a long time, and rebounds in the 13th FYP, showing that China's energy policy reform still has a long way to go.

The energy intensity effect (ΔCI^{ei}) plays the decisive role in decreasing carbon emission intensity during the periods of 11th FYP, 12th FYP and 13th FYP, with contributions in each period being 142.5% (−0.65 ton per 10^4 Yuan), 93.6% (−0.64 ton per 10^4 Yuan) and 99.8% (−0.24 ton per 10^4 Yuan), respectively. This conclusion is in line with studies such as Lin and Wang[61]. The energy intensity effect is a comprehensive reflection of the element substitution effect (ΔCI^{pech}), efficiency change effect (ΔCI^{eff}) and technical change effect (ΔCI^{tech}). The characteristics of different driving factors are different in each period.

The element substitution effect (ΔCI^{pech}) increases the carbon emission intensity in majority time. The element substitution effect describes the changes in carbon emission intensity due to substitution between energy and other elements, which reflects the ability of capital, labor and other factors to substitute energy. China's regional development policies may explain the negative effects of element substitution. Since the 11th FYP and the 12th FYP, the Chinese government has stepped up the development strategy of western development and the rise of the central region. The central and western regions have continuously undertaken industrial transfers from the eastern region, accelerating economic development in the central and western regions. Due to the limitation of development stage, the development of central and western regions is still characterized by severe resource dependence, so the element substitution effect is negative. With the proposal of ecological civilization construction, China pays more and more attention to the development of capital-dependent and technology-dependent industries. Therefore, this negative effect gradually decreases, and capital and labor resources can effectively substitute energy as well as reduce carbon emission intensity. This conclusion manifests that the differentiated policies for promoting the development of different regions positively affected Chinese carbon emissions. Simultaneously, replacing energy consumption with R&D and investment can proceed smoothly.

The overall efficiency change effect (ΔCI^{eff}) is related toa positive impact in most situations and decreases by 0.15 ton per 10^4 Yuan from 2006 to 2019 which could explain 11% of the total carbon emission intensity change. However, during the13th FYP, the change in efficiency is not satisfactory and even causes a rise in carbon emission intensity. One possible explanation is that during the rapid development phase of the Chinese economy, the production process was too dependent on factors such as capital and technology. Correspondingly, the management is relatively extensive, resulting in inefficient production and additional carbon emissions.

Input biased technical change, output biased technical change as well as magnitude of technical change diversely affected Chinese carbon emission intensity in different FYPs. That is, alternation and volatility are the typical characteristics of biased technological progress effect. Specifically, in 11th FYP, output biased technical change effect (ΔCI^{obtc}) and the effect of magnitude of technical change (ΔCI^{matc}) contributed significantly to the decline in carbon emission intensity, with an annual average decrease of 0.31 and 0.29 ton per 10^4 Yuan, respectively, while the input biased technical change effect (ΔCI^{ibtc}) impeded seriously the decline in carbon emissions. Since 2011, the trend has reversed that the input biased technical change effect turns to be positive. Yet the output biased technical change effect and the effect of magnitude of technical change transformed into increasing carbon emissions. The situation in the 13th FYP is similar to that in the 11th FYP. It should be noted that the impact of various components related to technical change is relatively strong compared with other factors, with a cyclical pattern, while gradually tends to be stable. This phenomenon can be seen as a microcosm of China's industrialization process. Since the 11th Five-Year Plan, with the continuous deepening of industrialization, China has actively carried out quantity of research and development (R&D) activities to achieve technical innovation and industrial upgrading, which has largely adjusted the production pattern. Due to the environmental regulation policies, the output of technical progress and magnitude of technical change show characteristics of emission reduction. Enterprises are more inclined to invest in input biased technology R&D to achieve savings on specific input factors, which leads to an increase in carbon emissions. As the acceleration of China's urbanization and industrialization the impact of the 2008–2009 Global Financial Crisis, China's economy entered a period of adjustment during the 12th FYP. During this period, the performance of each component of technical change is quite different from that of the previous stage. During the 13th FPY period, China's total economic output continued to rise to a new level. Meanwhile, the formation of China's strong domestic market has accelerated, the supply and demand structure have become more reasonable and the innovation ability has been continuously improved. China's economy has entered a high-quality development stage; output biased technical change effect and the effect of magnitude of technical change play a new role in reducing emissions.

▶ *Regional decomposition results*

In Section "Carbon emission intensity in different regions", we noticed that carbon emission intensity in the vast majority of regions has declined from 2005 to 2019, while the magnitude of the decline is different among regions. The heterogeneity of driving factors led to notably different changes in regional carbon emissions (Fig. 4). In order to explore this heterogeneity, the following part of this paper moves on to describe in greater detail the regional decomposition results. We will continue to use the groupings mentioned above, i.e. Cluster Ⅰ to Cluster Ⅳ.

Energy structure change effect in Cluster Ⅰ is most negative, with an average increment of 1.37 ton per 10^4 Yuan, followed by Cluster Ⅳ and Cluster Ⅱ, which have a relatively minor increment effect of 0.13 ton per 10^4 Yuan and 0.001 ton per 10^4 Yuan respectively. These results are generally in line with Lin and Wang[61], stating that the energy structure adjustment is behind the emission growth in majority China. Different from the other groups, energy structure changes in Cluster Ⅲ play positive role in reducing the carbon emission intensity. This inconsistency may be due to the stage of regional development. Majority regions in Cluster Ⅲ are developed provinces in China, taking Beijing and Chongqing as typical representatives; these regions have relatively developed economies and are affected by the energy structure reform policies. In recent years, the energy structure has been continuously optimized. Cluster Ⅰ, Cluster Ⅱ and Cluster Ⅳ are mostly located in western and central China (e.g. Inner Mongolia, Ningxia and Xinjiang). During the 11th FYP and 12th FYP, these areas experienced high-speed industrialization with high-carbon energy consumption. These regions are advancing energy structure reforms, but it is difficult to avoid the negative impact of high energy consumption.

The contribution of element substitution is similar to energy structure change in each group. As for element substitution effects, the average negative impacting extent was larger in Cluster Ⅰ with an average increment of 2.46 ton per 10^4 Yuan. Followed by Cluster Ⅳ and Cluster Ⅱ, the element substitution effect is 0.25 ton per 10^4 Yuan and 0.14 ton per 10^4 Yuan respectively. The only positive effect occurred in Cluster Ⅲ, with a reduction of 0.24 ton per 10^4 Yuan. With deep

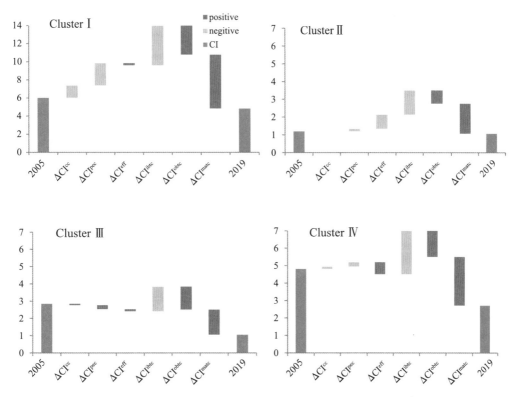

Fig. 4 The variation of driving factors in each group (ton per 10^4 Yuan)

implementation of Western Exploitation Strategy, the central and western regions have gained rapid development in industries by attracting more and more foreign manufacture and process orders. This process has effectively promoted the flow of factors, which promote regional economic development. Shao et al.[62] pointed out that the relationships between capital and energy, labor and energy, as well as capital and energy are not stable in China; the complementary relationships are not existing. Capital and labor input increase contribute to industrial output growth, rather than carbon emission reduction capacity.

Efficiency changes effect diversely affected the carbon emission intensity in different regions and positive in most regions. Its average effect in Cluster Ⅰ, Cluster Ⅲ and Cluster Ⅳ is −0.25 ton per 10^4 Yuan, −0.12 ton per 10^4 Yuan and −0.69 ton per 10^4 Yuan, respectively. The result indicates that China's management ability has been improved in recent years, especially in Cluster Ⅳ. Most of the regions involved in the Cluster Ⅳ are resourcebased provinces (such as Shanxi, Jilin, Hebei and Shandong) which have experienced varying degrees of industrial transformation and governance reform. The positive efficiency changes effect shows the rise of government governance level in these regions. In comparison, the efficiency changes effect in Cluster Ⅱ shows negative contributions resulting in an annual increase of the carbon emission intensity by 0.80 ton per 10^4 Yuan. Hence, to reach the maximum potential for carbon emission reduction, China should pay more attention to improve their efficiency.

On the whole, the comprehensive effect of technical change on reducing carbon emissions is positive in all groups. There are similar conclusions in the research of Zha et al.[45] as well as Yang and Li[44], that is, the important role of technical change in carbon emission reduction is clarified, and the credibility of this research conclusion is also confirmed. It should be noticed that the contribution of different biased components in the technical change to carbon emissions differs significantly among the four groups. For all groups, input biased technical change is estimated to be the main resistance to increase the carbon emission intensity. In particular, the increment of carbon emission intensity driven by input biased technical change is larger in Cluster Ⅰ with an annual average increase of 0.44 ton per 10^4 Yuan than others, followed by Cluster Ⅳ. Output biased technical change positively affects the carbon emission intensity in all groups, and especially in Cluster Ⅰ, which decreases 0.35 ton per 10^4 Yuan for each year. Simultaneously, the direction of magnitude of technical change effect is similar to output biased technical change. The sharp decrease in the magnitude of technical change largely positively affected carbon emission intensity in Cluster Ⅰ, followed by Cluster Ⅳ, Cluster Ⅱ and Cluster Ⅲ. The differentiated performance of each component of biased technical change in each group reflects the process of regional technical innovation in China. Generally speaking, the purpose of enterprise innovation is to obtain economic benefits. Correspondingly, technical innovation focuses on how to save scarce production factors, resulting in the input biased technical change. Yet the resource misallocation inhibits the positive effect of input biased technical change on carbon emission[63]. On the contrary, when the level of technology and economy reached a certain level, the government and the public began to realize the importance of environmental protection and urged enterprises to pay attention to their social responsibilities, especially environmental responsibilities[64].

Conclusions and Policy Recommendations

▶ *Conclusions*

From biased technical change view, this paper constructs a novel multidimensional decomposition approach for carbon emission intensity and interprets the drivers affecting the reduction of carbon emission intensity from six perspectives, namely, energy structure effect, element substitution effect, efficiency changes effect, magnitude of technical change effect, input biased technical change effect and output biased technical change effect. To explore the regional heterogeneity of the drivers, this paper gives the classification by the level of carbon emission intensity and its change. The main conclusions are as follows:

First, the carbon emission intensity has regional heterogeneity, and from the perspective of spatial distribution, it shows a downward trend from west to east. Second, there are obvious differences in the types and degrees of the effect of the six drivers in each period. During the 11th FYP and 13th FYP, output biased technical change and the magnitude of technical change are the critical factors in China's carbon emission reduction process, while the energy structure effect, the element substitution effect, efficiency changes effect and input biased technical change effect impeded the decline in carbon emissions. It is worth noting that in the 12th FYP, the role of six drivers has been reversed, which may be related to the industrialization process and the financial crisis. Third, from the perspective of regional heterogeneity, energy structure change and element substitution can only play a positive role in emission reduction in Cluster Ⅲ which has high economic development level and low carbon emission intensity. The efficiency changes effects play a positive role in Cluster Ⅰ, Cluster Ⅲ and Cluster Ⅳ, but negatively in the Cluster Ⅱ. Input biased technical change and output biased technical change have the most prominent effects and have opposite effects on carbon emissions, showing negative and positive effects, respectively. The degree of the biased technical change is directly proportional to the intensity of carbon emissions, that is, the emission reduction effect of input biased technical change and the negative emission-increasing effect of output biased technical change are inclined to be more prominent in Cluster Ⅰ and Cluster Ⅳ. In addition, the impact of the magnitude of technical change is similar to output biased technical change.

▶ *Policy recommendations*

Based on the above-mentioned empirical results, the major policy implications for regional carbon reduction as follows. The first suggestion aims to promote the rational flow of resources. The measurement manifests that element substitution effects in reducing carbon emissions are not satisfied, and input biased technical change also has seriously hindered the process of emission reduction, which concerns the regional misallocation of resources. Comprehensively restricted by historical and social reasons, regional protectionism is regarded as an important reason to hinder the rational flow of resources. The economically developed areas generally take measures to consolidate the advantages of local development, while other areas also tend to take measures to protect local development, resulting in market differentiation and fragmentation[65]. Hence, a feasible approach to overcome the impediment is removal of trade barriers and facilitating the orderly flow of personnel and capital among regions, to accelerate the formation of the unified national market.

The second suggestion aims to improve the efficiency. These experiments confirmed that promoting economic efficiency is an important approach to achieve "Double Carbon Target". At present, the efficiency of most provinces in China is relatively satisfactory, explaining 11% of the total carbon emission intensity change. However, in Cluster II (such as Zhejiang and Tianjin), the efficiency effect increased carbon emissions. Consequently, effective measures should be adopted to improve efficiency, such as identify production bottlenecks, find and eliminate wastefulness, as well as organize and standardize production processes across the board.

The third suggestion aims to promote low-carbon technologies. The output biased technical change and magnitude of technical change are the primary driving forces of the carbon reduction. A key policy priority should therefore be to plan for the long-term care of technical innovation. In order to fully express the inhibitory effect of technical change on carbon emission intensity, the government should facilitate the development of Clean Development Mechanism (CDM) projects, as well as support the new energy, novel material and other enterprises.

In addition, it is necessary to consider the periodically and heterogeneity of regional carbon emission. The heterogeneity of driving factors led to notably different changes in regional carbon emissions. Therefore, how to formulate differentiated emission reduction strategies has become a key. Cluster I, characterized by high carbon emission base and low emission reduction speed, such as Guangxi, Inner Mongolia, Ningxia and Xinjiang, is the key area to achieve China's "Double Carbon Target". The policy priorities of these regions are to undertake the industrial transfer in the eastern region and attach equal importance to economic development and environmental protection. As for Cluster II, it is characterized by low carbon emission intensity and slow emission reduction, mostly in eastern coastal provinces, such as Shanghai and Zhejiang. The economic efficiency of these regions restricts emission reduction, so the dominant policy aims to raise economic efficiency, to deepen the structural reform of the energy supply side and build a diversified and clean energy supply system. The low-carbon strategy is best implemented in Cluster III, which has the lower carbon emission intensity and significant emission reduction rate. Our policy suggestion is to focus on input biased technical innovation. Cluster IV mainly involves resource-based provinces with high carbon emission intensity and remarkable emission reduction, taking the Shanxi and Jilin as typical representatives. The occupation strategy of these areas is to pay attention to industrial upgrading and give full play to the advantages of clean technology.

A limitation of this study is that the paper is mainly based on the investigation at the regional level. In fact, carbon emissions from different industrial sectors also have significant differences; industrial scale may also affect the carbon emission. The role of sectoral heterogeneity, industrial scale and other factors toward carbon emission can be examined in the future studies, by using the multidimensional decomposition framework.

Availability of Data and Materials

The datasets generated and analyzed during the current study are available in the National Bureau of Statistics of China and Ministry of Ecology and Environment of the People's Republic of China repository, http://www.stats.gov.cn/english/Statisticaldata/AnnualData/ and http://english.mee.gov.cn/Resources/Reports/.

Funding

This work is supported by the National Natural Science Foundation of China (No. 71973132) and National Social Science Fund of China (No. 19VHQ002).

Author Contribution

Lili Ding designed the study and wrote and reviewed the manuscript; Kaixuan Zhang contributed to the writing of the final version of the manuscript; Ying Yang collected the data and analyzed the data and wrote the manuscript. All authors have read and approved the final version of the paper.

References

1. Zhao G, Liu C. Carbon emission intensity embodied in trade and its driving factors from the perspective of global value chain[J]. Environ Sci Pollut Res, 2020, 27: 32062-32075.
2. Tang K, Liu Y, Zhou D, et al. Urban carbon emission intensity under emission trading system in a developing economy: Evidence from 273 Chinese cities[J]. Environ Sci Pollut Res, 2021, 28(5): 5168-5179.
3. Cheng S, Fan W, Meng F, et al. Toward low-carbon development: assessing emissions-reduction pressure among Chinese cities[J]. J Environ Manag, 2020, 271: 111036.
4. Liu Z, Guan D, Moore S, et al. Climate policy: Steps to China's carbon peak[J]. Nature, 2015, 522(7556): 279-281.
5. Wang F, Wang G, Liu J, et al. How does urbanization affect carbon emission intensity under a hierarchical nesting structure? Empirical research on the China Yangtze River Delta urban agglomeration[J]. Environ Sci Pollut Res, 2019, 26(31): 31770-31785.
6. Lin B, Zhou Y. Does the Internet development affect energy and carbon emission performance?[J]. Sustain Prod Consum, 2021, 28: 1-10.
7. Song H, Zhang K, Piao S, et al. Soil organic carbon and nutrient losses resulted from spring dust emissions in Northern China[J]. Atmos Environ, 2019, 213: 585-596.
8. Zhang W, Li J, Li G, et al. Emission reduction effect and carbon market efficiency of carbon emissions trading policy in China[J]. Energy, 2020, 1961: 117117.
9. Peng H, Qi S, Cui J. The environmental and economic effects of the carbon emissions trading scheme in China: The role of alternative allowance allocation[J]. Sustain Prod Consum, 2021, 28: 105-115.
10. Xu X, Niu D, Xiao B, et al. Policy analysis for grid parity of wind power generation in China[J]. Energy Policy, 2020, 138: 111225.
11. Wu S, Zheng X, Khanna N, et al. Fighting coal—effectiveness of coal-replacement programs for residential heating in China: Empirical findings from a household survey[J]. Energy Sustain Dev, 2020, 55: 170-180.
12. Wu Y, Tam V W, Shuai C, et al. Decoupling China's economic growth from carbon emissions: Empirical studies from 30 Chinese provinces (2001–2015)[J]. Sci Total Environ, 2019, 656: 576-588.
13. Wang Y, Zheng Y. Spatial effects of carbon emission intensity and regional development in China[J]. Environ Sci Pollut Res, 2021, 28(11): 14131-14143.
14. Yang G, Zha D, Wang X, et al. Exploring the nonlinear association between environmental regulation and carbon intensity in China: The mediating effect of green technology[J]. Ecol Indic, 2020, 114: 106309.
15. Huang C, Zhang X, Liu K. Effects of human capital structural evolution on carbon emissions intensity in China: A dual perspective of spatial heterogeneity and nonlinear linkages[J]. Renew Sust Energ Rev, 2021, 135: 110258.
16. Cheng Z, Li L, Liu J. Industrial structure, technical change and carbon intensity in China's provinces[J]. Renew Sust Energ Rev, 2018, 81: 2935-2946.
17. Liang S, Zhao J, He S, et al. Spatial econometric analysis of carbon emission intensity in Chinese provinces from the perspective of innovation-driven[J]. Environ Sci Pollut Res, 2019, 26(14): 13878-13895.

18. Long R, Gan X, Chen H, et al. Spatial econometric analysis of foreign direct investment and carbon productivity in China: Twotier moderating roles of industrialization development[J]. Resour Conserv Recycl, 2020, 155: 104677.
19. Erdogan S. Dynamic nexus between technological innovation and buildings Sector's carbon emission in BRICS countries[J]. J Environ Manag, 2021, 293: 112780.
20. Nguyen D H, Chapman A J, Farabiasl H. Nation-wide emission trading model for economically feasible carbon reduction in Japan[J]. Appl Energy, 2019, 255: 113869.
21. Wang C, Guo Y, Shao S, et al. Regional carbon imbalance within China: An application of the Kaya-Zenga index[J]. J Environ Manag, 2020, 262: 110378.
22. Xiao H, Sun K, Tu X, et al. Diversified carbon intensity under global value chains: A measurement and decomposition analysis[J]. J Environ Manag, 2020, 272(15): 111076.
23. Jiang M, An H, Gao X, et al. Structural decomposition analysis of global carbon emissions: The contributions of domestic and international input changes[J]. J Environ Manag, 2021, 294: 112942.
24. Ninpanit P, Malik A, Wakiyama T, et al. Thailand's energy-related carbon dioxide emissions from production-based and consumption-based perspectives[J]. Energy Policy, 2019, 133: 110877.
25. Zhou X, Zhou D, Wang Q, et al. Who shapes China's carbon intensity and how? A demand-side decomposition analysis[J]. Energy Econ, 2020, 85: 104600.
26. Su B, Ang B. Multiplicative structural decomposition analysis of aggregate embodied energy and emission intensities[J]. Energy Econ, 2017, 65: 137-147.
27. Wang Z, Su B, Xie R, et al. China's aggregate embodied CO_2 emission intensity from 2007 to 2012: A multi-region multiplicative structural decomposition analysis[J]. Energy Econ, 2019, 85: 104568.
28. Su B, Ang B, Liu Y. Multi-region input-output analysis of embodied emissions and intensities: Spatial aggregation by linking regional and global datasets[J]. J Clean Prod, 2021, 313(1): 127894.
29. Dong F, Yu B, Hadachin T, et al. Drivers of carbon emission intensity change in China[J]. Resour Conserv Recycl, 2018, 129: 187–201.
30. Ma M, Yan R, Du Y, et al. A methodology to assess China's building energy savings at the national level: An IPAT–LMDI model approach[J]. J Clean Prod, 2017, 143(1): 784-793.
31. Ma M, Cai W, Cai W. Carbon abatement in China's commercial building sector: A bottom-up measurement model based on kaya-LMDI methods[J]. Energy, 2018, 165(Part A): 350-368.
32. Cao Q, Kang W, Xu S, et al. Estimation and decomposition analysis of carbon emissions from the entire production cycle for Chinese household consumption[J]. J Environ Manag, 2019, 247: 525-537.
33. Wang X, Wei Y, Shao Q. Decomposing the decoupling of CO_2 emissions and economic growth in China's iron and steel industry[J]. Resour Conserv Recycl, 2020, 152: 104509.
34. Xu S, He Z, Long R. Factors that influence carbon emissions due to energy consumption in China: Decomposition analysis using LMDI[J]. Appl Energy, 2014, 127: 182-193.
35. Chontanawat J, Wiboonchutikula P, Buddhivanich A. An LMDI decomposition analysis of carbon emissions in the Thai manufacturing sector[J]. Energy Rep, 2020, 6: 705-710.
36. Gu S, Fu B, Thriveni T, et al. Coupled LMDI and system dynamics model for estimating urban CO_2 emission mitigation potential in Shanghai, China[J]. J Clean Prod, 2019, 240(10): 118034.
37. Zhang M, Mu H, Ning Y. Accounting for energyrelated CO_2 emission in China, 1991-2006[J]. Energy Policy, 2009, 37(3): 767-773.
38. Zhou P, Ang B W. Decomposition of aggregate CO_2 emissions: A production theoretical approach[J]. Energy Econ, 2008, 30: 1054-1067.
39. Wang C. Decomposing energy productivity change: A distance function approach[J]. Energy, 2007, 32(8): 1326-1333.
40. Wang Q, Chiu Y, Chiu C. Driving factors behind carbon dioxide emissions in China: A modified production-theoretical decomposition analysis[J]. Energy Econ, 2015, 51: 252-260.
41. Du K, Xie C, Ouyang X. A comparison of carbon dioxide (CO_2) emission trends among provinces in China[J]. Renew Sust Energ Rev, 2017, 73: 19-25.
42. Wang M, Feng C. The impacts of technological gap and scale economy on the low-carbon development of China's industries: An extended decomposition analysis[J]. Technol Forecast Soc Chang, 2020, 157: 120050.
43. Wang M, Feng C. The consequences of industrial restructuring, regional balanced development, and market-oriented reform for China's carbon dioxide emissions: A multi-tier meta-frontier DEA-based decomposition analysis[J]. Technol Forecast Soc Chang, 2021, 164: 120507.
44. Yang L, Li Z. Technology advance and the carbon

dioxide emission in China—Empirical research based on the rebound effect[J]. Energy Policy, 2017, 101: 150-161.
45. Zha D, Yang G, Wang Q. Investigating the driving factors of regional CO_2 emissions in China using the IDA-PDA-MMI method[J]. Energy Econ, 2019, 84: 104521.
46. Zhang W, Tang X, Yang G, et al. Decomposition of CO_2 emission intensity in Chinese MIs through a development mode extended LMDI method combined with a production-theoretical approach[J]. Sci Total Environ, 2020, 702: 134787.
47. Yang J, Hao Y, Feng C. A race between economic growth and carbon emissions: What play important roles towards global lowcarbon development?[J]. Energy Econ, 2021, 100: 105327.
48. Li J, See K F, Chi J. Water resources and water pollution emissions in China's industrial sector: Agreen-biased technological progress analysis[J]. J Clean Prod, 2019, 229: 1412-1426.
49. Acemoglu D. Why do new technologies complement skills? Directed technical change and wage inequality[J]. Q J Econ, 1998, 113(4): 1055-1089.
50. Song M, Xie Q, Wang S, et al. Intensity of environmental regulation and environmentally biased technology in the employment market[J]. Omega, 2020, 100: 102201.
51. Essid H, Ouellette P, Vigeant S. Productivity, efficiency, and technical change of Tunisian schools: A bootstrapped Malmquist approach with quasi-fixed inputs[J]. Omega, 2014, 42(1): 88-97.
52. Ding L, Yang Y, Wang W et al. Regional carbon emission efficiency and its dynamic evolution in China: A novel cross efficiency-Malmquist productivity index[J]. J Clean Prod, 2019, 241: 118260.
53. Kaya Y. Impact of carbon dioxide emission control on GNP growth: Interpretation of proposed scenarios[R]. Intergovernmental Panel on Climate Change/Response Strategies Working Group May, 1989.
54. Chen L, Cai W, Ma M. Decoupling or delusion? Mapping carbon emission per capita based on the human development index in southwest China[J]. Sci Total Environ, 2020, 741: 138722.
55. Lin B, Du K. Decomposing energy intensity change: A combination of index decomposition analysis and production-theoretical decomposition analysis[J]. Appl Energy, 2014, 129: 158-165.
56. Acemoglu D. Directed technical change[J]. Rev Econ Stud, 2002, 69(4): 781-809.
57. Ding L, Yang Y, Zheng H, et al. Heterogeneity and the influencing factors of provincial green-biased technological progress in China: Based on a novel Malmquist-Luenberger multidimensional decomposition index[J]. China Popul Resources Environ, 2020, 41(9): 84-92.
58. Peng J, Xiao J, Wen L, et al. Energy industry investment influences total factor productivity of energy exploitation: A biased technical change analysis[J]. J Clean Prod, 2019, 237(10): 117847.
59. IPCC. 2006 IPCC Guidelines for National Greenhouse Gas Inventories[R]. Institute for Global Environmental Strategies (IGES) for the IPCC, Kanagawa, 2006.
60. Shan H. Re-estimating the capital stock of China: 1952–2006[J]. J Quant Techn Econ, 2008, 25(10): 17-31.
61. Lin B, Wang M. The role of socio-economic factors in China's CO_2 emissions from production activities[J]. Sustain Prod Consump, 2021, 27: 217-227.
62. Shao S, Yang Z, Yang L, et al. Synergetic conservation of water and energy in China's industrial sector: From the perspectivesof output and substitution elasticities[J]. J Environ Manag, 2020, 259(1): 110045.
63. Kong Q, Peng D. Resource misallocation, production efficiency and outward foreign direct investment decisions of Chinese enterprises[J]. Res Int Bus Financ, 2021, 55: 101343.
64. Du Y, Li Z, Du J, et al. Public environmental appeal and innovation of heavy-polluting enterprises[J]. J Clean Prod, 2019, 222: 1009-1022.
65. Wang Y, Liao M, Wang Y, et al. The impact of foreign direct investment on China's carbon emissions through energy intensity and emissions trading system[J]. Energy Econ, 2021, 971: 105212.
66. Wang M, Feng C. Decoupling economic growth from carbon dioxide emissions in China's metal industrial sectors: A technological and efficiency perspective[J]. Sci Total Environ, 2019, 691: 1173-1181.
67. Wang B, Yu M, Zhu Y, et al. Unveiling the driving factors of carbon emissions from industrial resource allocation in China: A spatial econometric perspective[J]. Energy Policy, 2021, 158: 112557.
68. Kim K, Kim Y. International comparison of industrial CO_2 emission trends and the energy efficiency paradox utilizing production-based decomposition[J]. Energy Econ, 2012, 34(5): 1724-1741.

Green finance and high-quality development of marine economy[①]

Sheng Xu[1], Ke Gao[2*]

1 Department of Finance, Ocean University of China, Qingdao, China
2 Ocean University of China – Laoshan Campus, Qingdao, China
* Corresponding author: Ke Gao (gao_ke97@163.com)

Abstract

Purpose: With the Chinese marine economy developing rapidly, the environmental problem has been occurring frequently, which needs green finance that supports energy conservation, environmental protection, and sustainable development to solve.

Design/methodology/approach: In this paper, the entropy method is used to measure the development level of green finance, the DEA-ML index is used to measure the green total factor productivity which is used to indicate the high-quality development level of the marine economy in 11 coastal provinces (cities), then the grey correlation degree between them whose result shows that there is a certain correlation between the two variables is calculated. The fixed-effect model was used to analyze the relationship between them.

Findings: The results show that the development level of green finance can promote the high-quality development of the marine economy, but there are still some problems in the process of green finance supporting the marine economy.

Originality/value: This paper seeks new growth drivers, green finance, for the high-quality development of the marine economy, which few scholars have studied.

Keywords: green finance; high-quality development of the marine economy; DEA-ML; fixed-effect model

Introduction

In the past, the Chinese marine economy which plays an increasingly important role in the national economy has paid too much attention to the speed of development in the context of excessive dependence on the input of factors of production, for which the marine economy develops at the expense of resource and environment. Therefore, Wang et al.[1] pointed out that assessing the level of high-quality development of the Chinese marine economy, seeking new momentum for economic growth, has become the focal point of marine economy development at the present stage.

Green finance that conforms to the trend of green development to protect the environment will help drive the transformation of marine industrial structure, promote innovation in marine science and technology and raise people's awareness of environmental protection to transform the marine economy from high-speed to high-quality development raised by Yu et al.[2]. However, the complicated approval procedures, small

---

① 本文于2022年7月发表在 *Marine Economics and Management* 第5卷，https://doi.org/10.1108/MAEM-01-2022-0001。

financing amount of green finance contrary to the characteristics of large financing demand, and the long cycle in the marine industry become the obstacle that promotes high-quality marine economy development. In addition, Xu et al.[3] indicated that marine industries account for only a small part of the industries supported by green finance because of the lack of green financial products related to the marine industry. So, it is important to study the relationship between green finance and the high-quality development of the marine economy while few scholars have studied the relationship between them.

Slightly different from other scholars, the green finance index system whose weight of specific indicators is determined by not only objective evaluation like the entropy method which Liu et al.[4] has used in their papers but also subjective assessment, such as expert scoring is established to measure the development level of green finance in coastal provinces in this paper. Meanwhile, the environmental pollution index is established to measure undesired outputs. When calculating the green total factor productivity, unlike other scholars who choose a single indicator, for example, Lin et al.[5] and Wu[6] used the quantities of industrial wastewater as unexpected output indicator to measure marine green total factor productivity.

In this paper, firstly, the relevant researches on green finance and high-quality development of marine economy are sorted out; secondly, method used in this paper, the entropy value method, DEA-ML index, the grey correlation analysis and the fixed-effect model, are introduced; then, relevant results are presented; at last, this paper looks into the future development of green finance and high-quality development of the marine economy, and puts forward relevant suggestions.

Literature Review and Mechanism Analysis

▶ *Literature related to green finance*

In the 1970s and 1990s, environmental problems arise for the rapid development of developed countries. With finance support for environmental protection arousing widespread concern, environment finance, another name of green finance, appeared. Scholars claimed that the purpose of green finance was to achieve environmental protection and sustainable economic development. With analyzing from finance institutions, green finance is considered as a behavior of financial institutions to reduce environmental risk and promote environmental quality. After the promulgation of the Equator Principles in 2003 and worldwide climate problems, green finance has taken off. Scholars turned their attention to solving environmental pollution and the greenhouse effect for sustainable economic development.

In China, Gao[7] discussed the connotation of green finance for the first time, he thought finance institutions could use green finance as a means for the coordinated development of the environment and economy. In the early days, green finance mainly concentrated on green credit, green bonds, green stock, green funds, green insurance, and so on. After the 18th CPC National Congress, Exploring the establishment of a green financial system has become a priority. Wang et al.[8] thought the construction of green finance index system has gradually become an important method to measure the development level of green finance. The entropy method is used by many scholars to determine the weight of each index; some scholars such as Zeng et al.[9] also determine the weight subjectively according to the actual operation of each index. Both approaches have their limitations, as for the entropy evaluation method, individual indicators may be particularly heavy for uneven development between regions, reflecting the unreal situation. The other approach is too subjective. This paper combines these two methods to measure the development level of green finance.

▶ *Literature related to the high-quality development of marine economy*

Jiang et al.[10] indicated since the 1990s, the Chinese marine economy has developed rapidly and become one of the important pillars of the national economy. The report to the 19th CPC National Congress pointed out that maritime power should be built rapidly, and China has taken the most seriously to the marine economy. But in the process of marine economic development, Luo et al.[11] considered that the seriously polluted marine environment, the shortage of marine resources, the unbalanced regional development, and other problem have been getting worse and worse. Under the requirement of high-quality economic transformation, it is important to seek indicators for evaluating high-quality marine development and new drivers. There are mainly two methods to evaluate the highquality development level of the marine economy by combing existing papers. The first one is to establish an indicator system for high-quality development of the marine economy, which uses the entropy evaluation method or other methods to calculate the weight of each. However, different results will be generated for scholars choosing indictors through different focuses. The other one is to measure the marine green total factor productivity which is deemed to be the source of high-quality development. This paper chooses the second method to measure the high-quality development level of the Chinese marine economy, and a comprehensive index system of environmental constraints is established to measure the undesired output, unlike most existing papers which use one single indicator, for example, Ren et al.[12] chose the volume of industrial wastewater discharged directly into the sea as undesirable output indicators.

▶ *Literature related to the relationship between finance and the high-quality development of marine economy*

In the context of the current economic slowdown, Dong[13] stated that new growth drivers should be sought and utilized to support the high-quality development transformation of the marine economy. Green finance, which could promote the high-quality development of the marine economy from many aspects, is one of them. Zhu[14] thought green finance can promote the transformation and upgrading of marine industrial structures. Upgrading the marine industrial structure can reduce energy consumption and promote technological innovation, as the marine industry after transformation is mostly clean, and technical industries, which were pointed out by Yang et al.[15]. Li and Gan[16] indicated that green finance can also promote international trade for higher quality products manufactured by marine enterprises after transformation, by which high-quality development of the marine economy can be actuated. Liu et al.[17] thought funds, an important factor of production, are reallocation from the high energy consumption marine industry to the green and energy saving industry through green finance. Unlike the emerging marine industry, such as the marine biopharmaceutics industry belonging to the environmental protection industry, lots of energy is consumed in the process of development by the traditional marine industry, which is represented by the marine communications and transportation industry, fishery industry. More capital is given to emerging marine industries by green finance, promoting the transformation of traditional marine industries into emerging industries, which conform to the requirements of high-quality development of the marine economy.

Technological innovation in the marine industry could be promoted by green finance, which was thought by Hong et al.[18]. Green finance will provide funds for technological innovation in the marine industry, which alleviates the bottleneck problem of insufficient funds in the process of marine technology innovation. Wang et al.[19] indicated that the long cycle, the high cost, and the high risk are the three characteristics of the marine science and technology innovation, which make it difficult to get financing, leading to the contradiction between the supply and demand of funds. With the development of green finance, Ling et al.[20] pointed out that money gradually flows into green finance, then financial institutions provide green financial products for marine enterprises, especially green credit that provides long-term funds for technological innovation. On the one hand, Wu and Liu[21] indicated that technological innovation promotes the high-quality development of the marine economy through the transformation of the traditional marine industry. Technological innovation in the marine industry makes the production process of enterprises more specialized. While improving labor productivity, the marine industry will be transformed from energy-consuming to technology-intensive. Breakthroughs and upgrades in marine technology stimulate new consumer demand and economic growth points, which fosters new marine industries. On the other hand, Wu et al.[22] also thought that technological innovation in the marine industry directly promotes high-quality economic development. Endogenous growth theory holds that technological progress is the decisive factor of economic growth, which makes products more efficient, improves the quality of the workforce, and promotes the improvement of the management system to drive economic growth. Technological innovations in marine industries can be directly applied to production, improve productivity and reduce environmental pollution, promoting high-quality development of the marine economy.

With the support of government policy, funds scattered among individuals could be gathered by finance to generate green investment through floating interest rate, differentiated credit, or other methods, which supports the green marine project. In addition, Peng and Zheng[23] pointed out that a green single is brought by green finance to the market. People pay more attention to the marine enterprises with green development when they invest, which gives these enterprises more opportunities. The green single also makes people establish the concept of energy conservation and environmental protection. It is shown below that the analysis figure of the mechanism of green finance promoting highquality development of the marine economy (Fig. 1).

In order to verify the relationship between green finance and high-quality development of marine economy to move forward a single step, firstly, the grey correlation analysis which was considered by Liu et al.[24] as a better model that needs less data, is used to study the relationship between the two. Then, the entity fixed-effect model is used to study the regression relationship between the two for the different development levels of marine economy in each coastal province which was pointed out by Liu et al.[25]

Methodology and Data

▶ *Methodology*

Entropy evaluation method

In this paper, the entropy method is used to confirm the green finance development level of 11 coastal provinces (cities). The entropy method, which is used to determine the weight coefficient according to the different degrees of each evaluation index, avoiding the interference of human factors and reflecting the importance of indicators in the comprehensive evaluation system objectively, belongs to the objective weighting method. In the entropy method, the greater the dispersion degree of data, the greater is the information entropy, which has a greater impact on comprehensive evaluation and greater weight. The obvious tendency and subjective judgment are overcome in the entropy method, which makes it widely used in various disciplines thought by Xu et al.[26]

The procedures of the entropy method are described as follows,

Standardize the raw data, if the indicator is positive,

$$y_{ij} = \frac{x_{ij} - \min(x_{ij})}{\max(x_{ij}) - \min(x_{ij})} \quad (1)$$

If the indicator is negative,

$$y_{ij} = \frac{\max(x_{ij}) - x_{ij}}{\max(x_{ij}) - \min(x_{ij})} \quad (2)$$

Normalization processing,

$$y'_{ij} = \frac{y_{ij}}{\sum_{i=1}^{n} y_{ij}} \quad (3)$$

Avoiding to calculate ln affected by 0, add 0.1^{-8} to the calculated result,

$$z_{ij} = y'_{ij} + 0.1^{-8} \quad (4)$$

Calculate the index weight

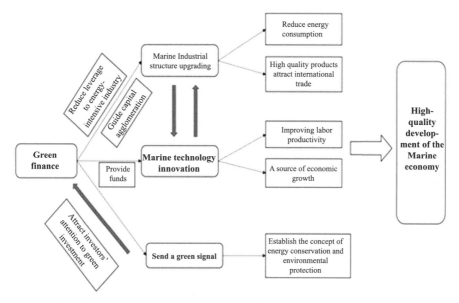

Fig. 1　Mechanism analysis diagram

$$\begin{cases} k = \dfrac{1}{\ln m} \\ e_j = -k \sum_{i=1}^{n} z_{ij}(\ln z_{ij}) \\ g_j = 1 - e_j \end{cases} \quad (5)$$

$$w_j = \dfrac{g_j}{\sum_{j=1}^{m} g_j} \quad (6)$$

DEA-Malmquist index

Under the assumption that returns to scale are constant, Malmquist-Leuenberger (ML), proposed by Fare, is chosen by this paper to measure green total factor productivity (GTFP), representing the level of high-quality development of the marine economy. GTFP is divided into technical efficiency (Effch) which is divided into pure technical efficiency (Pech) and economics of scale (Sech), technical progress (Tech).

$$M_t = \text{TFP} = \text{Effch} \times \text{Tech} = (\text{Pech} \times \text{Sech}) \times \text{Tech} \quad (7)$$

$$\begin{aligned} M_t(x^{t+1}, y^{t+1}, x^t, y^t) &= \sqrt{\dfrac{D^{t+1}(x^{t+1}, y^{t+1})}{D^{t+1}(x^t, y^t)} \times \dfrac{D^t(x^{t+1}, y^{t+1})}{D^t(x^t, y^t)}} \\ &= \dfrac{D^{t+1}(x^{t+1}, y^{t+1})}{D^t(x^t, y^t)} \times \\ &\quad \sqrt{\dfrac{D^t(x^{t+1}, y^{t+1})}{D^{t+1}(x^{t+1}, y^{t+1})} \times \dfrac{D^t(x^t, y^t)}{D^{t+1}(x^t, y^t)}} \\ &= \text{Effch} \times \text{Tech} \quad (8) \end{aligned}$$

(x^t, y^t) and (x^{t+1}, y^{t+1}) are t input and output vector at stage t and $t + 1$ stage respectively;

D^t and D^{t+1} are distance functions which use t and $t + 1$ stage as technical reference.

Grey correlation analysis

After years of development, the grey correlation proposed by Chinese professor Julong Deng, which judges the degree of correlation by the degree of sequence correlation within the system between variables theory has become mature. The specific calculation process is as follows.

First, the reference sequence and comparison sequence should be clarified.

The reference sequence: $X_0 = \{X_0(1), X_0(2), \cdots, X_0(n)\}$

The compare sequence: $X_i = \{X_i(1), X_i(2), \cdots, X_i(n)\}$

The standardized two sets of data using the methods of dividing by the average to avoid dimensional variance in estimation.

$$Y_i(n) = \dfrac{X_i(n)}{\dfrac{1}{K}\sum_{n=1}^{k} X_i(n)}, n = 1, 2, \cdots, k, i = 1, 2, \cdots \quad (9)$$

Then, calculate the difference between the reference sequence and the comparison sequence after being standardized.

$$\Delta_i(n) = |Y_0(n) - Y_i(n)|, n = 1, 2, \cdots, k, i = 1, 2, \cdots \quad (10)$$

$$\Delta_i = (\Delta_i(1), \Delta_i(2), \cdots, \Delta_i(n)); i = 1, 2, \cdots \quad (11)$$

Finally, the correlation coefficient and correlation degree are calculated.

$$\varepsilon_i(k) = \frac{\min\{\min|X_0(k) - X_i(k)|\} + \rho \cdot \max\{\max|X_0(k) - X_i(k)|\}}{|X_0(k) - X_i(k)| + \rho \cdot \max\{\max|X_0(k) - X_i(k)|\}} \quad (12)$$

Among them, $\rho = 0.5$

The grey correlation degree is

$$Y_i = \frac{1}{n}\sum_{k=1}^{n}\varepsilon_i(k) \quad (13)$$

Entity fixed-effect model

In general, the panel data model can be written in the following form.

$$y_{it} = \alpha_{it} + \beta_{it}X_{it} + u_{it}, i = 1, 2, \cdots, N; t = 1, 2, \cdots, T \quad (14)$$

If the intercept term of the model only varies with the cross-section, rather than time, this model is called the entity fixed-effects model. It is the basic assumption of the entity fixed effect model that $E(u_{it} | \alpha_i, X_{it}) = 0, i = 1, 2, \cdots, N$. The entity fixed effects model can be expressed in the following form.

$$y_{it} = \alpha_i + \beta_{it}X_{it} + u_{it}, i = 1, 2, \cdots, N, t = 1, 2, \cdots, T \quad (15)$$

In the above model, y_{it} is the explained variable, α_i is the intercept term whose number is equal to cross-section, and u_{it} is the stochastic error term.

▶ *Data*

Selection of green finance indicators

By referring to research of relevant scholars, green finance mainly includes four aspects: green credit, green securities, green insurance, and green investment. Relevant data come from China Industrial Statistics Yearbook and Insurance Yearbook and Wind database.

1. Green credit: There are two three-level indicators, the proportion of green credit and the proportion of interest expenditure of high energy consumption and high pollution industries. On the one hand, bank credit can positively support, while restraining the development of energy-intensive industries. On the other hand, contrarian indicators also need to be considered. Because of the lack of loan disclosure information on high energy consumption and high pollution industries, the proportion of interest expenditure of these two industries is chosen as one of the indicators of green credit.

2. Green securities: The proportion of the total market value of environmental protection enterprises in the total market value of regional stocks is selected to measure the index, which reflects the financing situation of environmental protection enterprises in the market.

3. Green insurance: The green insurance index includes the Proportion of agricultural insurance scale referring to the proportion of environmental liability insurance income, and the agricultural insurance loss rate referring to the ratio of agricultural insurance expenditure to income. China has introduced corporate environmental liability insurance since 2013, so systematic statistics are lacking. The development of green insurance can be replaced by agricultural insurance data, because of the deep relationship between agriculture and the natural environment.

4. Green investment: The green investment includes government spending on environmental protection and investment in environmental governance as a percentage of GDP.

To sum up, the established index system is shown in the following Table 1.

Selection of DEA-ML index indicators

First, input and output indicators need to be identified. Labor and capital are sources of economic growth, which is claimed by neoclassical economics. This paper chooses ocean-related employed personnel and marine capital stock which were calculated by using the following formula referring to Zhang[27] and Ding and Zhu[28] as input indicators.

$$K_t = (1-\delta)k_{t-1} + I_t/p_t \quad (16)$$

$$kt' = GOP/GDP \times k_t \quad (17)$$

Energy consumption which uses provincial data multiplied by the ratio of GOP to GDP is taken as an input index unlike other scholars choosing science and technology input index. Output indicators are divided into desirable output and undesirable output. Gross ocean production which is adjusted by price indices, is chosen as the desired output. A comprehensive index system of undesired output that contains industrial waste gas, effluents and solid waste is established. Specific indicators are listed in the following table, and then the entropy method introduced above was used for evaluation. The data of input, desirable output, and undesirable output come from China Marine Statistical Yearbook and China Environmental Statistics Yearbook (Table 2).

Table 1　The green finance indicator system

Level 1 indicators	Level 2 indicators	Level 3 indicators
Green finance indicator system	Green credit	The proportion of green credit
		The proportion of interest expenditure on high energy consumption and high pollution industries
	Green securities	The proportion of the total market value of environmental protection enterprises
	Green insurance	The proportion of agricultural insurance scale
		The agricultural insurance loss rate
	Green investment	The proportion of government spending on environmental protection
		The proportion of investment in environmental governance

Table 2　The input and output indicator

Composite indicator	Level 1 indicators	Level 2 indicators
Input indicators	Labor	Ocean-related employed personnel
	Capital	Marine capital stock
	Energy consumption	Total energy consumption
Desirable output indicators	Gross ocean production	GOP
Undesirable output indicators	Pollution index	Industrial waste gas
		Industrial effluents
		Industrial waste gas (CO_2, SO_2, and dust)

Variable selection of grey correlation analysis

This paper uses GTFP which is measured by DEA-ML, representing the development of high-quality development of the marine economy and the level of green finance development which is obtained by the entropy method above to perform grey correlation analysis.

Variable selection of fixed-effect regression

GTFP is selected as an explained variable, and the level of green finance development is used as a core explanatory variable. After reviewing other scholars' research, it is found that GTFP is also affected by the level of foreign direct investment (fdi), degree of openness (ope), and degree of affluence of residents (yd). So these three variables, measured by the ratio of foreign direct investment to GDP, the ratio of total exports to GDP, and the disposable income of urban residents, are selected as control variables. This paper constructs the following regression equation, which takes the natural logarithm of all the variables to alleviate the impact of heteroscedasticity on the estimation results.

$$\ln(GTFP)_{it} = u_i + \beta_1\ln(gre)_{it} + \beta_2\ln(fdi)_{it} \\ + \beta_3\ln(yd)_{it} + \beta_4\ln(op)_{it} + e_{it} \quad (18)$$

The Results

▶ *The result of the entropy evaluation method*

After calculation, the weight of each indicator is shown in the following Table 3 Weight 1.

Due to the data on green insurance varies greatly in time and space, green insurance accounts for a larger proportion, which is not consistent with the fact that green insurance develops slowly. In this paper, the subjective method is used to reconfirm the weight of indicators. The weight of green credit, green securities, green investment, and green insurance respectively are 50%, 15%, 10%, 15%, and evenly distributed within the secondary index referring to the research of relevant scholar Zeng et al.[9]. By averaging the weights obtained by the two methods, the heaviest weight can be obtained. The final results are shown in the following Table 3 Weight 2.

The development level of green finance in each region each year can be got by multiplying the normalized y'_{ij} by the weights in the table above. Then, replace individual outliers by trend function and draw the development level of green finance in each province each year on a map.

The following conclusions can be drawn from the above Fig. 2.

1. The development level of green finance in 11 coastal provinces shows an obvious wave-like uplift trend, Especially Shanghai, Guangdong, and Liaoning rising faster.

2. In 2008, the development of green finance in different regions is unbalanced. Jiangsu is at the peak of its development in the past decade for balanced development and higher green insurance. But in 2009, its level falls sharply for the decline in expenditure on energy conservation and environmental protection. Tianjin was at its lowest level for nearly a decade, and then it grows quickly.

3. In 2018, all coastal provinces (cities) have a high level of green finance development that is almost the highest over the past decade. Green finance in Hebei has been well developed, while Tianjin and Fujian are on the low side.

▶ *The result of the DEA-ML index*

The GTFP is calculated by MAX DEA software and drawn on a graph. When GTFP is greater than 1, it indicates that the high-quality development level of the marine economy has improved compared with last year. On the contrary, it indicates a decline in the level of development.

The following conclusions can be drawn from the above Fig. 3.

Table 3 The green finance index system and weight

Level 1 indicators	Level 2 indicators	Level 3 indicators	Weight 1	Weight 2
Green finance indicator system	Green credit	The proportion of green credit	0.0221	0.136
		The proportion of interest expenditure of high energy consumption and high pollution	0.0495	0.1497
	Green securities	The proportion of the total market value of environmental protection enterprises	0.037	0.0938
	Green insurance	The proportion of agricultural insurance scale	0.0301	0.065
		The agricultural insurance loss rate	0.0279	0.0639
	Green investment	The proportion of government spending on environmental protection	0.4577	0.2663
		The proportion of investment in environmental governance	0.3747	0.2248

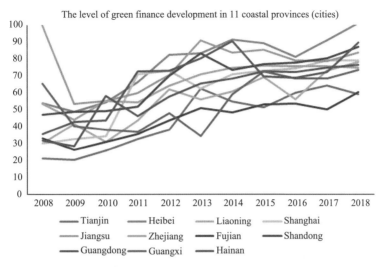

Fig. 2　Time series plot of green finance

1. From 2008 to 2018, the high-quality development level of the Tianjin marine economy has been rising for proper environmental governance.

2. The development level of Hebei province is poor, especially from 2008 to 2013 perhaps because of the undesired increase in output during development.

3. In 2008, Only Tianjin improves over last year, while other provinces are down from last year, Especially Hainan province.

4. In 2018, except for Liaoning province seeing a slight decline, the development levels of other provinces improve. With the implementation of green finance policies, the marine green total factor productivity of all provinces is basically greater than 1, which means the level of marine economic development has improved from 2013.

▶ *The result of the grey correlations*

The grey correlations between green finance and marine high-quality development which are shown in the following Table 4 are calculated by the above formula in 11 coastal provinces (cities).

Table 4　The results of grey correlation degree

Province (cities)	Tianjin	Hebei	Liaoning	Shanghai	Jiangsu	Hainan
Results	0.6908	0.6032	0.6168	0.5944	0.6535	0.7027
Province (cities)	Zhejiang	Fujian	Shandong	Guangdong	Guangxi	Average
Results	0.6463	0.5841	0.6847	0.6298	0.697	0.6458

As can be seen from the results in Table 4 above, the data of the grey correlation degree of 11 coastal provinces (cities) is basically greater than 0.6, at about 0.65, which shows that there is a certain correlation between green finance and the high-quality development of Marine economy. Hainan, Guangxi, and Tianjin have higher correlation degrees, while the correlation degree is low in Shanghai and Fujian.

▶ *The results of fixed-effect regression*

The benchmark return

Based on panel data of 11 coastal provinces from 2008 to 2018, this paper uses stata15 for estimation and analysis and chooses the method of stepwise regression to avoid the effect of multicollinearity

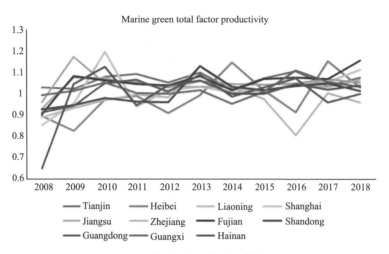

Fig. 3 Time series plot of GTFP

on the results. At the same time, the hypotheses that all data have the same variance were modified by using the robust command. The estimated results are shown in the following Table 5.

Judging from the regression results, as the control variables are added one by one, the core explanatory variable (gre), the level of green finance development, maintains a certain positive correlation with the explained variable (GTFP), high-quality development of the marine economy represented by green total factor productivity, while the significance declines. The value of DW is around 2, which indicates that residual sequences are unlikely to have sequence correlation. The value of the F statistic also rejects the null hypothesis that the independent variable of the equation is 0. In all overviews, it is suitable for the established regression which has guiding significance to the development of green finance and the marine economy.

The coefficient between the variables ln(GTFP) and ln(gre) is positive from the above table, which is shown that green finance will promote the high-quality development of the marine economy as expected.

Stability test

This paper adopts the method of reducing sample data to make regression again which are selected from 11 coastal provinces randomly to

Table 5 The results of fixed-effect regression

	FE(1)	FE(2)	FE(3)	FE(4)
ln(gre)	0.0443** (2.27)	0.0301* (2.01)	0.0340* (2.01)	0.0306* (1.89)
ln(yd)		0.7618*** (3.40)	1.1147*** (3.60)	0.9475** (3.84)
ln(fdi)			0.0469 (1.54)	0.0528* (1.88)
ln(ope)				−0.0745*** (−4.92)
Intercept item	0.2081** (2.33)	−1.6266** (−2.93)	−2.1658*** (−3.58)	−1.8952*** (−3.70)
DW value	2.0475	2.1071	2.1578	2.1561
F statistic	5.15	13.77	6.96	11.19

Note: *, **, and *** indicate significance at 10%, 5%, and 1% levels, respectively. The numbers in parentheses are t values.

test the stability, avoiding accidental factors for the erroneous conclusion. Besides, the data in the preceding part of the paper are winsorized and then regressed to eliminate estimation bias caused by outliers. The regression results are shown in the following table, Models 5 and 6 are the results of sample reduction, and Model 7 is the result of winsorization (Table 6).

After processing the data, the sign and significance level of variable lngre do not change significantly in the result of regression again, so this model which explains the relationship between green finance and high-quality development of the marine economy is basically stable.

Conclusion and Outlook

▶ *Research conclusion*

First, according to the grey correlation degree analysis, the grey correlation degree between green finance and GTFP at about 0.65 in 11 coastal provinces (cities), which can be considered preliminarily that there is a certain relation between green finance and high-quality development of the marine economy represented by GTFP. Hainan and Guangxi have the highest correlation probably because of smaller economies of scale for greater operational efficiency of green finance. Tianjin and Shandong followed while Shanghai and Fujian are at the bottom. May be marine industry is relatively small in the region's economic development; as a result, green finance does not support it enough.

In the empirical test, from the regression results, the regression coefficient of the core explanatory variable is 0.0306 and significant at the 10% level, which means that the development of green finance will promote the high-quality development of the marine economy. The coefficient of the variable gre is low, because it is small for the order of magnitude of green financial data obtained by the entropy method. With the addition of control variables, there was a slight decrease in the significance level of variable gre, it may be that the transmission mechanism of green finance to support high-quality development of the marine economy is not smooth enough. Marine industries are not the only ones, on the contrary, just a small part supported by green finance and there is no marine green finance product. It is complex to apply for green credit and the amount applied is small, which contraries the characteristics of the long cycle and a large amount of financing demand of Marine enterprises. In addition, the high-quality development of the marine economy has a certain substance. These reasons lead to that although green finance can promote the development of the marine economy, its significance level is low.

The coefficient of the control variable yd and fdi is positive, which means levels of affluence and foreign direct investment will promote the high-quality development of the marine economy in

Table 6 The results of the stability test

	FE(5)	FE(6)	FE(7)
ln(gre)	0.0476* (2.32)	0.0472* (1.97)	0.0293** (2.31)
ln(yd)	0.5599* (1.98)	1.0701** (3.10)	0.8866*** (3.91)
ln(fdi)	−0.0233 (−1.05)	0.0635 (1.07)	0.0563* (1.94)
ln(ope)	−0.0445** (−2.33)	−0.0623** (−2.51)	−0.007*** (−3.37)
Intercept item	−1.279 (−1.80)	−2.0310** (−3.03)	−1.7118*** (−3.67)
DW value	1.742	2.1711	1.9502
F statistic	11.01	675	10.64

Note: *, **, and *** indicate significance at 10%, 5%, and 1% levels, respectively. The numbers in parentheses are t values

line with expectations, while the control variable ope expressed as the ratio of exports to GDP whose empirical result is contrary to expectations is expected to develop into a positive relationship with the high-quality of marine economy. In the early days, Chinese exports were dominated by low-technology, highly labor-intensive goods, then there were energy products such as coal, which do not lead to the upgrading of industrial structure, nor produce technological innovation, and the massive exploitation of fossil fuels can cause serious environmental problems. Although in recent years, Chinese export products have been transformed to be technical, a time lag exists in the development of the high-quality marine economy. These reasons restrict the high-quality development of the marine economy, leading to the negative intercept of the variable ope.

▶ *Suggestions*

With China transforming from old growth drivers to new ones and the proposal of maritime power policy, green finance which promotes the high-quality development of marine economy from the above study, while the transmission mechanism between the two is not yet smooth, needs to be vigorously developed. This paper put forward the following suggestions for the current problems between green finance and marine economy development.

First, the roles of the government and financial institutions need to be fully played. The government should formulate reasonable, scientific, and effective policies, and strengthen enforcement and supervision based on understanding the characteristics of the marine economy and the connotation of green finance. The government should direct the flow of funds through financial incentives and tax breaks to guide financial institutions to support the marine economy. The existing state-owned financial institutions should be specialized in the marine business, which leads their business to be separate from the traditional financial business and be connected with the marine economy. Strengthen international cooperation, actively carry out cooperation with international financial institutions, which guides international green financial funds to flow into the marine industry, and strives for highquality capital supply from the international green bank, environmental fund, and other financial institutions.

Blue finance should be vigorously developed based on green finance market to solve the problem that only a small part of green finance supports marine industries. Blue bonds whose characteristics of long term, low redemption risk, and not changing the ownership structure of the enterprise can solve the financing problem of sea-related enterprises very well as an important tool of blue finance. In addition, attention should be paid to marine-themed investments, attracting more investors who have a long-term investment perspective to enter.

▶ *Outlook*

Limited by the lack of marine data in the writing process, for some data, this paper has to use other methods instead. For example, ocean data are replaced by using provincial data multiplied by the ratio of GOP to GDP, which, to some extent, has led to a bias in estimation.

From the theoretical analysis, there may be an intermediary effect that is not significant in empirical tests probably because of the bias in data collection between green finance and the high-quality development of the marine economy. In the future, we will continue to collect more accurate data for further research.

With the gradual development of green finance, people pay more attention to the efficiency of development rather than the absolute amount of development. In the future, we will measure the efficiency of green finance development in 11 coastal provinces (cities) to study the impact of green finance development efficiency on marine economy transformation.

Acknowledgements

This research is supported by the National Social

Science Fund Major Projects (18VHQ003): Study on the indicator system and path of maritime power in the new era.

References

1. Wang S H, Lu B B, Yin K D. Financial development, productivity, and high-quality development of the marine economy[J]. Marine Policy, 2021, 130: 74-90.
2. Yu C H, Wu X Q, Zhang D Y, et al. Demand for green finance: Resolving financing constraints on green innovation in China[J]. Energy Policy, 2021, 153: 134-150.
3. Xu Z L, Zhai S, Qian C. The impact of green financial agglomeration on the ecological efficiency of marine economy[J]. Journal of Coastal Research, 2019, 94: 988-991.
4. Liu S, Xu R, Chen X. Does green credit affect the green innovation performance of high-polluting and energy-intensive enterprises? Evidence from a quasi-natural experiment[J]. Environmental Science and Pollution Research, 2021, 27: 298-313.
5. Lin Q, Xu M, Juan T, Environmental regulation and green total factor productivity: Evidence from China's marine economy[J]. Polish Journal of Environmental Studies, 2021, 30: 5117-5131.
6. Wu D. Impact of green total factor productivity in marine economy based on entropy method[J]. Polish Maritime Research, 2018, 25: 141-146.
7. Gao J. Green Finance and sustainable development of finance[J]. Financial Theory and Teaching, 1998, 4: 20-22.
8. Wang X, Zhao H, Bi K. The measurement of green finance index and the development forecast of green finance in China[J]. Environmental and Ecological Statistics, 2021, 28: 263-285.
9. Zeng X W, Liu Y Q, Man M J, et al. Measurement analysis of the development degree of green finance in China[J]. Journal of China Executive Leadership Academy Yan'an, 2014, 7: 107-112.
10. Jiang X Z, Liu T Y, Su C W. China's marine economy and regional development[J]. Marine Policy, 2014, 50: 227-237.
11. Luo J, Meng B, Wang S, et al. Research on the development of China's marine economy from the perspective of the operational efficiency of shipping enterprises[J]. Journal of Coastal Research, 2014, 6: 12-15.
12. Ren W, Ji J, Chen L, et al. Evaluation of China's marine economic efficiency under environmental constraints-an empirical analysis of China's eleven coastal regions[J]. Journal of Cleaner Production, 2018, 184: 806-814.
13. Dong Z. A study on the transformation and upgrading of Zhejiang marine economic industrial structure based on the analysis and evaluation of scientific and technological innovation[J]. Journal of Coastal Research, 2019, 98: 231-234.
14. Zhu D. Exploring the impact of green financial derivatives on China's environmental protection[J]. Ekoloji, 2018, 27: 1857-1865.
15. Yang H, Peng C, Yang X, et al. Does change of industrial structure affect energy consumption structure: A study based on the perspective of energy grade calculation[J]. Energy Exploration and Exploitation, 2019, 37: 579-592.
16. Li C, Gan Y. The spatial spillover effects of green finance on ecological environmentempirical research based on spatial econometric model[J]. Environmental Science and Pollution Research, 2021, 28: 5651-5665.
17. Liu P D, Zhu B Y, Yang M. Has marine technology innovation promoted the highquality development of the marine economy?—Evidence from coastal regions in China[J]. Ocean and Coastal Management, 2021, 209: 256-274.
18. Hong M, Li Z, Drakeford B. Do the green credit guidelines affect corporate green technology innovation? Empirical research from China[J]. International Journal of Environmental Research and Public Health, 2021, 18: 532-549.
19. Wang J, Shi X, Du Y. Exploring the relationship among marine science and technology innovation, marine finance, and marine higher education using a coupling analysis: A case study of China's coastal areas[J]. Marine Policy, 2021, 132: 340-362.
20. Ling S, Han G, An D, et al. The impact of green credit policy on technological innovation of firms in pollution-intensive industries: Evidence from China[J]. Sustainability, 2020, 12: 254-274.
21. Wu N, Liu Z. Higher education development, technological innovation and industrial structure upgrade[J]. Technological Forecasting and Social Change, 2021, 162: 145-159.
22. Wu F, Wang X, Liu T. An empirical analysis of high-quality marine economic development driven by marine technological innovation[J]. Journal of Coastal Research, 2020, 25: 465-468.
23. Peng J, Zheng Y. Does environmental policy promote energy efficiency? Evidence from China in the context of developing green finance[J]. Frontiers in Environmental Science, 2021, 9: 126-150.
24. Liu S J, Cai H, Yang Y J. Research progress of grey relational analysis model"[J]. System Engineering

Theory and Practice, 2013, 33: 2041-2045.
25. Liu B Q, Xu M, Xie S M. Regional disparities in China's marine economy[J]. Marine Policy, 2017, 88: 1-7.
26. Xu Y, Gong B, Cheng W, et al. Persimmon fruit quality comprehensive evaluation method based on entropy weight TOPSIS model, involves determining evaluation index weight by entropy weight method, and determining optimal solution and worst solution in weighting matrix. 2018.
27. Zhang J. Estimation of provincial physical capital stock in China: 1952–2000[J]. Economic Research, 2004, 10: 35-44.
28. Ding L, Zhu L. Measurement and influencing factors of green total factor productivity of marine economy in China[J]. China Science and Technology Forum, 2015, 12: 78-90.

Further reading

Jiang L. The measurement of green finance development index and its poverty reduction effect: Dynamic panel analysis based on improved entropy method[J]. Discrete Dynamics in Nature and Society, 2020, 20: 340-362.

Liu R, Wang D, Zhang L, et al. Can green financial development promote regional ecological efficiency? A case study of China[J]. Natural Hazards, 2019, 95: 325-341.

Liu N N, Liu C Z, Xia Y F, et al. Examining the coordination between green finance and green economy aiming for sustainable development: A case study of China[J]. Sustainability, 2020, 12: 9-19.

Shao Q L, Guo J, Kang P. Environmental response to growth in the marine economy and urbanization: A heterogeneity analysis of 11 Chinese coastal regions using a panel vector autoregressive model[J]. Marine Policy, 2021, 124: 234-256.

Venturini F. Product variety, product quality, and evidence of endogenous growth[J]. Economics Letters, 2012, 117: 74-77.

Yan X F. Research on the action mechanism of circular economy development and green finance based on entropy method and big data[J]. Journal of Enterprise Information Management, 2021, 10: 156-170.

Yu X, Wang P. Economic effects analysis of environmental regulation policy in the process of industrial structure upgrading: Evidence from Chinese provincial panel data[J]. Science of the Total Environment, 2021, 753: 53-63.

Zhou X, Tang X, Zhang R. Impact of green finance on economic development and environmental quality: a study based on provincial panel data from China[J]. Environmental Science and Pollution Research, 2020, 27: 19915-19932.

Global trends and prospects of blue carbon sinks: A bibliometric analysis

Lu Jiang[1], Tang Yang[2], Jing Yu[1,3*]

1 College of Oceanic and Atmospheric Sciences, Ocean University of China, Qingdao 266100, China
2 School of Environmental and Municipal Engineering, Qingdao University of Technology, Qingdao 266033, China
3 Institute of Marine Development of Ocean University of China, Qingdao 266100, China
* Corresponding author: Jing Yu (by6801@ouc.edu.cn)

Abstract

Blue carbon sinks (mangroves, saltmarshes, and seagrasses) are considered an effective nature-based approach for climate change mitigation. Despite growing interest, a systematic review of this topic is still scarce. This study evaluated 1348 blue carbon sink related articles from 1990 to 2020 using bibliometric technology. The results from total of 85 countries, 1538 institutions, and 4492 authors indicated that blue carbon sink research shows the characteristics of rapid growth. The most active country, institution, and author were USA, Chinese Academy of Sciences, and Duarte CM, respectively. Relatively close academic collaboration has formed in blue carbon science. Environmental Sciences was the most popular category with 590 papers. The percentages of articles related to mangroves, saltmarshes and seagrasses were 63.87%, 40.36%, and 40.65%, respectively. Mangrove carbon sinks are the most popular topic, and stable isotope and remote sensing are the most researched technologies for mapping and quantifying blue carbon sinks. The threats to blue carbon sinks are complex and distinctive. Restoration, conservation, and management of blue carbon ecosystems aimed to improve their carbon sink capacity is becoming a hot issue and should be further investigated in the future. These findings provide a scientific roadmap for further research in this field and will enable stakeholders to identify the research trend.

Keywords: blue carbon sink; bibliometric analysis; hotspot; research trend; future direction

Introduction

A broad consensus has been reached among the scientific community that climate change is driven by the emissions of carbon dioxide (CO_2)[1,2]. The term 'blue carbon sink' was introduced to describe the great contribution of coastal vegetated ecosystems, consisting of mangroves, saltmarshes, and seagrasses, in sequestrating atmospheric CO_2[3]. As an emerging concept in the field of climate change mitigation, blue carbon sinks have drawn considerable attention from many policymakers, the general public, and scholars[4-6]. Due to their high CO_2 sequestration and flux rates[7-9], blue carbon sinks are regarded as the principal nature-based approach in carbon sequestration and climate change mitigation, despite their small global extent[4]. As the most intense carbon sink on earth, they account for only 0.05% of the plant biomass on land but store as much as 71% of all carbon storage in ocean sediments[6].

---

① 本文于2022年7月发表在 *Environmental Science and Pollution Research* 第29卷，http://doi.org/10.1007/s11356-022-22216-4。

However, global warming and anthropogenic activity have caused the degradation and loss of blue carbon ecosystems worldwide[10-13]. Approximately one-third of the blue carbon sinks have been lost in recent several decades[5]. Pervasive eutrophication[14], deforestation[15], aquaculture expansion[16,17], and land reclamation[18] were the main reasons for the significant decline in blue carbon sinks. Blue carbon sinks are vulnerable to climatic factors[19], and the degradation and loss of blue carbon sinks leads to the release of CO_2 into the atmosphere again[20]. Moreover, human society is threatened by a series of serious risks caused by climate change, such as sea level rise, flooding, and coastal erosion[21].

Many blue carbon sink studies have been conducted on the purpose of climate change mitigation in both the natural and social sciences[22-25]. Studies have reviewed the development of blue carbon science and the challenges in clarifying the scientific landscape of blue carbon sinks[5,26-28]. To gain deeper insights about blue carbon sinks, factors influencing them, such as biological invasions and habitat characteristics, have been discussed[15,29]. However, comprehensive reviews of the status and hotspots of blue carbon sink research are lacking. Bibliometric analysis is regarded as one of the most effective tools for quantitative and qualitative analyses of the academic literature[30] and can be used to explore the characteristics, structure, and academic collaborations in the field of scientific research[31]. Bibliometric analysis is also one of the most effective methods for assessing and predicting research trends in specific topics[32].

To understand the global trends and prospects of blue carbon sinks, a systematic review was conducted based on a bibliometric analysis. Scientific papers related to the blue carbon sink in the Web of Science Core Collection (WoSCC) database were chosen as raw data. Descriptive analysis was performed from the perspective of growth trends, geographical distribution of publications, active categories and journals, authorship, and highly cited references. Collaborative relationships among the most productive countries, institutions, and authors, were also identified. Basic characteristics of blue carbon sink related publications are analyzed, as well as the collaboration of big names. The research trends and future directions in the blue carbon sink field are explored by keyword co-occurrence analysis. The results will contribute valuable information for identifying key research questions and forecasting future trends of blue carbon sinks.

Methodology

▶ *Data source and search criteria*

WoSCC is one of the most widely used databases and includes a collection of thousands of journals covering more than 74.8 million scholarly data and 1.5 billion cited references across 254 categories[33]. In April 2021, this review searched the academic literature information in WoSCC from 1990 to 2020 using the following terms: Topics = (coast* or "tidal wetland*" or "mangrove*" or "salt marsh*" or "salt marsh*" or "saline wetland*" or seagrass* or "sea grass*") and ("carbon sink*" or "carbon sink*" or "carbon storage" or "sequester* carbon" or "carbon sequest*" or "blue carbon" or "carbon burial" or "carbon pool*"). Records from China included those from Hong Kong, Macao, and Taiwan. And the number of publications from the UK, represented as the total number of records from Wales, Scotland, Northern Ireland and England. A manual filter was then implemented based on the article's content, and papers unrelated to blue carbon sinks were deleted. In total, 1348 papers with the type of "article", "review" and "proceeding paper" were selected for analysis. A summary of the data is presented in Table S1, and the research framework was illustrated in Fig. 1.

▶ *Analysis methods*

Academic influence analysis

The academic influence of blue carbon sink studies was evaluated using the impact factor (IF), CiteScore, Source Normalized Impact per Paper

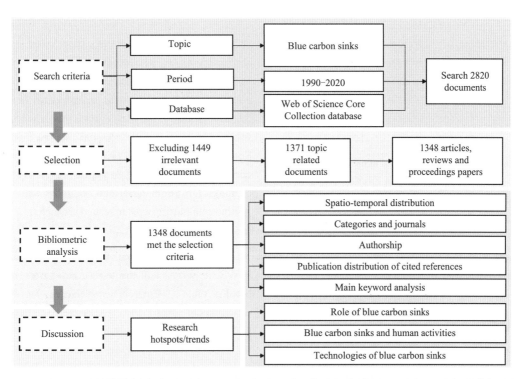

Fig. 1 Paper selection and flow chart of the research framework

(SNIP), Q ranking, h-index, citation count, and Altmetric Attention Score (AAS) indices. The IF, CiteScore, SNIP, and Q ranking are common evaluation indicators of the academic influence of a journal[34-38]. A journal's IF is calculated from the ratio of the total citation frequency to the total number of articles published in the previous two years. The h-index is defined as the number of publications co-authored by the researcher with at least h citations each[39], and it was selected to assess the academic influence of blue carbon sink research in terms of journals, countries, organizations, and authors. The IF and h-index indicators, which are calculated based on citation counts, were used to assess the influence of journals. The h-index is widely used in terms of countries, organizations, and authors. The citation count and AAS were chosen to measure the impact of highly cited articles in this study. Different degrees of attention from academia and the general public toward the same article, can be indicated by a comparison of the citation count and AAS value. The AAS is a weighted indicator representing the mentioned frequency of papers on social media platforms, such as Facebook, Twitter, research websites, and blogs[40-42]. The weights of the different sources depends on their potential to attract attention.

Hotspots analysis

Co-word analysis, a very well-established content analysis technique often used to reveal hotspots in scientific publications[43], was selected to identify popular topics of blue carbon sink research in this study. The extraction of keyword information and generation of data files used in the co-word analysis were performed using BibExcel. Different keyword expressions were manually combined before analysis. For example, the keywords of mangroves, mangrove forests, and mangrove were combined into mangrove. Further work with respect to visualization and cluster analysis, was conducted using VOSviewer software (version 1.6.16), which was also used to create an overlay map[44], in which the size of a node represents the frequency of keywords and the thickness of the line indicates the strength of the connection. Different nodes belonging to different hotspots were classified into different clusters using color as the dividing mark.

▶ *Evaluation metrics*

The methods of calculating the IF[45], CiteScore[35], SNIP[36], h-index[39], AAS[41], and Q ranking[34] have been reported in previous studies.

The IF, Q ranking, and citation count of the journals were obtained from the Journal Citation ReportsTM-2020, which is provided by the Web of Science. The CiteScore and SNIP of the top 10 most productive journals were obtained from the Scopus database. The h-index of journals, countries, institutions, and scholars was computed using document information downloaded from WoSCC. The AAS of the cited papers was obtained from https://www.scienceopen.com/ on the date June 27, 2021.

Results

▶ *Spatio-temporal distribution*

The growth trends and geographical distributions of the publications were analyzed, and the results are shown in Fig. 2. The number of articles related to blue carbon sinks increased exponentially from 1 in 1990 to 267 in 2020 (Fig. 2a). The evolution of papers can be divided into two stages. The first stage was from 1990 to 2009, during which the number of articles related to blue carbon sinks grew slowly, thus reflecting the fact that the blue carbon sink concept was in its infancy. The second stage was from 2010 to 2020, during which the number of articles significantly increased, thus indicating that the blue carbon sink concept was developing. The two stages were separated by the year 2009, when the term "blue carbon" was first introduced by United Nations Environment Programme (UNEP) [6]. A total of 1258 studies have been published in the last 10 years, and they account for 93.32% of the total number of blue carbon sink publications, which suggests a booming research trend in the field.

The geographical distribution of publications based on the information extracted from author addresses is illustrated in Fig. 2b. Blue carbon sink studies are widely distributed across all continents except Antarctica. The general characteristics of publications in these continents is displayed in Table S2. Asia published the most articles (694 articles), followed by North America (593) and Europe (560). South America and Africa ranked last among all continents, with 107 and 84 papers, respectively, although ten South American countries and 15 African countries engaged in related research.

▶ *Category and journal analysis*

Blue carbon sink related studies were published in 64 categories, and the top 10 categories are listed in Table S3. Environmental Sciences was the most popular category, with 590 papers. Ecology, and Marine and freshwater biology, occupied the second and third places with 315 and 274 papers, respectively. The details of the annual growth trends of the related categories are shown in Fig. 3a. Environmental Sciences, Ecology, Marine and Freshwater Biology, Oceanography, and Geosciences-Multidisciplinary were in the development stage from 1990 to 2009 and showed a significant increase after 2009. The development trend of blue carbon sink studies in Meteorology and Atmosphere Sciences has slowed in recent years. Scientific results for blue carbon sink research were published in 319 journals, and the top 10 most productive journals published 27.97% of the total publications (Table S4). *Estuarine, Coastal and Shelf Science* was the most productive journal, with 74 accepted papers, and it was also characterized by the highest h-index of 22. *Science of the Total Environment* and *PloS ONE* ranked second and third place, with 55 and 35 articles, respectively. *Global Change Biology* ranked 5th, although it had the highest IF value of 8.555. Most of the top 10 journals maintained a growing trend in the number of papers over time (Fig. 3b). Most of the top 10 productive journals had consistent rankings in terms of journal influence indicators. For example, *Global Change Biology* and *Science of the Total*

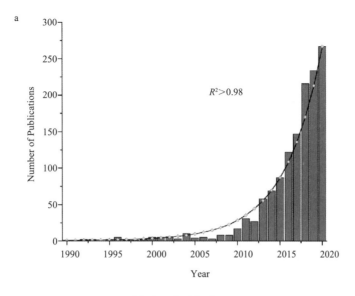

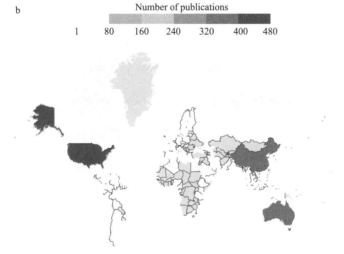

a. Temporal trends. b. Geographical distributions.

Fig. 2　Spatio-temporal distribution of blue carbon sink research

Environment, two Q1 level journals, ranked first and second in terms of IF and CiteScore, respectively. However, a few journals had large disparities in the rankings of the various indicators. *Frontiers in Marine Science* ranked third in terms of IF but had the seventh CiteScore value. This may be caused by the differences between WoS and Scopus in terms of the coverage of journals and categories. The Journal of Geophysical Research-Biogeosciences ranked fifth in terms of CiteScore but ninth in terms of SNIP. The difference in the categories may have led to the higher ranking of Journal of Geophysical Research-Biogeosciences in CiteScore because SNIP removes the effect of disciplinary differences in its calculations and can be used for cross-disciplinary comparisons.

▶ *Authorship analysis*

A total of 85 countries, 1538 institutions, and 4492 authors were involved in blue carbon sink research. Table S5 lists the basic information of the top 10 most prolific countries. The USA was

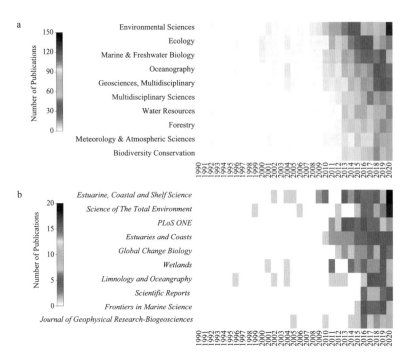

a. Annual growth trend of top 10 most productive categories. b. Annual growth trend of top 10 most productive journals.

Fig. 3　Category and journal analysis of blue carbon sink research from 1990 to 2020

the most productive country and had the highest h-index. In terms of the quantity of publications or h-index, Australia and China ranked second and third, respectively. European countries, such as Germany, Spain, and the UK, also performed well in blue carbon sink research. Developing countries, including China, Indonesia, and Brazil, have achieved relatively fruitful results in blue carbon sink research.

The basic conditions of the top 10 most productive institutions are listed in Table S6. The Chinese Academy of Sciences was the most productive institution, with 85 articles, followed by the University of Queensland (73) and Edith Cowan University (61). Except for the Chinese Academy of Sciences, all of the top 10 most productive institutions belong to developed countries. An obvious inconsistency in rank was found between publication quantity and h-index. For example, as the most productive institution, the Chinese Academy of Sciences ranked seventh in terms of

h-index. Developing countries should strengthen their abilities at the institutional level and focus on enhancing their academic influence.

As shown in Table S7, Duarte, C.M. from Saudi Arabia was the author with the highest yield. In terms of the h-index, Duarte simultaneously ranked first and was the only author with an h-index greater than 30. Australian authors also performed well, and six Australian researchers were involved among the top 10 authors. The h-index of Serrano was the highest among Australian researchers and ranked second among all authors.

As shown in Fig. 4, a close cooperative relationship was observed among the top 10 most productive countries, institutions, and authors. The USA, Edith Cowan University, and Duarte, C.M., were respectively the most active country, institution, and author in the top 10 entities in terms of publication quantity. An unconnected cooperative relationship was observed between high-yield developing countries, such as China,

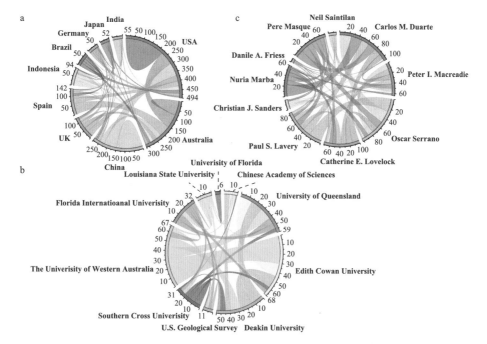

A chord diagram represents the connections between several entities (called nodes). Each entity is represented by a fragment on the outer part of the circular layout, and then arcs are drawn between each entities. The size of the arc is proportional to the importance of the connection. The numbers stand for the number of publication in terms of the cooperation of top 10 most productive countries, institutions and authors. The different colors were used to distinguish different countries.

Fig. 4 Collaboration of the top 10 (a) countries, (b) institutions, and (c) authors for blue carbon sink research

Indonesia, India, and Brazil. China, Indonesia, and Brazil all have the United States as their largest cooperative partners, and India's closest partner is Australia. A clear phenomenon of highly productive institutions belonging to developing countries was observed, such as the Chinese Academy of Science, which behaved inactive in international academic collaboration. Models of scientific cooperation between developing countries should be explored in the future. Further, more international cooperation should be conducted by institutions in developing countries to realize more achievements in the blue carbon sink field.

▶ *Citation analysis*

The top 10 most cited publications and the corresponding AAS are presented in Table S8. Citation count is a great indicator of academic concern, whereas AAS reflects social attention.

Mcleod et al.[5] was the most cited article, followed by Donato et al.[8] and Chmura et al.[7]. In terms of research content, the top 4 most cited publications involved a comprehensive review clarifying the importance of blue carbon sinks[5], and three articles discussed the carbon sink ability of mangroves[8], saltmarshes[7] and seagrasses[9]. Among the three articles related to carbon sink ability, Donato's article focusing on mangroves attracted the most attention from scholars. Based on the AAS, the publication of Fourqurean's seagrass ecosystem[9] attracted the most public attention, with 79% of the readers being the public and 19% of them being scientists. In summary, mangrove is the most concerned ecosystem in academia, although seagrass is the most focal ecosystem for the general public in the field.

The AAS ranking differed from that of the

citation count for the top 10 most cited papers (Table S8). For instance, Mcleod's paper ranked first in terms of citation count but only fifth in terms of AAS. The highest AAS was for Fourqurean's paper, which ranked fourth in terms of citation count. This phenomenon shows the different perspectives of the academic community and the general public on blue carbon sink research.

▶ *Keyword co-occurrence analysis*

As shown in Fig. 5, the keyword with the highest density was "blue carbon", which is at the center of the map. It is strongly linked to mangrove, saltmarsh and seagrass, which were widely thought to be blue carbon sinks in previous literature[5,7-9]. The percentages of articles related to mangrove, saltmarsh, and seagrass were 63.87%, 40.36%, and 40.65%, respectively. Mangrove carbon sink is the most popular topic. In addition, some emerging words, such as climate change, ecosystem service, and restoration, were also characterized by high density and closely centered on blue carbon. A two-stage process (Fig. S1) was used to display the keyword density from 1990 to 2010 and 2011 to 2020. A closer association of mangrove, saltmarsh, and seagrass can be found by comparing the density maps of the two phases.

As shown in Fig. 6a, most of the keywords appeared after 2015. This may reflect the emerging nature of the blue carbon sink field and its considerable development trend in recent years. A cluster visualization map of the keywords is illustrated in Fig. 6b. Five clusters were identified in the present study. Based on cluster 1, carbon cycling was a new research interest in the topic of saltmarsh carbon sinks. Cluster 2 mainly focused on mangrove carbon sinks, and some keywords related to human activities, such as land use and deforestation, were also involved. Cluster 4, which concentrates on technologies applied in blue carbon sink research, such as remote sensing (RS) and stable isotope, is also included in Cluster 2. This indicates that related technologies were mainly applied in mangrove carbon sink research. Related topics of The

The color of a node represents the quantity and weights of neighboring items. The larger the number of items in the neighborhood of a node and the higher the weights of the neighboring items, the closer the color of the point is to orange. The text size stand for the quantity and weights of neighboring items.

Fig. 5 Density visualization map of the co-occurrence keywords for blue carbon sink research from 1990 to 2020

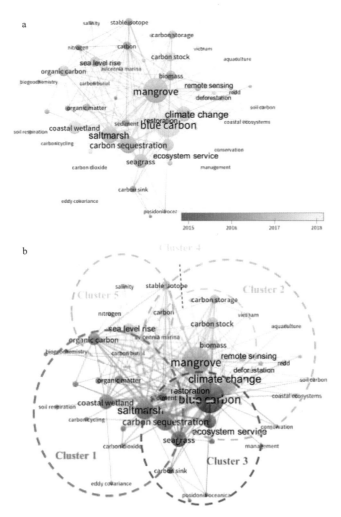

(a) Overlay visualization map of co-occurrence keywords. (b) Network visualization map of co-occurrence keywords. A node with greater yellow color in the Fig. 6a indicates that the keyword appeared later, while a node with more purple color indicates that the keyword appeared earlier. The frequency of keywords is indicated by the size of the node. The color of the nodes in Fig. 6b is used to distinguish different clusters.

Fig. 6 Keyword co-occurrence analysis for blue carbon sink research from 1990 to 2020

seagrass carbon sink were classified in Cluster 3. Primary production is an important form of carbon dioxide fixation in seagrass ecosystems[46]. Threats to seagrass habitats, such as eutrophication, and conservation and management, were also included in Cluster 3. Cluster 5 included the carbon stocks and climate change. Sea level rise, a significant disaster caused by climate change, was the core word in Cluster 5.

To clarify the research trends at specific time scales, this study divided the period 1990–2020 into the period 1990–2010 (Fig. S2) and the period 2011–2020 (Fig. S3). Strong continuity in the topics is observed for the two phases. Some topics, such as ecosystem service, carbon cycling, and stable isotope, exist in the keyword network of the two phases. However, richer topics emerged in the period 2011–2020 compared with the period 1990–2010. As the main reasons for the loss and degradation of blue carbon ecosystems, human activities, such as deforestation, aquaculture, and land use, have drawn great public interest in the period 2011–2020.

Because of the loss and degradation of blue carbon ecosystems, the conservation, restoration, and management of blue carbon sinks are emerging topics and future research directions.

Discussion

▶ *Growth trends of the publications*

The number of articles published for blue carbon sink research has shown exponential growth over the past several years (Fig. 2a). Before the term "blue carbon" was proposed forward by the UNEP, 65 publications for blue carbon sink research had been published in WoSCC, which indicates that the value of blue carbon sinks has received attention[47-49]. The number of blue carbon sink related papers has increased significantly since 2009. The approval of authoritative international organizations, and the concerns and collaboration of multidisciplinary scholars, may be the main driving forces for this research[5,50]. This phenomenon is also related to the support of related projects[51] and the attention to climate change related disasters[52]. In addition, blue carbon sink related technologies, such as remote sensing and stable isotope, are becoming increasingly mature, which is essential and necessary to conduct more in-depth research.

▶ *Geographical distribution of publications*

Climate change is a global issue and an important reason for the popularity of blue carbon sink research worldwide (Fig. 2b). However, the regional characteristics could still be identified. At the continental level, Asia ranked first in terms of publication quantity, followed by North America and Europe. The distribution of blue carbon sinks may be an important reason for these regional features (Fig. 7). Previous studies have suggested that blue carbon sinks are widely distributed in North America, Asia, and Europe[53-55], which have the largest proportion of seagrasses (69.6%), mangroves (28.6%) and saltmarshes (46.2%), respectively[28]. For blue carbon sink-rich areas, more studies are needed to promote adaptation to climate change[56,57].

The geographical distribution of publications in the field can also be explained by the level of scientific research expenditures. As shown in Fig. 8, North America, Western Europe, East Asia, and the Pacific, were characterized by a higher level than the world average in research and development expenditure[60], and they also represented the most productive areas in blue carbon sink research.

▶ *Category, published source, and authorship analysis*

The review identified 1348 papers belonging to 64 categories, indicating that blue carbon sink research is a popular multidisciplinary topic. Blue carbon science has a broad scope because it seeks to explore all potential opportunities for climate change mitigation and adaptation in coastal ecosystems[61]. However, more than half of the categories contained fewer than 10 articles. An obvious imbalance in the distribution of blue carbon sinks was identified in terms of category. Environmental Sciences, Ecology, and Marine and Freshwater Biology, were the top three categories of blue carbon sinks (Table S3). Studies on blue carbon sinks in economics and management are also urgently needed but still in their infancy[3].

A total of 319 journals have published blue carbon sink related papers, which indicate a wide interest in blue carbon sinks at the journal level. This phenomenon is positive and indicates that articles with the theme of blue carbon sink fit the subjects of a large number of journals. The broad acceptance of blue carbon sink related articles could promote the spread of scientific results in the field. However, specific journals that publish articles related to blue carbon sinks are lacking.

Although 319 journals were involved in the field, 158 journals only published one paper. This suggests that only a few journals have focused on the progress of blue carbon sink research. Global Change Biology, as the one with the highest IF in the top 10 most productive journals, ranked second and fifth in terms of *h*-index and publication number, respectively (Table S4). This phenomenon could be caused by the perception of authors that articles

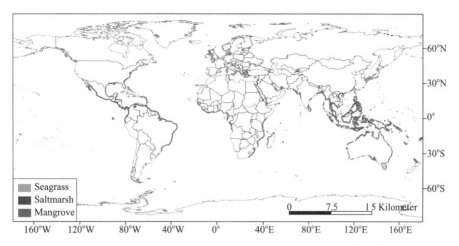

Fig. 7　Global distribution of blue carbon ecosystems[54,58,59]

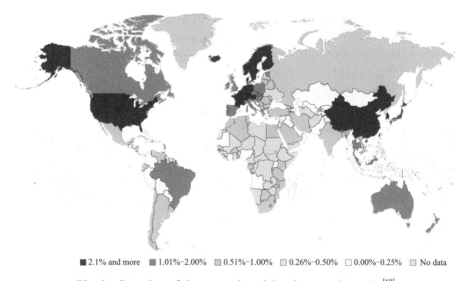

Fig. 8　Snapshot of the research and development intensity[60]

published in journals with relatively high IF have high quality or prestige[62].

Authors with a high h-index play an important role in drawing popular awareness to blue carbon sinks. For example, Carlos M. Duarte, who ranked first in the list of the top 10 most productive authors (Table S7), is also a co-author of the blue carbon report[6], which is considered as the first formal introduction of "blue carbon" to the public. There are several possible reasons for the importance of authors with a high h-index. Most scholars with a high h-index in the blue carbon sink field, such as the top 10 most productive authors listed in the paper, have global-level research experience. These authors should play leading roles in the global conservation of blue carbon ecosystems[63,64,98]. Scientists with a high h-index can more easily obtain support funding, which is essential to obtaining excellent research results, thus affecting the popularity of certain research directions[65].

Both developed and developing countries exhibited strong enthusiasm for blue carbon research (Table S5), which is related to the abundant blue carbon resources in these countries. Indonesia and

Brazil ranked first and second in terms of mangrove area in the world[66], and China is also rich in blue carbon resources[67]. In general, a relatively close academic collaboration in blue carbon research (Fig. 4) has been formed among high-yielding authorships. Academic collaboration is regarded as an important factor in promoting research progress[68].

▶ *Main keyword analysis*

Role of blue carbon sinks in climate change mitigation

The "blue carbon" concept was introduced as a metaphor to highlight that vegetated coastal ecosystems contribute significantly to carbon sequestration[5]. The close connection of "blue carbon" and "climate change" has been identified (Fig. 6b) in this study. Blue carbon sinks have been regarded as an effective nature-based solution for adapting to climate change[3,6]. Blue carbon sinks, including saltmarshes (Cluster 1), mangroves (Cluster 2) and seagrasses (Cluster 3), have attracted the attention of scholars. High primary productivity and a long carbon cycle are significant features of blue carbon sinks[5,28,69]. Because of their strong capacity and huge value, blue carbon sinks have been incorporated into climate mitigation strategies by international societies and some nations[61]. Strategies involving blue carbon sinks for climate change mitigation are most effective at the national scale[57].

Blue carbon sinks and human activities

The blue carbon sinks were affected by direct human activities (such as land reclamation, urban expansion, and deforestation) and indirect human activities (such as climate change) (Fig. 6b). Human activities have led to a worldwide loss of blue carbon sinks. These processes have not only caused a reduction in the carbon sink capacity of mangroves, seagrasses, and saltmarshes[10,11] but also led to an increase in carbon emissions[70].

The threats to blue carbon sinks are complicated and distinctive[28]. Mangroves are heavily affected by mangrove disturbances and conversion, seagrass ecosystems are affected by diverse factors, such as eutrophication, fisheries, and harbor activities[71,72], and saltmarshes are generally affected by reclamation, biological invasion, and sea level rise[73-75]. In addition, 3363 km^2 (2.1%) of the global mangrove area was lost between 2000 and 2016, with 62% caused by human activities[76]. The global loss of seagrass area between 1990 and the early 2000s was estimated to reach 7% per year[76], and 20%–50% of saltmarshes are predicted to be either lost or degraded during the current century[77,78]. Anthropic activity is widely regarded as one of the main reasons for the loss and degradation of blue carbon ecosystems[76-79,]; thus, activities focused on restoring, conserving, and managing blue carbon sinks is required.

The restoration, conservation, and management of blue carbon sinks has drawn the interest of researchers in recent years (Cluster 3 and Cluster 4) and has become a greater priority in marine management[80]. For example, considerable financial[81] and ecological efforts[82,83] have been directed towards restoring blue carbon sinks. A previous study indicated that appreciating interconnectivity between habitats is key to the management of blue carbon sinks[84], while another study indicated that payments for ecosystem services might contribute to protecting blue carbon sinks[85].

Ten fundamental questions regarding the development of blue carbon science have been listed in "The Future of Blue Carbon Science"[3]. The research hotspots summarized here were consistent with the questions listed in this article. According to our results, more fruitful topics are associated with natural science than social science in the blue carbon sink field (Fig. 6), which was also revealed by questions in Macreadie's paper[3]. The global extent and temporal distribution of blue carbon ecosystems have been widely studied, thus revealing the global loss and degradation of blue carbon ecosystems[76-79]. Factors that influence blue carbon sinks, such as human activities and climate change, as mentioned in Macreadie's paper[3], were also popular topics identified in our study (Fig. 6 and Fig. S3).

Development of blue carbon sink related technologies

RS and stable isotope are the two popular technologies in published papers on blue carbon sinks (Fig. 6b). RS is a useful technology for ecosystem change detection and has also been widely used in mapping and monitoring blue carbon ecosystems[86]. Furthermore, it has been applied to estimate the biomass and carbon stocks of blue carbon ecosystems in recent years[87-89]. The development of RS technologies, such as synthetic aperture radar and unmanned aerial vehicles, has promoted the progress of studies in blue carbon sinks[86,90]. A popular future direction for RS technologies applied in blue carbon sink research is to combine them with machine learning technologies[89]. With the gradual increase in knowledge of blue carbon sinks, more researchers have focused on the source of organic carbon sequestrated in blue carbon ecosystems. Stable isotope was the most frequently used tool in blue carbon studies to track the provenance and fate of organic carbon over the past 30 years[91]. Studies have revealed the key roles of saltmarshes, mangroves, and seagrasses[64] in carbon sequestration based on carbon isotopic analysis. New technologies, such as biomarkers, molecular properties, and environmental DNA, have also been developed to identify the source and quantities of different primary producers[91].

RS and stable isotope techniques have been applied to map and quantify blue carbon sinks over the last 30 years[92,93]. RS has been widely used to monitor blue carbon ecosystems[86] and provided information on the extent of global blue carbon ecosystems, which is essential in the blue carbon sink field[3]. In addition, RS has advantages in aboveground biomass modelling[94] and soil carbon estimation[95]. Stable isotope has been used to the identify carbon sources because of their special advantages in revealing carbon flow and storage[96].

Both RS and stable isotope are important techniques for quantifying blue carbon sinks, and the integration of these two technologies has become an emerging trend[92,93].

Differences are observed between RS and stable isotope. Owing to the wide coverage, easy availability, and low cost of RS images, RS technology is more often used in large-scale studies[89]. With the development of image accuracy, blue carbon estimation using RS technology has gradually improved. However, stable isotope is advantageous for identifying carbon sources and weighting the contribution[97]. Several studies have combined RS and the stable isotope to provide more information on blue carbon estimation[92,93].

▶ *Research gaps*

A review of the 1348 blue carbon sink related papers in the current work identified several gaps in the available literature. First, the economic valuation and management of blue carbon sinks, which are emerging concepts, are still in their infancy[3] (Fig. 6a). Second, the results of blue carbon content assessments are still controversial, although more scientists have studied the blue carbon content at different scales and obtained relevant results. Improvements in data accuracy and simulation models are necessary for global blue carbon estimation in the future. Furthermore, the use of RS and stable isotope in the field of blue carbon sinks is evolving, and more work needs to be combined with new technologies, such as machine learning. Future studies should focus on validating the effectiveness of these technologies using biomarkers, molecular properties, and environmental DNA. Finally, there is more space for promotion regarding scientific collaboration between inter-institution/country/author.

Conclusion

In recent years, blue carbon sinks have attracted the attention of researchers and the public owing to the threat of global climate change. The purpose of this study was to reflect the research status of the blue carbon sink field, identify key topics, and establish future directions for research through bibliometric analysis. The number of studies on blue carbon sinks has increased significantly with the growing trend in international interdisciplinary cooperation. Relatively close cooperation exists between developed and developing countries and among developed countries in the blue carbon sink field. In the future, cooperation among developing countries should be promoted. Big-name journals and influential scholars have played important roles in advancing research on blue carbon sinks. RS and stable isotope have been widely used to map and quantify blue carbon sinks. The integration of RS and stable isotope, a new attempt, has been used in studies. In addition, multidisciplinary technologies, such as machine learning, should be more widely applied to promote the ability of the RS and stable isotope in improving the accuracy of blue carbon sink estimation.

Many challenges remain to be solved, such as improving the development of blue carbon sink research and mitigating climate change. More basic research and appropriate frameworks are indispensable for increasing the accuracy and credibility of the assessment results, and universal methods and standards are imperative for the scientific evaluation of carbon sink capacity and economic value. Blue carbon ecosystems are still threatened globally; therefore, restoration, conservation, and management of blue carbon ecosystems should focus on improving the status of blue carbon sinks and their carbon sequestration capacity. Meanwhile, international campaigns to reduce emissions from deforestation and prevent forest degradation, among others, should make more efforts to protect blue carbon ecosystems. Several new technologies, such as biomarkers and environmental DNA, have emerged for organic carbon tracking. Therefore, researchers should pay more attention to validating these methods..

Funding

This work is supported by Major National Social Science Project Fund-Compensation Standard and System Design of Marine Ecological Damage (No. 16ZDA049).

Author Contribution

Lu Jiang: Data curation, formal analysis, investigation, methodology, writing(original draft). Tang yang: Visualization, writing(review and editing). Jing Yu: Conceptualization, Funding acquisition, methodology, project administration, supervision.

Conflict of Interest

The authors declare no competing interests.

References

1. IPCC. Climate Change 2007: Sysnthesis Report[R]. 2007. Available at: https://www.ipcc.ch/report/ar4/syr/.
2. Solomon S, Plattner G K, Knutti R, et al. Irreversible climate change due to carbon dioxide emissions[J]. Proc Natl Acad Sci, 2009, 106(6): 1704-1709.
3. Macreadie P I, Anton A, Raven J A, et al. The future of Blue Carbon science[J]. Nat Commun, 2019, 10: 3998.
4. Atwood T B, Connolly R M, Almahasheer H, et al. Global patterns in mangrove soil carbon stocks and losses[J]. Nat Clim Change, 2017, 7(7): 523-528.
5. Mcleod E, Chmura G L, Bouillon S, et al. A blueprint for blue carbon: Toward an improved understanding of the role of vegetated coastal habitats in sequestering CO_2[J]. Front Ecol Environ, 2011, 9(10): 552-560.
6. Nellemann C, Corcoran E, Duarte C M, et al. Blue Carbon. A Rapid Response Assessment[R/OL]. United Nations Environment Programme, GRID-Arendal, 2009. https://www.grida.no.
7. Chmura G L, Anisfeld S C, Cahoon D R, et al. Global carbon sequestration in tidal, saline wetland soils[J]. Global Biogeochem Cy, 2003, 17(4): 1111.
8. Donato D C, Kauffman J B, Murdiyarso D, et al. Mangroves among the most carbon-rich forests in the tropics[J]. Nat Geosci, 2011, 4(5): 293-297.
9. Fourqurean J W, Duarte C M, Kennedy H, et al. Seagrass ecosystems as a globally significant carbon stock[J]. Nat Geosci, 2012, 5(7): 505-509.
10. IPCC. Global Warming of 1.5 °C[R]. 2018. Available at: https://www.ipcc.ch/sr15/.
11. Siikamaki J, Sanchirico J N, Jardine S L. Global economic potential for reducing carbon dioxide emissions from mangrove loss[J]. Proc Natl Acad Sci,

2012, 109(36): 14369-14374.
12. Svensson C J, Hyndes G A, Lavery P S. Food web analysis in two permanently open temperate estuaries: Consequences of saltmarsh loss?[J]. Mar Environ Res, 2007, 64(3): 286-304.
13. Waycott M, Duarte C M, Carruthers T J B, et al. Accelerating loss of seagrasses across the globe threatens coastal ecosystems[J]. Proc Natl Acad Sci, 2009, 106(30): 12377-12381.
14. Jiang Z, Liu S, Zhang J, et al. Eutrophication indirectly reduced carbon sequestration in a tropical seagrass bed[J]. Plant Soil, 2018, 426(1-2): 135-152.
15. Davidson I C, Cott G M, Devaney J L, et al. Differential effects of biological invasions on coastal blue carbon: A global review and meta-analysis[J]. Global Change Biol, 2018, 24(11): 5218-5230.
16. Ahmed N, Glaser M. Coastal aquaculture, mangrove deforestation and blue carbon emissions: Is REDD plus a solution?[J]. Mar Policy, 2016, 66: 58-66.
17. Apostolaki E T, Holmer M, Marba N, et al. Reduced carbon sequestration in a Mediterranean seagrass (*Posidonia oceanica*) ecosystem impacted by fish farming[J]. Aquacult Env Interac, 2011, 2(1): 49-59.
18. Wu J, Zhang H, Pan Y, et al. Opportunities for blue carbon strategies in China[J]. Ocean Coast Manage, 2020, 194: 105241.
19. Lovelock C E, Reef R. Variable impacts of climate change on blue carbon[J]. One Earth, 2020, 3(2): 195-211.
20. Pendleton L, Donato D C, Murray B C, et al. Estimating global "Blue Carbon" emissions from conversion and degradation of vegetated coastal ecosystems[J]. PloS ONE, 2012, 7(9): e43542.
21. Nicholls R J, Cazenave A. Sea-level rise and its impact on coastal zones[J]. Science, 2010, 328(5985): 1517-1520.
22. Friess D A, Phelps J, Garmendia E, et al. Payments for Ecosystem Services (PES) in the face of external biophysical stressors[J]. Global Environ Change, 2015, 30: 31-42.
23. Macreadie P I, Nielsen D A, Kelleway J J, et al. Can we manage coastal ecosystems to sequester more blue carbon?[J]. Front Ecol Environ, 2017, 15(4): 206-213.
24. Moritsch M M, Young M, Carnell P, et al. Estimating blue carbon sequestration under coastal management scenarios[J]. Sci Total Environ, 2021, 777: 145962.
25. Schile L M, Kauffman J B, Crooks S, et al. Limits on carbon sequestration in arid blue carbon ecosystems[J]. Ecol Appl, 2017, 27(3): 859-874.
26. Alongi D M. Carbon sequestration in mangrove forests[J]. Carbon Manag, 2012, 3(3): 313-322.
27. Duarte C M, Krause-Jensen D. Export from seagrass meadows contributes to marine carbon sequestration[J]. Front Mar Sci, 2017, 4: 13.
28. Himes-Cornell A, Pendleton L, Atiyah P. Valuing ecosystem services from blue forests: A systematic review of the valuation of salt marshes, sea grass beds and mangrove forests[J]. Ecosyst Serv, 2018, 30: 36-48.
29. Mazarrasa I, Sarnper-Villarreal J, Serrano O, et al. Habitat characteristics provide insights of carbon storage in seagrass meadows[J]. Mar Pollut Bull, 2018, 134: 106-117.
30. Geissdoerfer M, Savaget P, Bocken N M P, et al. The circular economy a new sustainability paradigm?[J]. J Clean Prod, 2017, 143: 757-768.
31. Mao G, Huang N, Chen L, et al. Research on biomass energy and environment from the past to the future: A bibliometric analysis[J]. Sci Total Environ, 2018, 635: 1081-1090.
32. Zhang D, Zhang Z, Managi S. A bibliometric analysis on green finance: Current status, development, and future directions[J]. Financ Res Lett, 2019, 29: 425-430.
33. Singh V K, Singh P, Karmakar M, et al. The journal coverage of Web of Science, Scopus and Dimensions: A comparative analysis[J]. Scientometrics, 2021, 126(6): 5113-5142.
34. Asadi H, Mostafavi E. The productivity and characteristics of *Iranian Biomedical Journal* (IBJ): A scientometric analysis[J]. Iranian Biomedical Journal, 2018, 22(6): 362-366.
35. Atayero A A, Popoola S I, Egeonu J, et al. Citation analytics: data exploration and comparative analyses of CiteScores of open access and subscription-based publications indexed in Scopus (2014-2016)[J]. Data in Brief, 2018, 19: 198–213.
36. ELSEVIER. Source-Normalized Impact per Paper (SNIP)[Z/OL]. 2018. https://journalinsights.elsevier.com/journals/0969-806X/snip.
37. Garfield E. Citation indexes for science; a new dimension in documentation through association of ideas[J]. Science, 1955, 122(3159): 108-11.
38. Garfield E. The history and meaning of the journal impact factor[J]. Jama-J Am Med Assoc, 2006, 295(1): 90-93.
39. Hirsch J E. An index to quantify an individual's scientific research output[J]. Proc Natl Acad Sci, 2005, 102(46): 16569-16572.
40. Batooli Z, Mohamadloo A, Nadi-Ravandi S. Relationship between altmetric and bibliometric indicators across academic social sites in article-level:

the case of Iranian researchers' "Top Papers" in clinical medicine[J]. Library Hi Tech, 2021.
41. Dagar A, Falcone T. Altmetric scores analysis reveals a high demand for psychiatry research on social media[J]. Psychiatry Res, 2021, 298: 113782.
42. Thelwall W, Haustein S, Lariviere V, et al. Do altmetrics work? Twitter and ten other social web services[J]. PloS ONE, 2013, 8(5): e64841.
43. Callon M, Courtial J P, Turner W A, et al. From translations to problematic networks: An introduction to co-word analysis[J]. Soc Sci Inf, 1983, 22(2): 191-235.
44. van Eck N J, Waltman L. Software survey: VOSviewer, a computer program for bibliometric mapping[J]. Scientometrics, 2010, 84(2): 523-538.
45. Kumar V, Upadhyay S, Medhi B. Impact of the impact factor in biomedical research: Its use and misuse[J]. Singap Med J, 2009, 50(8): 752–755.
46. Duarte C M, Kennedy H, Marba N, et al. Assessing the capacity of seagrass meadows for carbon burial: Current limitations and future strategies[J]. Ocean Coast Manage, 2013, 83: 32-38.
47. Choi Y, Wang Y, Hsieh Y P, et al. Vegetation succession and carbon sequestration in a coastal wetland in northwest Florida: Evidence from carbon isotopes[J]. Global Biogeochem Cy, 2001, 15(2): 311-319.
48. Kaldy J E, Onuf C P, Eldridge P M, et al. Carbon budget for a subtropical seagrass dominated coastal lagoon: How important are seagrasses to total ecosystem net primary production?[J]. Estuaries, 2002, 25(4A): 528-539.
49. Twilley R R, Chen R, Hargis T. Carbon sinks in mangroves and their implications to carbon budget of tropical coastal ecosystems[J]. Water Air Soil Pollut, 1992, 64(1-2): 265-288.
50. Herr D, von Unger M, Laffoley D, et al. Pathways for implementation of blue carbon initiatives[J]. Aquat Conserv, 2017, 27: 116-129.
51. Wylie L, Sutton-Grier A E, Moore A. Keys to successful blue carbon projects: Lessons learned from global case studies[J]. Mar Policy, 2016, 65: 76-84.
52. Arnaud M, Baird A J, Morris P J, et al. Sensitivity of mangrove soil organic matter decay to warming and sea level change[J]. Global Ecol Biogeogr, 2020, 26(3): 1899-1907.
53. Giri C, Ochieng E, Tieszen L L, et al. Status and distribution of mangrove forests of the world using earth observation satellite data[J]. Global Ecol Biogeogr, 2011, 20(1): 154-159.
54. Mcowen C, Weatherdon L V, Bochove J, et al. A global map of saltmarshes (v6.1)[J]. Biodiver Data J, 2017, 5:e11764.
55. McKenzie L J, Nordlund L M, Jones B L, et al. The global distribution of seagrass meadows[J]. Environ Res Lett, 2020, 15(7): 074041.
56. Murdiyarso D, Purbopuspito J, Kauffman J B, et al. The potential of Indonesian mangrove forests for global climate change mitigation[J]. Nat Clim Change, 2015, 5(12): 1089-1092.
57. Taillardat P, Friess D A, Lupascu M. Mangrove blue carbon strategies for climate change mitigation are most effective at the national scale[J]. Biol Letters, 2018, 14(10): 20180251.
58. Bunting P, Rosenqvist A, Lucas R, et al. The Global Mangrove Watch: a New 2010 global baseline of mangrove extent[J]. Remote Sens., 2018, 10(10): 1669.
59. UNEP-WCMC Short F T. Global distribution of seagrasses (version 7.1). Seventh update to the data layer used in Green and Short (2003)[Z/OL]. Cambridge (UK), UN Environment World Conservation Monitoring Centre, 2021. https://doi.org/10.34892/x6r3-d211.
60. UNESCO. Global Investments in R&D[R/OL]. 2020. Available at: http://uis.unesco.org/sites/default/files/documents/fs59-global-investments-rd-2020-en.pdf.
61. Lovelock C E, Duarte C M. Dimensions of blue carbon and emerging perspectives[J]. Biol Letters, 2019, 15(3): 20180781.
62. Glanzel W, Moed H F. Journal impact measures in bibliometric research[J]. Scientometrics, 2002, 53(2): 171-193.
63. Lavery P S, Mateo M A, Serrano O, et al. Variability in the carbon storage of seagrass habitats and its implications for global estimates of blue carbon ecosystem service[J]. PLoS ONE, 2013, 8(9): e73748.
64. Kennedy H, Beggins J, Duarte C M, et al. Seagrass sediments as a global carbon sink: Isotopic constraints[J]. Global Biogeochem Cy, 2010, 24: GB4026.
65. Venable G T, Khan N R, Taylor D R, et al. A correlation between national institutes of health funding and bibliometrics in neurosurgery[J]. World Neurosurg, 2014, 81(3-4): 468-472.
66. Hamilton S E, Casey D. Creation of a high spatio-temporal resolution global database of continuous mangrove forest cover for the 21st century (CGMFC-21) [J]. Global Ecol Biogeogr, 2016, 25(6): 729-738.
67. Gao Y, Yu G, Yang T, et al. New insight into global blue carbon estimation under human activity in land-sea interaction area: A case study of China[J]. Earth-Sci Rev, 2016, 159: 36-46.

68. Song R, Xu H, Cai L. Academic collaboration in entrepreneurship research from 2009 to 2018: A multilevel collaboration network analysis[J]. Sustainability, 2009, 11(19): 5172.
69. Luisetti T, Jackson E L, Turner R K. Valuing the European 'coastal blue carbon' storage benefit[J]. Mar Pollut Bull, 2013, 71(1-2): 101-106.
70. Hamilton S E, Friess D A. Global carbon sinks and potential emissions due to mangrove deforestation from 2000 to 2012[J]. Nat Clim Chang, 2018, 8(3): 240-244.
71. Burkholder J M, Tomasko D A, Touchette B W. Seagrasses and eutrophication[J]. J Exp Mar Biol Ecol, 2007, 350(1-2): 46-72.
72. Nordlund L M, Gullstrom M. Biodiversity loss in seagrass meadows due to local invertebrate fisheries and harbour activities[J]. Estuar Coast Shelf S, 2013, 135: 231-240.
73. Bu N S, Qu J F, Li G, et al. Reclamation of coastal salt marshes promoted carbon loss from previously-sequestered soil carbon pool[J]. Ecol Eng, 2015, 81: 335-339.
74. Yuan J J, Ding W X, Liu D Y, et al. Exotic Spartina alterniflora invasion alters ecosystem-atmosphere exchange of CH_4 and N_2O and carbon sequestration in a coastal salt marsh in China[J]. Global Change Biol, 2015, 21(4): 1567-1580.
75. Ruiz-Fernandez A C, Carnero-Bravo V, Sanchez-Cabeza J A, et al. Carbon burial and storage in tropical salt marshes under the influence of sea level rise[J]. Sci Total Environ, 2018, 630: 1628-1640.
76. Salinas C, Duarte C M, Lavery P S, et al. Seagrass losses since mid-20th century fuelled CO_2 emissions from soil carbon stocks[J]. Global Change Biol, 2020, 26(9): 4772-4784.
77. Barbier E B, Hacker S D, Kennedy C, et al. The value of estuarine and coastal ecosystem services[J]. Ecol Monog, 2011, 81: 169-193.
78. Kirwan M L, Megonigal J P. Tidal wetland stability in the face of human impacts and sea-level rise[J]. Nature, 2013, 504: 53-60.
79. Goldberg L, Lagomasino D, Thomas N, et al. Global declines in human-driven mangrove loss[J]. Global Change Biol, 2020, 26(10): 5844-5855.
80. Howard J, Mcleod E, Thomas S, et al. The potential to integrate blue carbon into MPA design and management[J]. Aquat Conserv, 2017, 27: 100-115.
81. Vanderklift M A. Marcos-Martinez R, Butler J R A, et al. Constraints and opportunities for market-based finance for the restoration and protection of blue carbon ecosystems[J]. Mar Policy, 2019, 107: 103429.
82. Tang J, Ye S, Chen X, et al. Coastal blue carbon: Concept, study method, and the application to ecological restoration[J]. Sci China Earth Sci, 2018, 61(6): 637-646.
83. Thorhaug A, Poulos H M, Lopez-Portillo J, et al. Seagrass blue carbon dynamics in the Gulf of Mexico: Stocks, losses from anthropogenic disturbance, and gains through seagrass restoration[J]. Sci Total Environ, 2017, 605: 626-636.
84. Smale D A, Moore P J, Queiros A M, et al. Appreciating interconnectivity between habitats is key to blue carbon management[J]. Front Ecol Environ, 2018, 16(2): 71-73.
85. Locatelli T, Binet T, Kairo J G, et al. Turning the tide: How blue carbon and payments for ecosystem services (PES) might help save mangrove forests[J]. Ambio, 2014, 43(8): 981-995.
86. Pham T D, Xia J S, Ha N T, et al. A review of remote sensing approaches for monitoring blue carbon ecosystems: Mangroves, seagrasses and salt marshes during 2010-2018[J]. Sensors, 2019, 19(8): 1933.
87. Doughty C L, Ambrose R F, Okin G S, et al. Characterizing spatial variability in coastal wetland biomass across multiple scales using UAV and satellite imagery[J]. Remote Sens Ecol Con, 2021, 7(3): 411-429.
88. Lyons M, Roelfsema C, Kovacs E, et al. Rapid monitoring of seagrass biomass using a simple linear modelling approach, in the field and from space[J]. Mar Ecol Prog Ser, 2015, 530: 1-14.
89. Pham T D, Yokoya N, Bui D T, et al. Remote sensing approaches for monitoring mangrove species, structure, and biomass: Opportunities and challenges[J]. Remote Sens, 2019, 11(3): 230.
90. Jones A R, Segaran R R, Clarke K D, et al. Estimating mangrove tree biomass and carbon content: A comparison of forest inventory techniques and drone imagery[J]. Front Mar Sci, 2020, 6: 784.
91. Geraldi N R, Ortega A, Serrano O, et al. Fingerprinting blue ccarbon: Rationale and tools to determine the source of organic carbon in marine depositional environments[J]. Front Mar Sci, 2019, 6: 263.
92. Stankovic M, Hayashizaki K I, Tuntiprapas P, et al. Two decades of seagrass area change: Organic carbon sources and stock[J]. Mar Pollut Bull, 2021, 163: 111913.
93. Watson E B, Corona A H. Assessment of blue carbon storage by Baja California (Mexico) tidal wetlands and evidence for wetland stability in the face of anthropogenic and climatic impacts[J]. Sensors, 2018, 18(1): 32.
94. Wicaksono P, Danoedoro P, Hartono Nehren U.

Mangrove biomass carbon stock mapping of the Karimunjawa Islands using multispectral remote sensing[J]. Int J Remote Sens, 2016, 37(1): 26-52.

95. Le N N, Pham T D, Yokoya N, et al. Learning from multimodal and multisensor earth observation dataset for improving estimates of mangrove soil organic carbon in Vietnam[J]. Int J Remote Sens, 2021, 42(18): 6866-6890.

96. Kelleway J J, Trevathan-Tackett S M, Baldock J, et al. Plant litter composition and stable isotope signatures vary during decomposition in blue carbon ecosystems[J]. Biogeochemistry, 2022, 158(2): 147-165.

97. Chen G C, Azkab M H, Chmura G L, et al. Mangroves as a major source of soil carbon storage in adjacent seagrass meadows[J]. Sci Rep, 2017, 7: 42406.

98. Atwood T B, Connolly R M, Ritchie E G, et al. Predators help protect carbon sinks in blue carbon ecosystems[J]. Nat Clim Chang, 2015, 5(12): 1038-1045.

主编简介

李建平

中国海洋大学"筑峰工程"第一层次特聘教授,未来海洋学院院长,深海圈层与地球系统教育部前沿科学中心主任委员会副主任,海洋碳中和中心主任,青岛海洋科学与技术试点国家实验室"鳌山人才"卓越科学家,国家重点基础研究发展计划(973计划)和国家重大研究计划项目首席科学家,美国夏威夷大学兼职教授,国际大地测量学和地球物理学联合会(IUGG)会士,英国皇家气象学会(RMetS)会士,国际气候与环境变化委员会(CCEC)主席,国际气候学委员会(ICCL)主席,东亚气候国际计划共同主席,新世纪"百千万人才工程"国家级人选,*Climate Dynamics* 执行主编,*Advances in Climate Change Research* 副主编,*Theoretical and Applied Climatology*、*Scientific Reports*、*Advances in Atmospheric Sciences* 等期刊编委。

长期从事气候学基础理论与应用研究,主要研究方向为气候动力学与可预报性、东亚季风动力学及其预测、海-陆-气相互作用、环状模及其影响等,在这些领域取得了具有国际影响的系统性创新成果,并成功用于业务预测,对气候动力学的发展和推动业务预测做出了重要贡献。在气候动力学和可预报性方面:建立了非定常外强迫下气候的全局分析理论;创建了定量度量可预报性期限的非线性局部李雅普诺夫指数(NLLE)及其向量谱(NLLV)理论、可预报性期限归因的条件非线性局部李雅普诺夫指数(CNLLE)方法;提出了最优数值积分的逐步调整法。在东亚季风变化机理和预测方面:从中高低纬相互作用和海-陆-气相互作用动力学角度,系统揭示了环状模与热带协同影响中国气候的机理;系统建立了非均匀基流行星波传播新理论,提出了扰动位能(PPE)新理论并应用于季风与海气相互作用研究;提出了有物理基础的东亚季风预测模型,成功应用于业务预测,推动和深化了东亚季风动力学研究与业务应用,研究成果被中国气象局、美国国家大气海洋局(NOAA)等业务部门使用,为业务预测提供了重要参考。在海-气相互作用方面:提出了海-气耦合桥理论、年代际海-气相互作用的延迟振子理论,阐明了其在全球及区域年代际变化中的作用,揭示了北大西洋多年代际振荡(AMO)对热带西太平洋海温年代际变化、亚洲气候年代际变化的影响机制,以及大西洋厄尔尼诺的热动力学调控机制。

获国家杰出青年科学基金资助,获国务院政府特殊津贴,获首届全国优秀博士学位论文、国家自然科学奖二等奖(排名第3)等荣誉。入选路透社"气候变化研究领域全球最具影响力的1000位科学家"(2021年)、爱思唯尔"中国高被引作者"(2020、2021、2022年)、美国斯坦福大学"全球前2%顶尖科学家榜单"(2019、2020、2021年)、"全球顶尖前10万科学家排名"(2021年)、"全球最佳地球科学家"(2022、2023年)等。已发表论文500余篇(包括SCI论文400余篇),编译著作10余部,在Google Scholar上被引用19 000余次。在国际会议做特邀报告70余次,组织国际会议及分会80余次。

主要作者简介[①]

许博超
教授，博士生导师

2006年本科毕业于中国海洋大学海洋化学专业，2008—2009年参与美国佛罗里达州立大学博士联合培养项目，2011年获中国海洋大学海洋化学专业博士学位。现任职于中国海洋大学海洋化学理论与工程技术教育部重点实验室。主要研究方向为同位素海洋学，聚焦多核素技术对海洋沉积动力、水动力过程及其生物地球化学效应的示踪研究。主持国家自然科学基金区域联合重点项目等国家级项目，并作为主要成员参与国家自然科学基金重点项目、重点国际（地区）合作研究项目、国家重点研发计划项目等。担任中国海洋湖沼学会化学分会副理事长、国际SGD虚拟工作坊（virtual seminar series）第4届轮值主席、国家环境监测仪器产业计量测试联盟委员会专家委员会委员。担任 *Marine Chemistry* 期刊副主编，*Geosystems and Geoenvironment* 期刊编委，*Acta Oceanologica Sinica* 期刊客座编辑，*Geochimica et Cosmochimica Acta*、*Limnology and Oceanography*、*Water Research*、*Chemical Geology*、*Hydrology and Earth System Sciences*、*Journal of Geophysical Research: Oceans* 等40余种期刊的审稿人。在 *Nature Communications*、*Geophysical Research Letters*、*Geochimica et Cosmochimica Acta*、*Earth-Science Reviews*、*Limnology and Oceanography Letters* 等国际期刊发表论文50余篇。

获优秀青年科学基金项目资助。主要研究方向为气候变化和极端天气对大气污染影响解析，聚集高分辨率地球系统模式研发、气候变化对极端天气和大气污染的影响，在动力降尺度开发及全球区域模式耦合等方面取得一系列创新性成果。主持2项国家自然科学基金项目以及国家重点研发计划"政府间国际科技创新合作"重点专项，负责崂山实验室科技创新项目课题。发表SCI论文100余篇，其中以第一或通信作者在 *Nature Climate Change*、*National Science Review*、*Science Bulletin*、*PNAS* 等期刊发表论文40篇，论文SCI他引1500余次。获南京大学紫金全兴环境奖青年学者奖、清华大学-浪潮集团计算地球科学青年人才奖、中国环境科学学会青年科学家优秀奖等。讲授本科生、研究生课程5门。担任SCI期刊 *Frontiers in Environmental Science* 大气与气候变化方向副主编、*Journal of Environmental Science* 大气环境领域青年编委，*Nature*、*Nature Climate Change*、*Science Advances*、*Journal of Geophysical Research* 等期刊审稿人。多次受国内外研究机构（如美国加利福尼亚大学伯克利分校、芬兰赫尔辛基大学、日本环境研究所）邀请做报告和交流访问，并作为会场召集人、特邀报告人等多次参加国内外气候变化与大气污染相互作用方面的会议。

高阳
教授，博士生导师

[①] 依正文作品顺序排序。

于蒙
副教授，硕士生导师

2011年本科毕业于中国海洋大学化学专业，2015—2016年参与瑞士苏黎世联邦理工学院地球科学学院地质系博士联合培养项目，2018年获中国海洋大学海洋化学专业博士学位。入选中国海洋大学青年英才工程第三层次。现任职于中国海洋大学海洋化学理论与工程技术教育部重点实验室。主要研究方向为海洋有机地球化学／海洋化学，应用多参数有机地球化学分析手段，包括用总有机质、生物标志物参数及其碳同位素（^{13}C、^{14}C）等方法进行河流－河口－边缘海连续体有机碳循环研究，聚焦有机碳源汇格局、碳埋藏的关键过程和控制机制研究，及单体分子碳－氢同位素地球化学研究。主持国家自然科学基金青年基金项目、青岛市博士后应用研究项目和校级科研项目；参与国家自然科学基金国际（地区）合作与交流项目、面上项目，科技部重点研发计划等国家级科研项目。在 *Environmental Science & Technology*、*Geochimica et Cosmochimica Acta*、*Earth and Planetary Science Letters*、*Chemical Geology*、*Science of the Total Environment*、*Journal of Geophysical Research: Biogeosciences* 等期刊发表多篇论文。

获瑞士苏黎世联邦理工学院博士学位，曾在美国哈佛大学从事博士后研究。担任中国海洋大学海洋化学理论与工程技术教育部重点实验室副主任、海洋碳中和中心执行主任、中国海洋大学首届青年创新交叉团队负责人。获评中组部海外高水平人才引进计划青年学者、山东省杰出青年。获瑞典自然科学基金 Starting Grant 项目基金资助。研究方向为 ^{14}C 生物地球化学、海洋有机地球化学、陆地－海洋碳汇过程、深海海沟碳循环、海洋环境重建等。主持国家自然科学基金重大研究计划重点资助项目、面上项目、专项项目。以第一或通信作者在 *Nature Communications*、*Geology*、*Earth and Planetary Science Letters*、*Global Biogeochemical Cycles*、*Geophysical Research Letters* 等期刊发表论文 20 余篇。

包锐
教授，博士生导师

徐晓峰
教授，博士生导师

山东省泰山学者青年专家。2006 年本科毕业于湖北大学材料科学与工程学院，2011 年获华南理工大学材料学专业博士学位，2011—2014 年、2014—2017 年、2017—2018 年分别在新加坡南洋理工大学、瑞典查尔姆斯理工大学、瑞典林雪平大学进行博士后研究。入选中国海洋大学青年英才工程第一层次。科研工作围绕高分子基光/电、光/热、光/化学转换材料与器件领域，以期实现对"太阳能"的高效综合利用。近 5 年在能源高分子材料设计与合成、器件制备与界面修饰、大面积器件制备技术与多功能集成等方向开展了系统的研究工作。发表 SCI 论文 80 余篇，SCI 引用超过 3500 次，h 因子为 35。获授权国家发明专利 4 件。担任 *Nature Sustainability*、*Advanced Materials*、*Advanced Energy Materials*、*Nano Energy*、*Chemistry of Materials*、*Journal of Materials Chemistry A*、*ACS Applied Materials & Interfaces* 等期刊审稿人。

2009 年本科毕业于大连理工大学水利水电工程专业，2014 年获天津大学水利工程专业博士学位。研究方向为波浪能、风能等海上新能源开发利用，海岸结构物地基冲刷与水动力分析。主持国家自然科学基金项目 2 项、联合基金项目课题 1 项、国家重点研发计划子课题 1 项、山东省重点研发计划项目 2 项及其他多项纵横向课题，参与国家自然科学基金重点项目、中国工程院重大咨询项目、山东省自然科学基金重大基础研究项目等多项课题。以第一作者或通信作者发表 SCI、EI 论文近 30 篇。论文得到了中国工程院唐洪武院士、杜修力院士，美国工程院 Ahsan Kareem 院士等多位知名专家的引用肯定。相关科技成果应用于我国沿海多个风电场的基础设计及波浪能装置的海试保障。

于通顺
教授，博士生导师

王栋
教授，博士生导师

获国家杰出青年科学基金资助，获评山东省有突出贡献的中青年专家。获中国海洋工程科学技术二等奖、茅以升土力学及岩土工程青年奖等奖项。研究方向为海洋岩土工程与工程地质。主持中国和澳大利亚的国家级基金项目7项，负责中国海洋石油集团有限公司、荷兰辉固（Fugro）国际集团、埃克森美孚公司等单位委托的攻关课题30余项。发表SCI论文88篇。研究成果应用于南海滑坡灾害评价、我国深海可燃冰试采井口基础设计、国内外多个海域的自升式平台插桩分析、西非海域平板锚设计等，被国际标准化组织（ISO）和挪威船级社（DNV）等国际权威机构的规范采纳。

曹飞飞
副教授，硕士生导师

全国黄大年教师团队（团队名称：绿色与智慧海岸工程教师团队）及山东省泰山学者创新团队（科研方向：海洋可再生能源利用）骨干。2009年本科毕业于中国海洋大学港口航道与海岸工程专业，2016年获中国海洋大学港口、海岸及近海工程专业博士学位。研究方向包括波浪能开发与利用、风-浪联合开发与利用、海洋环境水动力。先后主持了国家自然科学基金青年基金、国家自然科学基金面上基金、山东省自然科学基金面上基金、山东省海洋工程重点实验室开放基金等项目，作为课题负责人承担了国家自然科学基金联合基金项目、山东省自然科学基金重大基础研究项目，作为主要人员参加了国家重点研发计划项目、国家高技术研究发展计划（863计划）项目、海洋可再生能源资金项目、山东省重大基础研究项目等海洋能领域重点课题。发表论文20余篇，获授权国家发明专利10余项、实用新型专利17项以及日本发明专利1项，获软件著作权16项，获中国专利优秀奖。

董双林
教授

山东省泰山产业领军人才，1997年获国家杰出青年科学基金资助。现兼任中国海洋学会海洋经济分会主任委员、中国海洋湖沼学会养殖生态学分会理事长等职。1992年7月于青岛海洋大学获得博士学位。曾任中国海洋大学副校长（2003—2015年）、国务院学位委员会学科评议组水产学科组召集人（2014—2019年）、中国水产学会副理事长（2005—2012年）等职。长期从事生态养殖理论与技术研究。研发了低洼盐碱地池塘安全养殖技术和滩涂海水池塘清洁养殖技术，推动了我国盐碱荒地渔业利用和滩涂池塘清洁生产的发展，并分别于2006年和2012年获得国家科技进步奖二等奖（首位）。近几年开拓了我国深远海鱼类绿色养殖领域，实现了温暖海域冷水鱼类养殖的世界性突破。出版专著1部、教材2部。以第一或通信作者发表论文250篇，其中SCI收录100余篇。获授权发明专利11项。培养博士研究生59名。

董云伟
教授，博士生导师

2020年获国家杰出青年基金资助。主要从事海洋生态学和养殖生态学研究。获中国海洋与湖沼学会"张福绥贝类学奖"青年创新奖、中国动物学会"长隆奖"新星奖。兼任中国海洋湖沼学会养殖生态学分会副理事长、中国海洋湖沼学会棘皮动物分会副理事长、中国生态学会海洋生态专业委员会常务委员、中国动物学会生理生态专业委员会常务委员、中国野生动物保护协会科学技术委员会委员、青岛市生态学会副理事长等。在 *PNAS*、*Global Change Biology*、*Proceedings of Royal Society B*、*Global Ecology and Biogeography* 等期刊发表论文100余篇，文章累计被引5000余次。担任 *Marine Life Science & Technology* 期刊学科编委，*Diversity and Distributions*、*Journal of Experimental Marine Biology and Ecology*、*Anthropocene Coasts* 等期刊副主编，*Marine Environmental Research*、*Journal of Thermal Biology*、《渔业科学进展》等期刊编委。

张潮
副教授，硕士生导师

山东省青年创新团队负责人，青岛－香港海洋环境与生态联合研究中心青年工作组委员。2011年本科毕业于天津工业大学环境工程专业。2017年获中国海洋大学环境科学专业博士学位。入选中国海洋大学青年英才工程第三层次。主要从事大气沉降的海洋生态效应、海洋生源要素（氮、磷等）的循环过程、海洋生物与环境的相互作用等方面的研究。主持和参与多项国家自然科学基金项目。曾多次赴中国近海及西北太平洋开阔海域开展科学研究工作。发表学术论文20余篇，包括中科院一、二区论文15篇。担任 Journal of Geophysical Research、Environmental Pollution、Global Change Biology、Science of the Total Environment 等环境领域国际主流期刊审稿人。

高会旺
教授，博士生导师

享受国务院政府特殊津贴专家，山东省教学名师，山东省优秀研究生指导教师，第7届、第8届国务院学位委员会学科评议组（环境科学与工程）成员，国家教材委专家委员会成员。主要从事大气物质输送、沉降及其对海洋生态系统影响的研究。主持国家自然科学基金重大项目课题、国家重大科学研究计划、国家科技重大专项、中外合作研究项目等。曾任973计划项目首席科学家、上层海洋－低层大气国际研究计划（SOLAS）科学指导委员会委员，引领多国参与的"亚洲沙尘与海洋生态系统"研究计划。发表论文200余篇，获省部级奖励5项。在教学方面，积极探索教改方法和培养模式，承担多项省级、校级教改项目。主讲的环境海洋学课程已建设成为国家级精品课程及国家级精品资源共享课、国家一流课程，并获省级教学成果特等奖1项、一等奖2项、二等奖1项。

郭金家
教授级高工,博士生导师

中国光学工程学会海洋光学专家委员会委员,山东激光学会理事,《光学精密工程》编委。主要从事激光光谱水下原位探测关键技术和系统研发工作,在水下原位探测光谱传感器研制方面具有较为丰富的经验和良好的研究基础,发表 SCI、EI 论文 70 余篇。作为负责人先后主持国家自然科学基金项目、863 计划项目、国家重点研发计划项目和山东省重大创新工程项目等,研发了多种类型的水下光谱传感器。主持研发了我国首台深海原位激光诱导击穿光谱(LIBS)探测系统,设备整体性能居于世界领先水平,该设备入选国家"十三五"科技创新成就展。针对水中二氧化碳等溶存气体和天然气水合物测量,研制了基于可调谐半导体激光吸收光谱术(TDLAS)的甲烷、二氧化碳浓度及同位素传感器,开发完成探头式深海拉曼光谱探测系统,并在东海和南海对水合物、碳酸盐岩等进行了原位测量。

获英国剑桥大学博士学位。入选山东省泰山学者青年专家、中国海洋大学青年英才工程第一层次。兼任中国成本研究会理事、广州市社会组织专家库成员、*Sustainability Accounting, Management and Policy Journal* 副主编。主要研究方向为高管激励、国资国企改革、政企关系、可持续发展会计。先后主持和参与国家级科研项目 4 项和多项省部级课题。在 *Entrepreneurship, Theory and Practice*、*Journal of Corporate Finance*、*Emerging Markets Review*、*Accounting & Finance*、《管理科学学报》等期刊发表论文 10 余篇。

谢素娟
教授,博士生导师

丁黎黎

教授，博士生导师

中国海洋大学经济学院副院长，董氏国际海洋可持续发展研究中心副主任，海洋经济发展研究中心副主任；山东省泰山学者特聘专家。主要研究方向为海洋经济与绿色发展、金融风险管理。作为国家社会科学基金重大项目首席专家，主持国家自然科学基金及国家社会科学基金4项，在《管理世界》《中国管理科学》等国内外刊物上发表80余篇学术论文。作为海洋经济领域智库专家，取得的30余项成果被中央海洋权益工作领导小组办公室、自然资源部海洋战略规划与经济司、国家海洋信息中心、山东省海洋局等部门采纳，并应用于海洋经济高质量发展的模块化管理工作，助力海洋经济治理能力提升。

中国海洋大学海洋发展研究院高级研究员，海洋碳中和中心主任委员，美国加州大学伯克利分校访问学者，美国纽约州立大学访问学者。山东大学经济学学士，复旦大学经济学硕士，中国海洋大学工学博士，对外经济贸易大学应用经济学博士后。研究方向为经济结构转型与绿色发展。担任国家社会科学基金重大项目首席专家，主持国家社会科学基金项目、教育部人文社会科学研究项目、教育部人文社会科学重点研究基地重大项目、国家海洋软科学基金项目等20余项。在SCI、SSCI、CSSCI、EI等国内外学术期刊发表论文100余篇，出版著作9部。荣获山东省、青岛市优秀科研成果奖及各类荣誉称号20余项。担任多个政府机构专家委员会及智库专家。多项研究成果被国家部委及相关部门采纳。担任教育部"长江学者奖励计划"通讯评审专家、国家社会科学基金项目评委及国内外多种学术期刊审稿人。

徐胜

教授，博士生导师

主要作者简介

余静
副教授,博士生导师

2006年获中国海洋大学环境工程博士学位。现任中国海洋大学中澳海岸带管理研究中心(中方)副主任、海洋发展研究院海洋空间规划与管理研究所副所长。兼任全国海洋标准化技术委员会海域使用及海洋能开发利用分技术委员会委员、中国海洋工程咨询协会海洋论证评估质量管理分会理事、中国太平洋学会海洋标准化分会理事、自然资源学会海洋资源专业委员会委员、山东环境科学学会海洋环境专业委员会秘书长等职。担任自然资源部国家级海域使用论证评审专家库和山东省环境影响评价评审专家库入库专家。主要研究方向为海洋区划与规划、海岸带综合管理、海洋环境影响评价与海域使用论证。主持或参与包括国家重点研发计划项目课题、国家自然科学基金联合基金项目、国家社会科学基金重大项目等10余项。在SCI、SSCI、CSSCI等国内外期刊发表论文70余篇,参编学术专著4部、国家标准2项、行业标准2项。